The Genesis of Grammar

Studies in the Evolution of Language

General Editors
Kathleen R. Gibson, *University of Texas at Houston,*
and James R. Hurford, *University of Edinburgh*

PUBLISHED

1
The Origins of Vowel Systems
Bart de Boer

2
The Transition to Language
Edited by Alison Wray

3
Language Evolution
Edited by Morten H. Christiansen and Simon Kirby

4
Language Origins
Evolutionary Perspectives
Edited by Maggie Tallerman

5
The Talking Ape
How Language Evolved
Robbins Burling

6
The Emergence of Speech
Pierre-Yves Oudeyer
translated by James R. Hurford

7
Why we Talk
The Evolutionary Origins of Human Communication
Jean-Louis Dessalles
translated by James Grieve

8
The Origins of Meaning
Language in the Light of Evolution 1
James R. Hurford

9
The Genesis of Grammar
Bernd Heine and Tania Kuteva

IN PREPARATION
The Evolution of Linguistic Form
Language in the Light of Evolution 2
James R. Hurford

The Invisible Miracle
The Evolutionary Origins of Speech
Peter MacNeilage

PUBLISHED IN ASSOCIATION WITH THE SERIES
Language Diversity
Daniel Nettle

Function, Selection, and Innateness
The Emergence of Language Universals
Simon Kirby

The Origins of Complex Language
An Inquiry into the Evolutionary Beginnings of Sentences, Syllables, and Truth
Andrew Carstairs McCarthy

The Genesis of Grammar

A Reconstruction

Bernd Heine and Tania Kuteva

OXFORD
UNIVERSITY PRESS

Great Clarendon Street, Oxford OX2 6DP

Oxford University Press is a department of the University of Oxford.
It furthers the University's objective of excellence in research, scholarship,
and education by publishing worldwide in

Oxford New York

Auckland Cape Town Dar es Salaam Hong Kong Karachi
Kuala Lumpur Madrid Melbourne Mexico City Nairobi
New Delhi Shanghai Taipei Toronto

With offices in

Argentina Austria Brazil Chile Czech Republic France Greece
Guatemala Hungary Italy Japan Poland Portugal Singapore
South Korea Switzerland Thailand Turkey Ukraine Vietnam

Oxford is a registered trade mark of Oxford University Press
in the UK and in certain other countries

Published in the United States
by Oxford University Press Inc., New York

© Bernd Heine and Tania Kuteva, 2007

The moral rights of the authors have been asserted
Database right Oxford University Press (maker)

First published 2007

All rights reserved. No part of this publication may be reproduced,
stored in a retrieval system, or transmitted, in any form or by any means,
without the prior permission in writing of Oxford University Press,
or as expressly permitted by law, or under terms agreed with the appropriate
reprographics rights organization. Enquiries concerning reproduction
outside the scope of the above should be sent to the Rights Department,
Oxford University Press, at the address above

You must not circulate this book in any other binding or cover
and you must impose the same condition on any acquirer

British Library Cataloguing in Publication Data

Data available

Library of Congress Cataloging in Publication Data

Data available

Typeset by SPI Publisher Services, Pondicherry, India
Printed in Great Britain
on acid-free paper by
Biddles Ltd., King's Lynn, Norfolk

ISBN: 978–0–19–922776–1   978–0–19–922777–8

1 3 5 7 9 10 8 6 4 2

Contents

Preface xi
List of abbreviations xiii

1 Introduction 1
1.1 Questions and approaches 1
   1.1.1 Previous work 4
   1.1.2 Assumptions 14
   1.1.3 The present approach 20
   1.1.4 On uniformitarianism 28
1.2 Grammaticalization 32
   1.2.1 Methodology 33
      1.2.1.1 The parameters 33
      1.2.1.2 Extension 35
      1.2.1.3 Desemanticization 39
      1.2.1.4 Decategorialization 40
      1.2.1.5 Erosion 42
      1.2.1.6 Discussion 45
   1.2.2 Problems 46
1.3 The present volume 53

2 An outline of grammatical evolution 57
2.1 Introduction 57
2.2 Layers 58
   2.2.1 Nouns and verbs 59
      2.2.1.1 The first layer: nouns 60

		2.2.1.2 The second layer: verbs	71
	2.2.2	The third layer: adjectives and adverbs	82
	2.2.3	The fourth layer: demonstratives, adpositions, aspects, and negation	87
	2.2.4	The fifth layer	93
	2.2.5	The final stages	98
	2.2.6	Treating events like objects	100
2.3	Evidence from signed languages		108
2.4	A scenario of evolution		110
2.5	Conclusions		114

3	Some cognitive abilities of animals		121
3.1	Introduction		121
3.2	What linguistic abilities do animals have?		125
	3.2.1	Communicative intentions	126
	3.2.2	Concepts	128
	3.2.3	"Lexicon"	135
	3.2.4	Functional items	138
	3.2.5	Compositionality	143
	3.2.6	Argument structure	144
	3.2.7	Linear arrangement	146
	3.2.8	Coordination	148
	3.2.9	Taxonomic concepts	150
3.3	Discussion		159
	3.3.1	Problems	160
	3.3.2	Language-like abilities in animals	162
	3.3.3	Grammaticalization in animals?	163
3.4	Conclusion		164

4	On pidgins and other restricted linguistic systems	166
4.1	Introduction	167
4.2	Kenya Pidgin Swahili (KPS)	169

4.3	The rise of new functional categories	175
4.4	Discussion	184
4.5	Grammaticalization in other pidgins	187
4.6	A pidgin window on early language?	193
4.7	Other restricted systems	198
4.8	An elementary linguistic system?	205
4.9	Conclusion	208
5	**Clause subordination**	**210**
5.1	Introduction	211
5.2	Expansion	216
5.3	Integration	224
	5.3.1 Relative clauses	224
	5.3.1.1 The demonstrative channel	225
	5.3.1.2 The interrogative channel	229
	5.3.2 Complement clauses	229
	5.3.2.1 Introduction	230
	5.3.2.2 The noun channel	230
	5.3.2.3 The verb channel	236
	5.3.2.4 The demonstrative channel	240
	5.3.2.5 The interrogative channel	242
	5.3.3 Adverbial clauses	244
	5.3.3.1 Introduction	244
	5.3.3.2 The noun channel	245
	5.3.3.3 The verb channel	248
	5.3.3.4 The demonstrative channel	250
	5.3.3.5 The adverb channel	250
	5.3.4 From complementizer or relativizer to adverbial clause subordinator	251
5.4	Discussion	254
5.5	Conclusions	260

6	**On the rise of recursion**	**262**
6.1	What is recursion?	262
	6.1.1 A definition	264
	6.1.2 Manifestations	266
	6.1.3 Simple vs. productive recursion	268
	6.1.4 Embedding, iteration, and succession	270
	6.1.5 Treatment of recursion in linguistic description	271
	6.1.6 Are there languages without recursion?	272
	6.1.7 Discussion	273
6.2	Animal cognition	276
6.3	The noun phrase	279
	6.3.1 Attributive possession	280
	6.3.2 Modifying compounding	283
	6.3.3 Adjectival modification	286
	6.3.4 Conclusion	287
6.4	Clause subordination	287
	6.4.1 Case studies	288
	6.4.1.1 The rise of a relative clause construction	288
	6.4.1.2 The rise of complement and adverbial clauses	291
6.5	Loss of recursion	293
6.6	Conclusions	294
7	**Early language**	**298**
7.1	Grammatical evolution	298
	7.1.1 Layers	298
	7.1.2 From non-language to language	311
	7.1.3 Lexicon before syntax	313
	7.1.4 Word order	315
	7.1.5 Functions of early language	318
	7.1.5.1 Cognition or communication?	318
	7.1.5.2 Motivations underlying grammaticalization	323

	7.1.5.3 Discussion	329
	7.1.6 Who were the creators of early language?	331
	7.1.7 Did language arise abruptly?	338
7.2	Grammaticalization—a human faculty?	342
7.3	Looking for answers	345
7.4	Conclusions	354
References		357
Index		401

Preface

The present volume grew out of our work within the research project on Restricted Linguistic Systems as a Window on Language Genesis at the Netherlands Institute for Advanced Study (NIAS) in the academic year 2005–6. We wish to express our gratitude to the other team members Anne Baker, Sandra Benazzo, Adrienne de Bruyn, Christa König, Pieter Muysken, Bencie Woll, and Henriette de Swaart, as well as to Beppie van den Bogaerde, and most of all to the project coordinator Rudie Botha for his cooperation and assistance, and to Wim Blockmans, the Rector of NIAS for his extraordinary hospitality and understanding. Without the initiative of Rudie Botha and the support of NIAS, this book would never have been written.

A number of other people have also contributed to this book. Our thanks are due in particular to Philip Baker, Peter Bakker, Ulrike Claudi, Elly van Gelderen, Tijs Goldsmith, Casper de Groot, John Haiman, Zelealem Leyew, Hiroyuki Miyashita, Fritz Newmeyer, Irene Pepperberg, Roland Pfau, and Klaus Zuberbühler for all their readiness to assist in the project, as well as to two anonymous referees of this work, to whom we owe a wealth of critical comments and suggestions. We also want to thank Barbara Sevenich, Birgit Wolandt-Pfeiffer, Birte Giebfried, and Anne Hartmann for their invaluable typographical help. Finally, a word of special thanks is due to Tom Givón, whose work was a constant source of inspiration—but also of frustration because time and again we discovered that what we have to say was in most cases not much different from what he had already said and written.

Preliminary work leading to the book was carried out during our stay at various research institutions. We are also indebted to the following institutions which provided us with opportunities to work on various aspects of the book, especially the following: Center for Advanced Study in the Behavioral Sciences, Stanford, the Institute for Advanced Study at LaTrobe University, and the Research Centre for Linguistic Typology, also at LaTrobe, whose Directors Bob Dixon and Sasha Aikhenvald offered the first-named author hospitality to work on the book. The second-named author thanks the Max Planck Institute for Evolutionary Anthropology, Leipzig, and in particular the Director of its Linguistics Department, Bernard Comrie, for providing both a most stimulating working atmosphere and financial support. Furthermore, our colleagues Lenore Grenoble and Lindsay Whaley of Dartmouth College offered the first-named author academic hospitality in 2002.

January 7, 2007

Bernd Heine & Tania Kuteva

Abbreviations

A	agent
A.DCL	affirmative declarative
ABL	ablative
ABS	absolute, absolutive
ACC	accusative
ADP	adposition
AFF	affix
AGR	agreement marker
ANA	anaphor
AND	andative
a.n.	authors' note
ASP	aspect
C1, C2, etc.	noun class 1, 2, etc.
CAS	case marker
CE	English-based creole
CF	French-based creole
CL	classifier
CLASS	classifier
COM	comitative
COMPL	complementizer
CONJ	conjunction
CONN	connective
COP	copula
CP	Portuguese-based creole
CPL	completive aspect, complementizer
DAT	dative
DI	distal demonstrative
DIR	directional
DP	dummy pronoun
DU	dual
E	ergative

EMPH	emphatic
ENC	enclitic
END	verbal ending
ERG	ergative
EXCL	exclusive
F	feminine
FOC	focus
FUT	future tense
GEN	genitive
GL	goal
HAB	habitual
HON	honorific marker
IMM	immediate tense–aspect marker
IMP	imperative
IMPFV	imperfective
INCE	inceptive
INCL	inclusive
IND	indicative
INF	infinitive
INT	intensifier
INTR	intransitive
IPS	impersonal
JUNC	juncture
KPS	Kenya Pidgin Swahili
LINK	linker
LOC	locative
LOGO	logophoric (pronoun)
M	masculine
N	noun; non-past prefix
N1, N2, etc.	noun class 1, 2, etc.
NAR	narrative
NEG	negative, negation marker
NF	non-future tense
NOM	nominative
NP	noun phrase
NT	non-topic
O	object
OBL	oblique case

OED	*Oxford English Dictionary*
OM	object marker
OR	originative (case)
PART	particle
PASS, PAS	passive
PAST	past tense
PERF	perfect
PFV	perfective
PL	plural
PNT	proper name topic marker
POL	politeness marker
POST	postposition
PR	proximal demonstrative
PRES	present tense
PRN	pronoun
PROG	progressive marker
PRON	pronoun
PST	past (perfective)
PTCPL	past participle
PURP	purpose marker
QUOT	quotative marker
R	relational suffix
REC	reciprocal
REF	reflexive
REFL	reflexive
REL	relative clause marker
S	subject; subject marker; sentence
SBJ	subjunctive
SBR	adverbial clause subordinator
SG	singular
SI	subject index
SUB	clause subordinator
SUBJ	subjunctive
SUFF	suffix
T	transitive marker
TAM	tense-aspect-modality
TERM	terminative aspect
TNS	tense marker

TOP	topic marker
TPA	today past
TR	transitive preposition
TRS	transitivizer
V	verb
VEN	venitive, ventive
VN	verbal noun
VP	verb phrase
1, 2, 3	first, second, third person
I, II	juncture I, II

Dedicated to Anne Buxton and Maya Pencheva

1 Introduction

> Like other biological phenomena, language cannot be fully understood without reference to its evolution, whether proven or hypothesized.
>
> (Givón 2002b: 39)

It is the above quotation that inspired us to embark on this book project. The claim is important; it is intuitively appealing—at the same time, it is hard to verify. It therefore provided a real challenge for the research which was subsequently carried out, and whose findings are presented here. But there was yet another factor which was accompanying us throughout the work on the present book, namely to test and apply to a topic as fascinating, challenging, and ambitious as that of language evolution the potential of grammaticalization theory as a tool for reconstructing the evolution of human language.

1.1 Questions and approaches

Work based on classical methods of historical linguistics has brought about a wealth of insights into the substance and structure of earlier forms of human languages. But this work gives us access to only a fairly small phase in the history of human languages: Linguistic reconstruction becomes notoriously fuzzy and conjectural once we are dealing with a time depth exceeding 8,000 years. While there exists a wide array of opinions on how far back reconstruction can be pushed and on how the genetic and areal relationship patterns among earlier languages may have been, there appears to be general agreement on the following point: The languages that were spoken 8,000 ago were typologically not dramatically different from what they are today. Thus, compared to the biologist, the linguist is in a deplorable situation:

In other words, the central aspects of language—syntax and phonology—have no evident homologs. In that sense, language is an emergent trait (or "key innovation") and poses, along with all such traits, particularly difficult problems for evolutionary biology. Likewise, there are no archaeological digs turning up specimens of the language of 100,000 years ago. While the fossil record has given us a reasonably clear picture of the evolution of the vocal tract, grammatical structure, needless to say, is not preserved in geological strata. (Newmeyer 2003: 61)

What we factually know about the neurology, cognitive capacity, and sociocultural environment shaping early language development is severely limited. Fossil bone findings—for example on the enlargement of the thoracic vertebral canal, the descent of the larynx, and an increase in encephalization—have been used in linguistic reconstruction work. But the exact significance of such findings remains largely unclear, and some of these findings, such as the descent of the larynx, have recently been shown to be entirely irrelevant (Fitch 2002). Davidson (2003: 144) therefore concludes that anatomical evidence has proved a poor guide to speech abilities, and has contributed little to the understanding of the emergence of language.

That we know essentially nothing about the evolution of language in pre-modern times is certainly not because not enough research has been done on it, and also not because there is a lack of plausible hypotheses on this issue: on the contrary, scholars have been working on it for at least three centuries, if not longer, and have come up with many fascinating hypotheses. Still, in spite of all this work, no empirically really satisfactory information has surfaced from this research on any of the following questions that we consider to be important to understand language genesis and evolution:

(1) Open questions
   a. Why did human language evolve, and what purpose did it serve?
   b. When and where did it evolve?
   c. Who were the creators of early language?
   d. Was its origin mono-genetic or poly-genetic, that is, do the modern languages derive from one ancestral language or from more than one?
   e. Were the forms and structures characterizing early language motivated or arbitrary?
   f. Did language originate as a vocal or a gestural system?

g. Can language genesis be related to the behavior of non-human animals?
h. Was language evolution abrupt or gradual?
i. Which is older—the lexicon or grammar?
j. What was the structure of language like when it first evolved?
k. How did language change from its genesis to now?
l. How long did it take to develop a structure that corresponds to what we find in modern languages?
m. How did phonology evolve?
n. How did the properties believed to be restricted to modern human languages arise, in particular syntax and the recursive use of language structures?

Not surprisingly therefore, many linguists believe that these questions are beyond the scope of their discipline, or of any discipline, and are therefore—in the wording of Chomsky (1988: 183)—"a complete waste of time". This view has a long tradition: and in much the same spirit, the Linguistic Society of Paris and the British Academy urged their members in the 1860s to refrain from discussing the origin of language, as a reaction to the speculation that had dominated discussions on this issue in linguistics and philosophy.

But the situation has changed in the course of the last decades: The question of how human language evolved has become a field of intense research. First, observations on present-day languages and cultures have come to be used by contemporary linguists as indirect evidence in favor of different hypotheses of how human language could have originated. For example, in a number of highly non-trivial works, Haiman (1994, 1998) argues that linguistic signs could have gradually—via repetition and ritualization—originated from motivated symptoms. Accordingly, all linguistic structures can be placed on a functional spectrum, whereby the more iconic a structure, the more likely it is to be at the expressive/ directive end; the more arbitrary a structure, the more likely it is to be at the phatic end of that spectrum (Haiman 1998: 166; see also Givón 1979b, 1979c). Second, new technologies offered by computer sciences, findings in biology, psychology, neuroscience, brain imaging, genetics, palaeo-anthropology and archaeology, together with more detailed information on restricted linguistic systems such as pidgin languages, home-sign systems invented by deaf children of non-signing parents, systems arising in specific conditions of first and second language acquisition, etc., have encouraged scholars to make a fresh attempt to look for answers to questions such as the ones listed in (1), and a plethora of books and articles have appeared, offering a wealth of stimulating hypotheses.

The problem now facing research on language evolution is not that there are no answers to the questions listed in (1); on the contrary, there are perhaps too many of them, with the effect that the novice in this research field may find it hard to decide which of the answers—many of them mutually contradictory—is the most credible. And there is also a problem with the question that has worried linguists over centuries, namely whether it will ever be possible not only to provide answers but to provide answers based on empirically sound knowledge rather than on conjecture. Even for a question like (1a), which appears to be readily accessible to empirical research, there is no uncontroversial answer, leading Fitch, Hauser and Chomsky (in press: 5) to the conclusion that "from an empirical perspective, there are not and probably never will be data capable of discriminating among the many plausible speculations that have been offered about the original function(s) of language."

1.1.1 Previous work

The kinds of linguistic systems that we are concerned with here have been referred to with a range of different terms. For our purposes, a twofold distinction is of interest: On the one hand there are the **modern languages**, characterized by the following features: (a) they consist of the languages spoken today; (b) they are immediately accessible to reconstruction by means of established methods of historical linguistics; and (c) they relate to linguistic developments of roughly the last eight millennia. On the other hand there is what we will call **early language**,[1] which lacks all these features, that is, (a) it is not available today; (b) it is not accessible via orthodox historical methodology; (c) it is clearly older than 8,000 years and covers the timespan from the genesis of human language to the beginning of modern languages; and, consequently, (d) all we know about it remains of necessity hypothetical. Among the many works that

---

[1] Early language must not be confused with "protolanguage" (Bickerton 1990), a term that will not be used in this work except when referring to works where it appears.

have been written and the stimulating views that have been expressed it would be hard to single out any that deserve particular mention, or to provide an overview of this field of research that would do justice to all of them. We will therefore be satisfied in what follows with highlighting a few major orientations of scholarly thinking and themes that surface from this research; for a detailed and insightful discussion, see Johansson (2005).

A perhaps noteworthy observation is that quite a number of these works use what one may call integrating approaches. These approaches are based on the assumption that the more data on different and unrelated phenomena can be combined, the stronger is the hypothesis on language genesis and evolution that can be built; Givón (2002a: 151) describes the logic of these approaches thus: "Many of the arguments for this are either conjectural, convergent or recapitulationist, so that they only make sense if taken *in toto.*" The kind of phenomena analyzed is largely the same from one author to another, including most of the following: (a) child language; (b) late second language acquisition; (c) pidgin languages; (d) agrammatic aphasics; (e) "Genie", the woman isolated from human contact from age three to thirteen (see Section 4.7); (f) the behavior of non-human primates; and (g) regularities in linguistic change. Paradigm examples of such approaches are Bickerton (1990), Jackendoff (1999, 2002: 237), Comrie (2000), Calvin and Bickerton (2000), Givón (2002a, 2002b, 2005), and Li and Hombert (2002). Givón's and Comrie's approaches stand out against most others in that they take (g), that is, generalizations on linguistic diachrony, as an important parameter of reconstruction; this parameter is ignored by other authors, whose integrating approach can be characterized as being essentially achronic in nature.

The procedure adopted by Givón and Comrie is mainly to look for common denominators among the structural properties of the phenomena listed above in order to propose a credible hypothesis on language genesis and/or language evolution. For example, Givón (2002a: 151–2) concludes that grammatical concepts are more abstract than the lexicon and, given the overwhelming concreteness of primate and early childhood lexicons, it is unlikely that the evolution of the grammatical code could have preceded the evolution of at least some rudimentary well-coded lexicon in time.

Among the phenomena that we listed above, one has found particular attention, namely child language or ontogeny, for two different reasons.

First, the assumption implicit in some of the integrating and other approaches that have been proposed to reconstruct language evolution is that ontogeny recapitulates phylogeny—in other words, the way young children acquire language offers a possible analog of the way humans acquired early language. And second, it is argued by some authors that young children play a crucial role in language change, hence they also must have been the agents in language genesis and/or language evolution (e.g. Piattelli-Palmarini 1989). Neither of these assumptions is, however, really uncontroversial. While the ontogeny-recapitulates-phylogeny hypothesis appears to have considerable potential for reconstruction, it is not unproblematic; as has been demonstrated by Slobin (2002), there is substantial counter-evidence to this hypothesis. And that young children were instrumentally involved in language genesis is also an assumption that is in need of reconsideration (see Section 7.1.6).

Models taken from evolutionary biology While there is a long and respectable tradition of research in evolutionary biology, there is no corresponding field of evolutionary linguistics. It is therefore not surprising that students of language evolution have drawn on models from biology in framing or justifying their hypotheses or for proposing explanatory accounts. A number of controversies on what are essentially linguistic issues have been based on biological notions and, accordingly, the hypotheses proposed have been shaped by the respective biological models (see Botha 2003a for detailed discussion).

One major line of research is based on Charles Darwin's ([1859] 1964) concept of adaptation, according to which a fitness-enhancing character is shaped or built for its present function by natural selection. In the same way as a complex design like the eye of a vertebrate can be accounted for by adaptation and gradual evolution, it has been argued since Pinker and Bloom (1990) that the complex design of human language can be explained as a biological phenomenon in the Darwinian sense, having arisen via a series of adaptive processes (Pinker 2003; Jackendoff and Pinker 2005; see also Haspelmath 1999c; Croft 2000; Mufwene 2001; Bierwisch 2001; Givón 2002a, 2005).

But what exactly was the nature of selection pressure leading to the emergence and evolution of human language? The following collection of views proposed within the Darwinian adaptation framework of research on language evolution may give an impression of the kinds of reasoning that have been invoked. Thus, it is argued that language would have allowed our ancestors to acquire information and pass it on far faster than biological evolution could achieve, giving them a decisive advantage in competition with other species. But whereas many agree that "language cannot have come cheap"[2] there is no real consensus on what could have been the reason for all the expensive changes that evolution has brought about in order to give us speech. On one view, let us call it the "hunting" view, early man was a great hunter, who needed to communicate plans for herding prey or trapping them in particular places. Our ancestors needed to communicate about places to hunt, new sorts of traps, locations of water, good caves, techniques for making tools, and ways to make and keep fire. The bottom line is: Men needed to speak in order to hunt better. But there is also a "foraging" view on which, at the dawn of human language, communication was necessary in order to exchange information about locations, the nutritional value of available foodstuffs, and the safety of available foodstuffs. On the "narrative" view, language evolved to enable the exchange of stories about the supernatural or the tribe's origins. According to the "Machiavellian Intelligence" view, social primates are believed to understand matters such as alliances, familial relationships, dominance hierarchies, and the trustworthiness of individual members of the group. The assumption therefore is that our hominid ancestors were social animals, that they could recognize and compare different social relationships, and that they could respond appropriately.

Social life is complex: it requires a lot of brain power to remember who is who, who has done what to whom, etc., and language therefore is said to have emerged in order to enable and maintain the complexity of social life, and on the "gossip" view, language evolved to allow us to gossip (Dunbar 1996). Dunbar argues that most of our talking is gossip, which enables us to cement and maintain our human relationships, in short: Gossip does for humans what grooming does for many other primates, language makes it possible for us to gossip. It acts as "a cheap and ultra-efficient form of grooming" (Dunbar 1996: 97; see also Dunbar 2006): We can talk to more than one person at once, pass on information about cheaters and scoundrels, and tell stories about who makes a reliable friend. Central to the

---

[2] The whole of our vocal apparatus had to be evolved, which meant (a) complex changes in the neck, mouth, and throat, and (b) the impossibility to drink and breathe at the same time, increasing the risk of choking.

"memetic" view is the notion of meme as an element of culture that may be considered to be passed on by non-genetic means, especially imitation (Blackmore 1999). Blackmore considers the meme to be an entity that plays the role of gene in the transmission of words, ideas, faiths, mannerisms, fashions. Humans have the unique ability to imitate, and so to copy from one another ideas, habits, skills, behaviors, inventions, songs, stories. Memes, like genes, are assumed to be selfish replicators; they compete to find space in our minds and cultures, for the sake of their own replication. The standpoint taken in memetic theory is that language evolved for the spread of memes: "When the environment changes, a species that can speak, and pass on new ways of copying, can adapt faster than one that can adapt only by genetic change" (Blackmore 1999: 99; see above).

The main problem associated with the above approaches having a Darwinian adaptation orientation concerns appropriate empirical evidence on the nature of selection pressures that can be held responsible for adaptation. Whereas evolutionary biologists have come up with sufficiently detailed reconstructions of the evolution of organisms, there is so far no really convincing hypothesis on the nature of the adaptive processes that ultimately led to the emergence of modern languages in all their phonological, morphological, syntactic, and semantic complexity (see Johansson 2002: 94–108, 2005).

An alternative line of research uses exaptation as an explanatory construct. Gould and Vrba (1982: 6–7) proposed the notion exaptation for biological characteristics that enhance fitness in their present role (or effect) but were not built by natural selection for this role. On this view, certain fundamental features of language originated like birds' wings and tail feathers, which evolved initially as an adaptation for thermoregulation and were later exapted for an entirely different function, namely for flight (e.g. Lieberman 1991; Gould 1991; Wilkins and Wakefield 1995, 1996; Carstairs-McCarthy 1999; Calvin and Bickerton 2000).

A related major line of research relies on the notion of spandrel co-optation, a variety of exaptation where characteristics that did not originate by the direct action of natural selection were later co-opted for their current utility or function. Characteristics belonging to this subcategory have been called "spandrels", a term borrowed from architecture: Spandrels are forms and spaces that arise as necessary by-products of another decision in design rather than as adaptations for direct utility in themselves (Gould 1997: 10751). On this view, the structures governing language are biological traits which originally served no function at all but acquired their present function via exaptation or "reappropriation" (cf. Piattelli-Palmarini 1989: 19; Wilkins and Wakefield 1995).

A problem with the last two kinds of hypotheses on language evolution is that they claim that something has or may have happened, but they do not show *how* it happened—that is, they do not offer any coherent account of the process leading from non-language to early language, or from early language to modern languages. While biologists have come up with fairy appropriate descriptions of exaptive processes, it remains unclear how exactly exaptation may have proceeded in the evolution of language structure.

As these remarks suggest, there is an enormous diversity in views that have been held in understanding language evolution, and it comes as no surprise that some of these views have given rise to controversies. In the following paragraphs we highlight some of these controversies that are particularly relevant for an understanding of the subject matter covered in this book.

From ape communication to language One controversy relates to the question of whether language can immediately be derived from forms of communication or cognitive abilities to be found in non-human animals, most of all in apes, or whether it is a phenomenon *sui generis* that has no homolog in animal behavior, having evolved in a discontinuous fashion without any links to pre-human forms of communication and/or cognition. In accordance with the former position, Jackendoff (2002: 236) argues that one can reconstruct from modern human language a sequence of distinct innovations over primate calls, with each being an improvement in communicative expressiveness and precision. The latter position, emphasizing the uniqueness of human languages, asserts that specific properties of modern languages, most of all the syntactic mechanism of recursion, do not appear to have any homolog in non-human animals (Hauser, Chomsky, and Fitch 2002). But there are also compromising positions that try to reconcile these contrasting views. A more differential perspective of continuity is propagated in particular by Aitchison (1998: 19–22): Distinguishing three constituents of language, namely (a) auditory mechanisms, (b) articulatory mechanisms, and (c) the brain, she maintains that continuity was strong in the case of (a), weak in the case of (c), while (b) shows the greatest discontinuity.

Discontinuity vs. continuity Irrespective of whether a bridge can be built between non-human animal communication and human language, there remains another question, namely whether the rise and early evolution of human language was gradual or abrupt (see Botha 2003a: 36–41 for discussion); let us refer to the two main claims that have been made on this issue as the gradualist hypothesis and the leap (or discontinuist) hypothesis, respectively. Proponents of both hypotheses draw on evidence from evolutionary biology and on biological notions such as adaptation, mutation, and exaptation in search of support for their respective positions.

The gradualist hypothesis[3] is strongly associated with, though not restricted to, the Darwinian (or Neo-Darwinian) paradigm of natural selection and adaptation. Arguments in support of the gradualist hypothesis tend to emphasize, first, that language evolution should not be viewed as being drastically different from other linguistic, socio-cultural, and biological phenomena. And second, on account of their enormous complexity, human languages—just like the vertebrate eye—cannot have arisen overnight but must have involved a series of developments, with one incrementally building on the other (e.g. Pinker and Bloom 1990; Newmeyer 1998b; Bichakjian 1999; Jackendoff 1999, 2002; Givón 2002a, 2002b, 2005; Heine and Kuteva 2002a; Tomasello 2003a). For example, Newmeyer (1998b: 311) holds that syntax is made up of various subsystems each of which is governed by a distinct set of principles which cannot be derived from a more fundamental principle or property. He concludes that these principles must have evolved in an incremental way, thus suggesting a gradualist scenario.

The leap hypothesis capitalizes on what is argued to be the unique nature of language (Kirby 2002: 173). Supporters of this hypothesis include Chomsky (1988) and Gould (1997), who argue that language may have evolved all at once as the product of a kind of macromutation (see Pinker 2003: 25), and roughly the same position is maintained by Bickerton (1990, 1998), for whom there was a single mutation, coincident with the transition from *Homo erectus* to *Homo sapiens*, that created true language out of what he calls "protolanguage." The major linguistic consequence of this mutation was the imposition of recursive hierarchical structure on pre-existing thematic structure, transforming "protolanguage" in one swoop to true modern human language:

---

[3] Bickerton (2005) refers to this hypothesis as "genre continuism."

the linkage of theta analysis with other elements involved in protolanguage would not merely have put in place the basic structure of syntax, but would also have led directly to a cascade of consequences that would, in one rapid and continuous sequence, have transformed protolanguage into language substantially as we know it today. (Bickerton 1998: 41)

Evidence for this hypothesis is seen in particular in the syntax of human languages whose nature is claimed to preclude the possibility of a gradual evolution (Berwick 1998). The following arguments in particular have been volunteered in support of the leap hypothesis or—more appropriately—against the gradualist hypothesis (see Botha 2003a: 79 ff.): First, there is no evidence of a gradual accumulation of linguistic capabilities over a long period. If indeed there had been a gradual process, then either we should find some level of "stable syntax" somewhere between early language and modern languages—showing up in atypical manifestations of human languages (such as aphasic and dysphasic syndromes), cryptolects, stages of first and second language acquisition—or there should be synchronic varieties of language bearing witness of intermediate stages (Bickerton 1998: 354–7; Crow 2002). Second, the syntax of modern languages is such that it could not have evolved in an incremental fashion: Berwick (1998: 338–9), for example, argues that many syntactic relations and constraints, such as basic skeletal tree structures, movement rules, etc., can be derived from "Merge", a combinatorial operation used in generative grammar for the derivation of sentences. Accordingly, it is argued, if there had been evolution of syntax, then it must have involved a kind of leap from non-combinatorial syntax to "Merge."

Both hypotheses highlight relevant issues of language evolution. But there is a difference. Whereas work based on the gradualist hypothesis has come up with scenarios on how the transition from non-language to modern languages can be accounted for (e.g. Jackendoff 2002; Heine and Kuteva 2002b; Davidson 2003), there appear to be no appropriate scenarios in support of the leap hypothesis. It remains largely unclear how, all of a sudden, the complex morphological and syntactic structures characterizing modern languages came into being. We will return to this discussion in Section 7.1.7.

A language-specific faculty? There is a wide range of further issues that have occupied students of language genesis and evolution over the past decades. The reader is referred in particular to the many contributions to be found in the volumes edited by Gibson and Ingold (1993), Hurford, Studdert-Kennedy and Knight (1998), Givón and Malle (2002), Wray (2002), or Christiansen and Kirby (2003), Tallerman (2005), and the many books that are available (e.g. Aitchison 1996; Bickerton 1990; Calvin and Bickerton 2000; Botha 2003a; Givón 2002a, 2005; Johansson 2005). Mention should be made in particular of one line of recent work which relates language and language evolution to animal communication systems on the one hand, and to non-language-specific human cognitive abilities on the other. Modern languages have a number of properties that contrast sharply with anything that we find in non-human primates and other animals, and that have puzzled researchers trying to establish links between forms of communication to be observed in apes and what we find in languages today. Properties that have been mentioned in particular are multi-propositional discourse coherence, spatially or temporally displaced reference, syntactic displacement, that is, the ability to move constituents from their natural argument positions and place them in other slots in the sentence, locality constraints which prevent displacement from acting over unbounded domains, or structural dependence (Hauser and Fitch 2003; Givón 2005; see also the "design features" of Hockett 1960).

But there is one property that has found particular attention. Hauser, Chomsky and Fitch (2002) propose distinguishing between a faculty of language in the broad sense (FLB) and a faculty of language in the narrow sense (FLN). According to these authors, most, if not all, of FLB is based on mechanisms shared with non-human animals, while FLN is specific to humans and human languages; FLN, these authors argue, comprises only the computational mechanisms of recursion as they appear in narrow syntax and the mappings to the interfaces. In short, recursion, that is, the embedding of a constituent within another constituent of the same type,[4] is claimed to be essentially the only uniquely human component of the faculty of language, being specific both to language and to humans. So how did this property arise in human languages? We will return to this issue in Chapters 6 and 7.

New directions Some of the problems that we dealt with above may perhaps be solved by future research. In the course of the last decades a number of new lines of research have been developed that might shed light

---

[4] As we will see in Chapter 6, this is only one way in which the term recursion is or has been used.

on problems that previously were not accessible via empirical research. The following remarks are far from presenting an exhaustive account of all the new directions that have surfaced in the recent literature; rather, we are restricted to mentioning a few salient research areas (for more detailed discussion, see Johansson 2005).

One promising direction of research is provided by work on computer simulation. Common to many of the approaches that have been devised is the theory that there are agents, intending to communicate and produce a self-generated and self-organized language-like system (e.g. Batali 1998; Steels et al. 2002; Steels 2003; Steels and Belpaeme 2005; Briscoe 2002; Tonkes and Wiles 2002; Kirby 2000, 2002).

Another promising line of research can be seen in the study of prehistoric human behavior as it manifests itself in particular in ritual burials, sea-crossings, and artefacts such as rock engravings, cave paintings, statuettes, and other pieces of art, where creativity accelerated dramatically roughly between 85,000 and 50,000 years ago (e.g. Davidson and Noble 1993; McBrearty and Brooks 2000; Klein and Edgar 2002; Mellars 2006; d'Errico et al. 2005; Henshilwood 2006; Conard 2006; Wurz 2006). Abstract geometrical designs, perforated shell beads, bone tools, and other products of early human behavior that were found in southern African sites—for example Blombos Cave, Klasies River, and Diepkloof—are taken by some to provide strong evidence for the presence of language during that period. So far, however, no compelling correlations between these findings and language evolution have been established (cf. Coleman 2006).

Furthermore, there are fields such as neurobiology and genetics, including epigenetics, which have a large potential for providing new insights into how early language arose and evolved. The discovery of mirror neurons in the brains of macaque monkeys allow us to view basic abilities required for communication, such as imitation and speech control, in a new light, and research on FOXP2, a gene on Chromosome 7, might lead to a better understanding, for example, of the contribution made by nonhuman primates and of possible selection pressures that could have influenced language genesis. Both neurology and genetics are dynamic research areas and it is hard to predict how they will enrich our knowledge on early language in the years to come.

Finally, another line of work can be seen in the search for "linguistic fossils". This search, which has its parallels, for example, in observations on archaic features in the physical structure of present-day organisms, is concerned on the one hand with the study of restricted linguistic systems (Botha 2003b, 2005, 2005/6); on the other hand it is based on the claim that certain design features of modern languages exhibit an evolutionarily "more primitive character", or resemble fossils of earlier evolutionary stages (Fónagy 1988; Bickerton 1990; Jackendoff 1999: 276, 2002: 264). Using "fossil" phenomena for reconstruction is not new in linguistics, it is essentially part of the method of internal reconstruction and other diachronic approaches (Givón 2000), and it also has some potential for the reconstruction of early language. A problem that needs further attention—in this case as in other cases that we discussed in this section—concerns the basis of comparison between two kinds of phenomena. For example, when Jackendoff (2002: 264) hypothesizes that a certain design feature X of modern languages resembles (or is related to) some feature Y of early language then this raises the question of what one actually knows about Y or—more precisely—what the methodological basis for one's hypothesis on Y is. Jackendoff does not elaborate on this issue, other than alluding to defective lexical items, items that have no syntax, etc., or noting that some of the defective lexical items "have almost the flavor of primate alarm calls," etc. In other words: Can we relate X and Y meaningfully to one another without having an appropriate theory that allows us to set up Y?

These and many other new directions of research suggest that, after a history of more than three centuries, work on language genesis and evolution, while still faced with fundamental problems, might now be entering a new phase—one that is determined more by empirical observations than speculation, and whose outcome is promising, though hard to predict.

1.1.2 *Assumptions*

We saw above that there are quite a number of problems, which are due mainly to the fact that language genesis and evolution are not immediately accessible to empirical analysis, and that any attempt at accessing them involves some "poking in the dark." Perhaps the main problem facing research on language genesis and development is methodological: What kind of method and evidence is required to produce a plausible hypothesis on language genesis and/or evolution (see Botha 2003a)?

Bickerton (2005) asserts that language evolution is an interdisciplinary field and that "it is at best incautious to frame hypotheses on the basis of evidence from a single discipline." While we are fully aware of the potential offered by interdisciplinary work and the integrating approaches sketched in Section 1.1.1, especially considering the complex subject matter at stake, Bickerton's position is not the one taken in this book. Rather, we believe that much more disciplinary research is required before it is possible to construct a viable overall theory of language evolution. Quite a number of hypotheses that are on the market have been based on interdisciplinary research—but many of them are of what in evolutionary psychology and biology tends to be referred to as the "just-so story" type (Eldredge 1995). That is to say, they are hypotheses that lack the ability to be testable at least in principle (see (11) below) on the basis of empirical evidence as it can be obtained from research provided by individual disciplines—Bickerton's (1990, 2005) own hypothesis being a case in point. Indeed, hypotheses based on evidence and the methodology of a single discipline are severely limited in scope; but they do at least allow some particular phenomenon to be considered in a consistent way, without having to rely on more or less conjectural reasoning across disciplines.

The goal of the present book is a modest one. By using one specific method of linguistic reconstruction, namely that of grammaticalization theory, we aim to find answers to at least some of the questions listed in (1) (see Section 7.3). This methodology is based on the following observations:

(2) Observations
    a. The development from early language to the modern languages is about linguistic change. Accordingly, in order to reconstruct it we need to know what is a possible linguistic change and what is not.
    b. An important driving force of linguistic change is creativity.
    c. Linguistic forms and structures have not necessarily been designed for the functions they presently serve.
    d. Context is an important factor determining grammatical change.[5]
    e. Grammatical change is directional.

---

[5] This concerns both the linguistic and the extra-linguistic context.

Much of the past and present work on the genesis of grammar relies on generalizations on synchronic language structure and does not take into account findings on how languages change, in particular which changes are possible ones and which are not. There is no indication that the principles of language change in early language were significantly different from the ones we observe in modern languages; hence, in accordance with (2a) we assume that early language can be studied on the basis of the same principles as modern languages. Conversely, a hypothesis on language evolution that is not in accordance with observations on change in modern languages is less plausible than one that is; we will return to this issue in Section 1.1.3.

Linguistic change may be viewed on the one hand as leading from one type of language to another, for example (with reference to some discussions that have been led on language evolution) from non-syntactic to syntactic language. On the other hand, it is taken to refer to a modification of individual properties or structures of languages. Our concern in this book is with the latter kind of change; however, given that there is an appropriate number or cluster of modifications of individual properties in some language or group of languages, it is possible to phrase such modifications in terms of linguistic change leading from one type of language to another.

A considerable portion of the recent literature on grammatical change highlights activities relating to language acquisition by children, learning, parsing, productivity, etc. While such work captures important aspects of linguistic behavior, it tends to underrate the significance of a factor that we consider to be central to many kinds of grammatical change, namely creativity, and (2b) constitutes a cornerstone of the methodology used in the present work. Creativity, as we understand it here, must not be confused with productivity, that is, with the use of a limited set of taxa and rules to produce a theoretically unlimited number of taxonomic combinations or structures. Rather than with conforming with rules or constraints, creativity in this sense is about modifying rules or constraints by using and combining the existing means in novel ways, proposing new meanings and structures. The perspective adopted is therefore of a different kind than that underlying the approaches alluded to above, and so is the methodology that will be used and, consequently, so also are the linguistic data that we will draw on when looking for answers to the questions in (1).

In a nutshell, the main thesis of the present book is that the emergence and development of human language is the result of a strategy whereby means that are readily available are used for novel purposes. Conceivably, this strategy is not an innovation of *Homo sapiens*, a manifestation of it might be seen, for example, in the tool-making abilities of some non-human primates. What is new, however, is that the strategy has been further developed, increasingly refined, and extended to a new domain of human behavior, namely linguistic communication.

Our concern here is less with the current utility of words or constructions but rather with what these entities have been designed for. Statement (2c) concerns an issue that has received considerable attention in evolutionary biology: Do categories serve the purpose for which they were designed? Answering this question is faced with the problem that there is not always agreement among scholars on how the functions of a given category are to be defined. However, there are some data that suggest an answer. To begin with, there is the general observation made independently in a number of studies that language change is a by-product of communicative intentions not aimed at changing language (see especially Keller 1994; Haspelmath 1999a). In fact, when a new functional category is created, there is nothing to suggest that this is what the speakers who are involved in this process really *intend* to happen.

The following example, which is characteristic of the way new auxiliaries for tense, aspect, or modality arise (see Section 2.2.3), may illustrate this (Heine and Miyashita 2006). In German, the item *drohen* 'to threaten' occurs both as a lexical verb, cf. (3a), and as an auxiliary having the syntax of a raising verb but the aspectual-modal meaning 'something undesirable is about to happen' (3b).

(3) German
    a. Karl   droht   seinem   Chef,   ihn   zu   verklagen.
       Karl   threatens to.his   boss   him   to   sue
       'Karl threatens his boss to take him to court.'
    b. Sein   Haus   droht   einzustürzen.
       his   house   threatens to.collapse
       'His house is about to collapse.'

Historical records show that (3a) represents the earlier situation, attested in Old and Middle High German, while (3b) is a later development of (3a) arising in Early New High German after 1700. But at no stage in the process from lexical verb to auxiliary can speakers of German be assumed to have intended to "design" a functional category or else a new auxiliary construction. The process started in the first half of the sixteenth century when instead of human agents, abstract nouns such as *Sünde* 'sin', *Urteil* 'verdict', *Gesetz* 'law', or *Tod* 'death' could be used productively as subject referents—in other words, when the use of the action verb *drohen* was extended metaphorically to inanimate concepts conceived as threatening agents. Subsequently, the use of the verb with inanimate subjects was generalized, where the lexical meaning 'someone acting intentionally' made no more sense. Consequently, the verb was desemanticized in such contexts and the aspectual-modal meaning surfaced. This triggered a number of decategorialization processes,[6] in that the erstwhile lexical verb lost most of its verbal properties, such as the ability to take auxiliaries or to control its subject—with the effect that by the mid-eighteenth century there evolved a functional category of the kind exemplified in (3b). In sum, the current utility of the auxiliary does not reflect the purpose for which it was designed, which was simply to treat threatening forces metaphorically like human agents.

Another, much discussed example concerns the French negation marker *pas*. In a sentence such as *Je (ne) sais pas* (I (NEG) know NEG) 'I don't know' it is a negation marker, but this is not what it used to be, or what it was designed for: *pas* was a noun meaning 'step' which was introduced as a device not specifically for negation but rather for efficiently supporting a negative predication[7] (cf. *He didn't move a step*), and it is only with the gradual decline of the erstwhile negation marker *ne* that *pas* assumed its present-day function as the primary or the only marker of negation. To conclude, the French negation marker serves a function that it was not designed for by earlier speakers of French. Once again, the original function—that of reinforcing another word—has little in common with the current utility of the item concerned. Even if one were to assume that there are cases where the product of grammatical change is exactly what the speakers concerned had intended to achieve, there are other cases showing that functional categories may change in

---

[6] Concerning the terms desemanticization and decategorialization, see Section 1.2.1.

[7] In the wording of an anonymous referee of an earlier version of this book, "the grammaticalization of Old French negation markers like *pas* < lat. *passu(m)* 'step', *mie* < lat. *mica(m)* 'crumb', *point* < lat. *punctu(m)* 'point' etc. started when they were used in context of scalar argumentation ('I will not move a step', 'I will not eat a crumb', 'I cannot even see a point', etc.). Here, they already serve the purpose of efficiently denying/negating".

such a way that they bear little resemblance to their original design. A paradigm case is provided by the grammaticalization of demonstratives as described by Greenberg (1978): The first step in this process is from demonstrative to definite article; subsequently the element may develop further to be used for indefinite reference, and in a final stage the erstwhile demonstrative may turn into a semantically (largely empty) marker of nominalization. To conclude, the evidence available suggests that the current utility of functional categories need not have anything to do with the motivations that speakers had when they "designed" them.

The example of the French negation marker *pas* alluded to above also illustrates (2d): It shows that it may be, and frequently is, the contextual or co-textual environment—in short, the context—which determines semantic and syntactic change: There is nothing in the meaning of the French noun *pas* 'step' that would suggest the meaning of negation. It was the use of this noun in one particular context that shaped the development from noun to "emphasizer" and finally to negation marker; otherwise, *pas* has remained until today what it used to be, namely a noun for 'step'. Another example to illustrate (2d) is the following. In many languages there is a grammatical distinction between an indicative and a subjunctive mood. While these two categories express contrasting meanings, it may happen that there is a change from indicative to subjunctive mood. The way this happens can be described as follows: In many languages there is a progressive aspect category expressing ongoing processes. Now, progressives tend to spread to contexts where they denote present tense and habitual events, eventually ousting an already existing present indicative tense construction in main clauses. But this old present tense construction can survive in subordinate clauses, and since subordinate clauses tend to be associated with subjunctive functions, the old construction can become a canonical marker of a new subjunctive mood. Languages that have been reported to have undergone a development along these lines include Modern Armenian, Tsakonian Greek, Modern Indic, Persian, and Cairene Arabic (Bybee, Perkins, and Pagliuca 1994: 230–6; Haspelmath 1998b: 41–5; Croft 2000: 127–8). This kind of change, which may also lead from present to future tense constructions, is not motivated by semantic or syntactic forces but rather by the pragmatic factor of context extension (see Section 1.2.1.2).[8]

---

[8] As pointed out by an anonymous referee of an earlier version of this book, this example concerns creativity and innovation only in a very indirect fashion.

Statement (2e) is a cornerstone of grammaticalization theory, which underlies the present work. For example, as we saw above, lexical verbs commonly develop into auxiliaries for tense, aspect, or modality, while it is unlikely that a tense auxiliary develops into a lexical verb. Similarly, demonstratives give rise to definite articles, and numerals for 'one' to indefinite articles, or body part nouns may give rise to adpositions (prepositions or postpositions), while it is unlikely that articles develop into demonstratives, adpositions into nouns, or auxiliaries into lexical verbs (but see Section 1.2.2).

1.1.3 The present approach

The procedure of reconstruction adopted here is the one summarized in (4).

(4) Reconstructing language evolution
    a. X and Y are phenomena that are related in some way.
    b. Hypothesis 1: X existed prior to Y.
    c. Hypothesis 2: There was a change X > Y (but X continues to exist parallel to Y).
    d. There is evidence in support of (4c).
    e. There are specific factors that explain (4c).

Examples (4a) through (4c) are in some form or other part of many approaches in the historical sciences. As has been demonstrated by Botha (2003a), (4d) is the most controversial component in studies on language evolution, but it is also the most crucial one. The factors that have been invoked to deal with (4e) have in many cases been taken from evolutionary biology, relating in particular to notions such as natural selection, adaptation, exaptation, mutation, etc. (see Section 1.1.1). For example, in their seminal paper on language evolution, Pinker and Bloom (1990) see adaptation by natural selection as the special factor of (4e):

> Language shows signs of complex design for the communication of propositional structures, and the only explanation for the origin of organs with complex design is the process of natural selection. (Pinker and Bloom 1990: 726)

As we noted in Section 1.1.1, however, the way these biological notions have been employed in studies on language evolution leaves a number of questions unanswered—we are not aware of any convincing way to account for natural selection in language evolution.

Over the past decades, the study of language genesis has been approached from a wide range of different angles and disciplines, many of which are not primarily linguistic in orientation. The approach used here on the basis of (4), that is grammaticalization theory, is linguistic in nature. It relies on regularities in the development of linguistic forms and constructions. Heine and Kuteva (2002b) claim that it is possible to push linguistic reconstruction back to earlier phases of linguistic evolution by using this theory. The purpose of the present book is to substantiate the claim made there by providing a wider range of crosslinguistic evidence, to propose a framework for reconstructing grammar and to relate this framework to some of the issues on early language that have been raised over the last decades.

We may illustrate this approach with an example from English. In the sentences of (5) there are two instances of the item *used*: In (5a), it has the function of a physical action verb, that is, of a lexical verb, while in (5b) it serves as an auxiliary verb expressing the aspectual notion of past habitual action.[9]

(5) English
    a. He used all the money.
    b. He used to come on Tuesdays.

From the history of English we know that the lexical use of *used* as in (5a) is older than the auxiliary *used* in (5b), and that the latter has in some way developed historically out of the former, in accordance with (4a) through (4c). This means that at some earlier stage in the history of English there was a lexical item *use* but no habitual marker *used to*.

That this is not an isolated case can be shown with two more examples, (6) and (7): In the (a)-sentences, *kept* and *is going* have the function of action verbs while in the (b)-sentences they serve as auxiliaries expressing aspect or tense functions: in the case of *kept* this is a durative aspect, and in the case of *is going* that of a future tense. Furthermore, as the history of the English language tells us, the (a)-uses of these items existed before the

---

[9] We are ignoring here the fact that the two *useds* are phonetically distinct, viz. [juzd] in (5a) vs. [just] in (5b) (Philip Baker, p.c.). This difference is a largely predictable product of the grammaticalization parameter of erosion, as we will see in Section 1.2.1.5. Furthermore, we are also ignoring the fact that there were various intermediate stages in the processes sketched here.

(b)-uses, and the latter can be traced back to the former, once again in accordance with the reconstruction model in (4).

(6) English
    a. He kept the money.
    b. He kept complaining.

(7) English
    a. He is going to town.
    b. He is going to come soon.

Such examples are not restricted to English; example (8) from French is strikingly similar to the English example (7): Both have a motion verb 'go' in (a), and in (b) there is a homophonous item denoting future tense, and in both cases, (b) can be shown to be historically derived from (a).

(8) French
    a. Il   va   à   la   maison.
       he   goes to the house
       'He's going home.'
    b. Il   va   venir   bientôt.
       he   goes to.come soon
       'He is going to come soon.'

But it is not only English and French that exhibit such examples; more examples can be found in hundreds of genetically unrelated languages[10] all over the world (see especially Bybee, Perkins, and Pagliuca 1994; Heine 1993; Kuteva 2001; see also Section 2.2.6 (b)), exhibiting the properties mentioned above as well as a number of others. A list of such properties is found in (9).[11]

(9) Generalizations
    a. There are two homophonous items A and B in language L, where A serves as a lexical verb and B as an auxiliary marking grammatical functions such as tense, aspect, or modality.
    b. While A has a noun as the nucleus of its complement, B has a non-finite verb instead.
    c. While A is typically (though not necessarily) an action verb, B is an auxiliary expressing concepts of tense, aspect, or modality.

---

[10] The expression "genetically unrelated languages" must be taken with care; it is a shorthand for "languages which so far have not been proven to be genetically related."

[11] Our reconstruction is based on attested cases of grammatical change, where we have written records and where these generalizations seem to hold. So we assume that these generalizations can be used for reconstructing cases where written records are missing.

d. B is historically derived from A.[12]
e. The process from A to B is unidirectional; that is, it is unlikely that there is a language where A is derived from B.
f. In accordance with (9d) and (9e), there was an earlier situation in language L where there was A but not B.

While properties (9a) through (9c) specifically refer to the data presented in the preceding examples, (9d) through (9f) are central to grammaticalization theory in general, and of the methodology used here in particular: They allow us to use a technique of linguistic reconstruction using the situation depicted in Figure 1.1 as a basis. In accordance with this technique, we will say that given a present situation where there are phenomena A and B, such as the ones presented in (9a–9e), we hypothesize that (9f) applies.

With reference to our English example in (5), (6), and (7) this means that in Modern English there are functional categories, exemplified by *used to*, *keep* V-*ing*, and *be going to*, that were not there at some earlier stage in the history of English, and that it is possible to reconstruct such an earlier stage on the basis of (9) even in the absence of any historical evidence.

This technique is not really new, it has been exploited in particular in another method of historical linguistics, namely that of internal reconstruction. However, unlike internal reconstruction, the technique as used here is not restricted to the analysis of language-internal processes; rather, it is comparative in nature and allows for reconstructions across languages[13] (see Section 1.2.1 below).

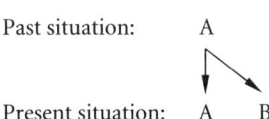

FIG. 1.1. Basis of reconstruction

---

[12] This generalization is in need of qualification. It hypothesizes a process from A to B that happened at some stage in the past, but it is possible that A is no longer exactly what it used to be at that stage.

[13] An argumentation that is in line with the present approach is found in some of Greenberg's works (Greenberg 1992: 154; see also Croft 1991).

More relevant for our purposes are works where the technique has been recruited—implicitly or explicitly—to deal with earlier situations in language evolution (e.g. Sankoff 1979; Comrie 1992; Aitchison 1996). Common to these works is the assumption that languages reveal layers of past changes in their present structure. Thus, Comrie (1992, 2002) argues that certain kinds of present linguistic alternation can be reconstructed back to earlier states without that alternation.[14]

This technique, and the methodology of which it is a part, will be described in Section 1.2. In the spirit of the works just cited we will endeavor to apply it to stages in the development of human language that are not accessible to other methods of historical linguistics. Our reconstruction work will be based on the following observations:

(10) Observations underlying reconstruction
 a. Grammaticalization theory offers a tool for reconstructing the rise and development of grammatical forms and constructions. It rests on generalizations about language change that happened in modern languages (see Section 1.2).
 b. There is no intrinsic reason to doubt that language change and the functional motivations underlying it were of the same kind in early language as what we observe in modern languages.[15]
 c. Accordingly, grammaticalization theory can be extended from modern languages to early language by extrapolating from the known to the unknown.
 d. Human language was structurally less complex at its earliest stage of evolution than modern languages are.

With the exception of (10a), which is supported by massive evidence (see Chapter 2), none of these observations is really unproblematic—they are bluntly speaking of the inferential-jump type. Statement (10b), and consequently also (10c), may be intuitively plausible and there does not appear to be any contradicting evidence, but there is no hard-core evidence in their

---

[14] It might be argued that this approach works in cases where appropriate historical evidence exists, but not necessarily in other cases. In other words, our claim that the presence of two structures A and B can be traced back to an earlier situation where there was A but no B cannot be generalized. It is a common analytic procedure, we argue, to describe the unknown in terms of the known given appropriate correlations between the two, and not much would be gained by rejecting such a generalization.

[15] In the wording of Comrie (2002: 256): "We propose no processes that are not attested in the historical period." We will return to this issue in Section 1.1.4.

support, either. And the same applies to (10d): While it echoes what has been claimed in much of the literature on language evolution within the last decades, there is no substantive evidence available to support it.

Another problem characterizing (10) is as follows: Whereas (10b) asserts sameness between early language and modern languages, (10d) maintains that there is a difference. That (10b) refers to "early language" and (10d) to the "earliest stage of evolution" does not solve this problem, since the latter is part of the former in accordance with our definition in Section 1.1.1. One may argue that there is a difference, in that (10b) concerns a process while (10d) concerns a state. Still, this would not be an entirely satisfactory solution either; we will return to this issue in Section 1.1.4.

Statement (10b) rests on the observation that the motivations underlying language use are crosslinguistically essentially the same, and that at any stage in the development of languages, the same processes of change were at work. For example, we observe that in modern languages there is a widespread phonetic change from plosive to fricative consonants (e.g. from p to f), from voiceless to voiced intervocalic consonants, or a morphosyntactic change from lexical verb to auxiliary, while a change in the opposite direction is clearly less likely. Accordingly, we argue that this justifies the assumption that such changes were also possible in early language. Recent research has shown that grammaticalization is regular across different kinds of linguistic communication systems, including signed languages (Janzen 1995, 1998, 1999; Janzen and Shaffer 2002; Sexton 1999; Wilbur 1999; Pfau 2004; Pfau and Steinbach 2005a, 2005b; Shaffer 2000), pidgin languages (see Chapter 4), as well as in situations of language contact (Heine and Kuteva 2003, 2005, 2006). Thus, we argue that hypotheses on language evolution that do not take (10b) into account (i.e. generalizations on diachronic processes to be observed in modern languages) are empirically less plausible, and we will ignore such hypotheses in the remainder of this work.

For example, there is little evidence in attested morphosyntactic changes for hypotheses that have been proposed on the basis of what may be called the holistic (or analytic) hypothesis: On this hypothesis, early language was characterized by holistic, monomorphemic linguistic signals conveying propositional contents,[16] and these signals are believed

---

[16] "From the viewpoint of present-day languages, every sentence of this early language would be fully suppletive" (Comrie 2002: 253).

to have undergone a segmentation process whereby complex but unanalyzable signals were broken down into words and syntactic structures (Wray 1998, 2000; Kirby 2002; Arbib 2005: 119). While we concur with the proponents of this hypothesis (e.g. Callanan 2006) that pragmatics is a crucial factor in language change, we are not aware of any diachronic evidence to the effect that such a segmentation process can commonly be found in language change: New grammatical categories do not normally arise via the reinterpretation of complex, unanalyzed propositions; accordingly, we consider this hypothesis to be less convincing for reconstructing language evolution (see also Section 7.1.1).

To be sure, it may happen that unanalyzable lexemes, such as *Watergate*, *Hamburger*, or *lemonade* in English, are segmented and give rise to new productive morphemes (*-gate*, *-burger*, *ade*), and folk etymology also provides examples to show that segmentation is a valid process of linguistic behavior. But such a process is, first, fairly rare (i.e. statistically negligible) and, second, it lacks the kind of regularity to be found in grammaticalization processes. Looking at the history of functional categories in English or other Indo-European languages it is hard to find grammatical elements that arose via segmentation of holistic expressions, while there is an abundance of examples of grammaticalization. Furthermore, the rise of new lexemes via compounding is a highly productive process in English and many other languages, while there does not appear to be a productive process in any language whereby mono-morphemic lexemes are split up into two or more new lexemes.[17] On account of such diachronic observations we side with Tallerman (2007) in assuming that the holistic hypothesis does not provide a convincing basis for reconstructing linguistic evolution.

In a similar fashion there are a number of other stimulating hypotheses that (10b) will prevent us doing justice to. One of them concerns what may be called in short the phonology-to-syntax hypothesis, which has been proposed to account for the rise of syntax: On this hypothesis, it was

---

[17] An anonymous reader points out that there are processes such as that from expressions for one-word utterances (e.g. German *ja!*, *doch!*, *schon!*) to modal particles (Waltereit 2001), which can be taken to lend support to the holistic hypothesis. While this is in fact a semantic process involving grammaticalization, it does not relate to what appears to be at the core of this hypothesis, namely that there is a segmentation process whereby complex but unanalyzable signals are broken down into words and syntactic structures. We do not wish to exclude the possibility that there may have been a holistic stage preceding early language, but those phases of language evolution that are accessible via grammaticalization theory are not really compatible with this hypothesis.

the neural organization underlying syllable structure, or phonology, that was co-opted to provide the syntax for strings of words (Lieberman 1984, 1998; Carstairs-McCarthy 1999: 147–8). As far as we are aware, there is no substantial evidence for this hypothesis in the history of modern languages, hence we will also ignore this hypothesis in the present work (for other problems with this hypothesis, see Bickerton 2005).

And much the same applies to hypotheses that invoke extra-linguistic forms of human behavior to account for language evolution, in particular the "Singing-Neanderthals" hypothesis, according to which there was a co-evolution of language and music (Mithen 2005). On this view, language evolved as a communication system distinct from music fairly late, at the *Homo sapiens* stage some 200,000 years ago, when language specialized in the transmission of information and music turned into a communication system for the expression of emotion. While there are in fact a number of interesting parallels shared by language and music, there is no convincing evidence to show that language change was affected by or can be related to musical behavior.

Circumstantial evidence for (10d) can be seen in the following observations: (a) As has been argued for by many authors on this subject, human language is the result of a direct transition from conceptualization and communication patterns of pre-human primates. It therefore must have passed through a stage where it was less complex than it is today (cf. Hurford 2003: 45, 49). (b) For good reasons, historical linguists tend to hold a "uniformitarian" position, according to which languages, as far back as they are recoverable via the methodology of historical linguistics, do not differ essentially in their degree of complexity. Still, there is at least some rudimentary evidence for a "non-uniformitarian" view according to which modern language was not always as complex as it is now; we will return to this issue in Section 1.1.4.

As the remarks just made suggest, the observations in (10) do not all have the same empirical status. Pinker and Bloom (1990) draw a distinction between two notions which are crucial for our approach: testability-in-principle and testability-in-practice (see Botha 2003a: 122):

(11) Two ways of testing a theory
    a. A theory (or hypothesis) is testable-in-practice if (a) it is testable-in-principle and (b) empirical data are in fact available with which its test-implications can be confronted.

b. A theory (or hypothesis) is testable-in-principle if (a) precise test-implications can be derived from it and (b) the empirical data can be specified which, if they are available, would indicate that the test-implications are false.

On the basis of this distinction, we can say that observation (10a) is in accordance with (11a), in that grammaticalization theory is testable-in-practice: It rests on generalizations on linguistic change which is immediately accessible and hence can be verified or falsified. But the remaining observations (10b) through (10d) are not: They concern hypotheses on early language, which is not immediately accessible in terms of historical analysis. It is possible to specify the kind of empirical data that, if they should become available, would indicate that the test implications can be falsified, but these hypotheses are not testable in practice.

Since the "inferential jump" from modern languages to early language that we alluded to above is not testable-in-practice, we require a hypothesis (or theory) responding to (11b)—one which licences inferential jumps on the relationship between two distinct phenomena, or, in other words, one which relates phenomena belonging to different ontological domains (namely those of modern languages and of early language) to one another in a principled way—that is, we need a **bridge hypothesis**. Following Botha (2003a: 147) we will say that such a hypothesis consists of an internally coherent set of assumptions which explicitly interlinks properties of entities of one ontological domain with properties of entities of another ontological domain. The hypothesis that we use rests on the observations proposed in (10).

In accordance with this hypothesis, the evidence that we will rely on is of a specific kind: It is based on observations made on the development of modern languages and our methodology rests exclusively on such observations. In other words, we will ignore observations that are not immediately accessible via this methodology. Underlying this methodology there is observation (10b) in that, with regard to language change, human behavior was essentially the same at the stage when early language arose as it can be observed to be today.

1.1.4 *On uniformitarianism*

It has been argued, implicitly or explicitly, that early language evolution was different in kind from modern language change, and that the study of the former therefore requires assumptions of a type that are not necessarily relevant to the study of the latter. Such discussions relate in particular to a notion that we consider relevant to any methodology employed for reconstructing early language, namely that of uniformitarianism. It is thanks to Newmeyer (2002, 2003) that this notion has received some attention in works on early language. The term has been used in a number of different senses, and these senses are not necessarily compatible with one another. To our knowledge, the following are senses of uniformitarianism that have been proposed in linguistics, which may be referred to as the U1, U2, and U3 assumptions.

(12) Assumptions of uniformitarianism
    U1 All modern languages are in some important sense equal.
    U2 Since the general structure of human languages of 5000 years back was about the same as it is today, it must also have been the same in early language.[18]
    U3 Linguistic change in early language was of the same kind as we observe in modern languages.

U1 is fairly widely accepted in contemporary linguistics. It is also the assumption adopted here, even if there are some languages that are claimed to lack salient properties of human languages, such as the Brazilian language Pirahã, which is said to show a lack of color words, numbers and counting, and recursion (Everett 2005; but see Section 6.1.6), or Riau Indonesian, which is said to lack syntactic categories (Gil 2001).

As we observed in (10) above, U3 is central to our framework (see also Christy 1983: 2). While assuming that early language differed in structure from modern languages, we argue that the processes of grammatical change were the same as those characterizing modern languages, both being shaped by parameters of grammaticalization.

U2 requires more detailed attention. Newmeyer (2002: 358) observes that most research into language origin and evolution has taken a position in accordance with U2, but he concludes that this position should be rejected:

---

[18] Comrie (2002: 252) provides the following analogy to question the logic underlying U2: "[F]rom the fact that my parents were humans, and that their parents were humans, and that their parents were humans, etc., one cannot logically conclude that there have been humans around from the beginning of time".

In other words, we have no reason to believe, and every reason to doubt, that the functionally motivated aspects of grammar have remained constant over time. (Newmeyer 2003: 75)

We side with Newmeyer's (2002, 2003) position on U2, in accordance with the approach described in Section 1.1.3. As we saw there, the perspective on language change that we adopt in this book is in accordance with U3, that is, it is based on the assumption that the processes of *language change* were the same in the past as they are in the present—hence, that it is possible to use generalizations on documented cases of language change to reconstruct earlier language states. And with reference to *language structure*, that is, with reference to U2, we also agree with Newmeyer in adopting a distinctly non-uniformitarian position: We hypothesize that early language had a structure that was different from the one characterizing modern languages—thereby following Comrie (2002: 257), who argues "that it is a valid exercise to reconstruct stages in the development of human language typologically different from, in particular less complex than, attested human languages".

That early language was typologically less complex than modern languages has in fact been maintained by many students of language evolution, and a number of scenarios have been proposed to account for this fact (see especially Givón 2002a, 2005; Jackendoff 2002; Johansson 2005, 2006). But this raises the question of how to reconcile this hypothesis with U3, according to which linguistic change was of the same kind throughout language evolution.

A possible piece of an answer is volunteered by Newmeyer (2002: 368–9), who argues that overall pre-literate societies make less use of clause subordination than literate societies. There is in fact some evidence to show that there are significant differences between spoken and written language use and reasons have been provided to account for this difference (e.g. Tannen 1982b: 3, 8; Chafe 1982: 39–45; Romaine 1992a: 147, 159; Harris and Campbell 1995: 310; Croft 2000: 83–4). Such observations might be taken to indicate that language use in pre-literate societies and/or in spoken communication reflects an earlier stage of language evolution. However, there are problems with such a view. First, the quantitative evidence in support of this view is still not entirely satisfactory. Second, as also acknowledged by Newmeyer (2002: 369), there are many pre-literate societies having languages with a remarkable amount of subordination and other forms of recursive structures (see e.g. König in prep.). In fact, a number of languages of traditional hunter-gatherer societies do not differ markedly with regard to subordination from languages of societies having had a long history of literacy—speakers of the Central Khoisan languages of southwestern Africa, for example, probably make more use of subordination than speakers of Italian or Chinese. And third, while it seems plausible that written discourse is more likely to invite clause subordination than spoken discourse,[19] this does not mean that written languages necessarily dispose of richer inventories of subordinating categories.

Our reason for rejecting assumption U2 is of a different nature. When introducing our methodology in Section 1.1.3 we observed that there are auxiliaries in English or other languages that can be historically derived in a principled way from lexical verbs. We therefore proposed a reconstruction procedure in (9), leading to the hypothesis that at some stage in early language there were lexical verbs but no auxiliaries. On the basis of this hypothesis we are led to conclude that early language at that stage was structurally less complex than modern languages since the former is hypothesized to have lacked a grammatical distinction that is present in modern languages.

But the situation is more complex. It is not for the first time that the process from verb to auxiliary that we discussed in Section 1.1.3 happened in English, or in other languages for that matter. New auxiliaries and other grammatical categories arise at all times in the history of a given language while old ones may co-exist with the new ones, or develop further into clitics and affixes, or disappear altogether, frequently—though not necessarily—being replaced by other new categories (see Section 2.5). And much the same applies to the many other grammatical categories that we are going to discuss in Chapter 2; accordingly, grammatical development in the modern languages is, at least to some extent, cyclical. The result is that modern languages are replete with grammatical categories, some of them less grammaticalized, others more strongly grammaticalized, and modern languages therefore exhibit a fairly high degree of grammatical complexity. Our concern in this book, however, is exclusively

---

[19] For example, Harris and Campbell (1995: 310) assert: "It is clear that hypotaxis is well suited to written language, for it provides a means for packing a great deal of information into a few words and has the potential of making the relationships among ideas specific. Complex hypotaxis is less appropriate in spoken language, since it places a greater burden on memory and processing ability".

with the situation that we hypothesize to have characterized early language, when these processes took place *for the first time*, that is, when there were, for example, verbs but no auxiliaries—hence, when human language was less complex than it is today. On the basis of this hypothesis, we see no reason to adopt assumption U2.

1.2 Grammaticalization

In his survey of studies on language evolution, Bickerton (2005: 6) claims that grammatical structures, being biologically based, cannot be added, changed, or deleted. Looking at the literature on historical linguistics that has accumulated over the last centuries, such a view would seem to be somewhat naïve.[20] To take a simple example: Indo-European languages are historically derived from a common, hypothetical ancestor, Proto-Indo-European. If Bickerton were right in his claim that grammatical structures cannot be changed then all Indo-European languages would have the same grammatical structure—that is, English would be structurally indistinguishable from Latin, Kurdish, or Hindi. Studies in grammaticalization show that change in grammatical structure is not only possible but is a predictable property of human languages and can be accounted for in a principled way. The present work is based on the application of grammaticalization theory, which relies on regularities in the change of linguistic forms and constructions, especially on the unidirectionality principle and the implications this principle has for the reconstruction of earlier language states.[21]

Grammaticalization is defined as the development from lexical to grammatical forms, and from grammatical to even more grammatical forms.[22] Since the development of grammatical forms is not independent of the constructions to which they belong, the study of grammaticalization is in the same way concerned with constructions, and with even larger discourse segments (see Traugott and Heine 1991a, 1991b; Heine, Claudi,

---

[20] Presumably Bickerton's notion of "grammatical structure" is not the same as the one commonly used in linguistic description. We are ignoring here the assumption that grammatical structure is biologically determined—an issue that is also not uncontroversial.
[21] For a critical assessment of this theory, see Section 1.2.2 below.
[22] For a fairly comprehensive list of definitions that have been proposed for grammaticalization, see Campbell and Janda (2001).

and Hünnemeyer 1991; Hopper and Traugott 1993; Bybee, Perkins, and Pagliuca 1994; Lehmann 1982; Heine 1997b; Kuteva 2001; Heine and Kuteva 2002a for details). In accordance with this definition, grammaticalization theory is concerned with the genesis and development of grammatical forms. Its primary goal is to describe how grammatical forms and constructions arise and develop through space and time, and to explain why they are structured the way they are. One main motivation for grammaticalization consists in using linguistic forms for meanings that are concrete, easily accessible, and/or clearly delineated to also express less concrete, less easily accessible and less clearly delineated meaning contents. To this end, lexical or less grammaticalized linguistic expressions are pressed into service for the expression of more grammatical functions.

That it is possible to push linguistic reconstruction back to earlier phases of linguistic evolution, that is, to phases where human language or languages can be assumed to have been different in structure from what we find today has been hypothesized in Section 1.1 (see also Heine and Kuteva 2002b; Smith 2006), and this will be the major concern of this book.

1.2.1 *Methodology*

Grammaticalization theory is concerned with regularities in language use as they can be observed in spoken and written linguistic discourse on the one hand and in language change on the other. It does not presuppose any formal theoretical constructs, such as a distinction between an E-language and an I-language, nor does it require assumptions to the effect that "language"—however this notion may be defined—is or should be conceived of as a system.

1.2.1.1 The parameters
Grammatical change has been described in terms of a wide variety of different models. In works on grammaticalization, the emphasis has been on two aspects of change. One aspect concerns semantics, in that this process is primarily one that leads from less grammaticalized to more grammatical meanings. The second aspect concerns pragmatics, and in particular the role of the frequency of use and context. In the handbook-like treatments mentioned above, attempts are made to reconcile these two aspects in some way or other. The methodology employed here (see

also Heine and Kuteva 2002a, 2005) rests on the assumption that grammaticalization is based on the interaction of pragmatic, semantic, morphosyntactic, and phonetic factors. There is a wide range of criteria that have been proposed (see e.g. Lehmann 1982; Heine, Claudi, and Hünnemeyer 1991; Hopper and Traugott 2003; Bybee, Perkins, and Pagliuca 1994); in our model it is the four parameters listed in (13). A wide range of alternative criteria have been proposed, such as syntacticization, morphologization, obligatorification,[23] subjectification, etc. We argue that they can be accounted for essentially with reference to these four parameters. Henceforth we will rely on these parameters, using them as a tool for identifying and describing instances of grammaticalization.

(13) Parameters of grammaticalization[24]
 a. extension, i.e. the rise of new grammatical meanings when linguistic expressions are extended to new contexts (context-induced reinterpretation)
 b. desemanticization (or "semantic bleaching"), i.e. loss (or generalization) in meaning content
 c. decategorialization, i.e. loss in morphosyntactic properties characteristic of lexical or other less grammaticalized forms
 d. erosion ("phonetic reduction"), i.e. loss in phonetic substance.

Each of these parameters concerns a different aspect of language structure or language use: (13a) is pragmatic in nature, (13b) relates to semantics, (13c) to morphosyntax, and (13d) to phonetics. Except for (13a), these parameters involve a loss in properties. But the process cannot be reduced to one of structural "degeneration"; there are also gains. In the same way as linguistic items undergoing grammaticalization lose semantic, morphosyntactic, and phonetic substance, they also gain properties characteristic

---

[23] Some students of this paradigm of linguistics argue that obligatorification, whereby the use of linguistic structures becomes increasingly more obligatory in the process of grammaticalization, should be taken as a definitional property of this process. As important as obligatorification is (see Lehmann 1982), it is neither a *sine qua non* for grammaticalization to take place, nor is it restricted to this process, occurring also in other kinds of linguistic change, such as lexicalization. Within the present framework, obligatorification—as far as it relates to grammaticalization—is a predictable by-product of decategorialization.

[24] Our use of the term "parameter" must not be confused with that found in some formal models of linguistics. For an alternative account of grammaticalization within a Chomskyan Minimalist framework, see van Gelderen (2004), where this process is described in terms of economy principles, entailing in particular a syntactic shift from specifier to head, e.g. from main verb to auxiliary, from demonstrative to definite article, etc.

of their uses in new contexts—to the extent that in some cases their meaning and syntactic functions may show little resemblance to their original use.

The ordering of these parameters reflects the diachronic sequence in which they typically apply: Grammaticalization tends to begin with extension, which triggers desemanticization, and subsequently decategorialization and erosion. Erosion is the last parameter to be involved; as we will see below, in many of the examples to be presented in this book, erosion is not (or not yet) a relevant parameter.

1.2.1.2 Extension

Of all the parameters, extension is the most complex, for two reasons. First, it has a sociolinguistic, a text-pragmatic, and a semantic component. The sociolinguistic component concerns the fact that grammaticalization starts with innovation (or activation) as an individual act, whereby some speaker (or a small group of speakers) proposes a new use for an existing form or construction, which is subsequently adopted by other speakers, ideally diffusing throughout an entire speech community (propagation; see e.g. Croft 2000: 4–5). The text-pragmatic component involves the extension from a usual context to a new context or set of contexts, and the gradual spread to more general paradigms of contexts (for an example of the dramatic effects that this process may have, see Section 1.1.3). The semantic component finally leads from an existing meaning to another meaning that is evoked or supported by the new context. Thus, text-pragmatic and semantic extension are Janusian sides of one and the same general process characterizing the emergence of new grammatical structures.

Second, the term extension has been used in a variety of different ways and in different frameworks. The most relevant use of it as a technical term is that by Harris and Campbell (1995), who refer to it as what they consider to be one of the three basic mechanisms of syntactic change.[25] In their model, extension has the effect that a condition on an existing rule is

---

[25] The other two mechanisms are reanalysis and borrowing. In more general terms, extension is defined by these authors "as change in the surface manifestation of a syntactic pattern that does not involve immediate or intrinsic modification of underlying structure" (Harris and Campbell 1995: 97). The use of "reanalysis" in the model of these authors overlaps with that of grammaticalization but should not be confused with the latter. In order to avoid confusion, we will not use the term "reanalysis" in this work.

removed. While this use differs in a number of respects from ours, there is considerable overlap, as is suggested by the examples given by these authors, some of which also satisfy our criteria of extension. What distinguishes the two mainly is, first, that in our model extension is framed in terms of pragmatic notions rather than of syntactic rules, second, that we restrict the term to changes where a given linguistic expression acquires new contexts of use while Harris and Campbell use extension in a more general sense, and third, we argue that extension entails some change in meaning, however minute this change may be (see under semantic generalization below), while meaning does not play a central role in Harris and Campbell's model.

A number of approaches have been used to deal with the phenomena relating to extension (see e.g. Bybee, Perkins, and Pagliuca 1994; Traugott and Dasher 2002: 34–9); in the present work we will use the four-stage model of context-induced reinterpretation depicted in Table 1.1 to describe the most salient characteristics of extension (see Heine 2002b for more details).

Table 1.1 suggests that the transition from a less grammatical (e.g. lexical) meaning of stage I to a more grammatical meaning of stage IV does not proceed directly; rather, it involves two intermediate stages, viz. stages II and III. A central question that arises is the following: What is the factor that is responsible for a new context triggering a new meaning (stage II) or backgrounds an existing meaning (stage III)? Available research findings suggest that there are two possible factors: (a) semantic generalization, whereby new contexts entail a more general meaning, and (b) invited inferencing, in that new contexts suggest new meanings.

Semantic generalization is the one most generally involved (Bybee, Perkins, and Pagliuca 1994). We may illustrate this with the following example. In (14a), the item *in* serves its canonical function as a spatial preposition, and much the same applies to (14b). But when *in* is extended to contexts such as (14c),[26] its meaning can no longer be reduced to that of a spatial preposition, that is, it has become more general—in other words, the more contexts of use a linguistic expression acquires, the more it tends to lose in semantic specificity and to undergo semantic generalization (that

---

[26] These contexts do not exhaust the number of invited inferences associated with *in*; for example, in the following sentence there is a kind of manner inference (Fritz Newmeyer, p.c.): *John died in pain.*

TABLE 1.1. *A model of extension (context-induced reinterpretation)*

Stage	Context	Resulting meaning	Type of inference
I Initial stage	Unconstrained	Source meaning	—
II Bridging context	There is a new context triggering a new meaning	Target meaning foregrounded	Invited (cancellable)
III Switch context	There is a new context which is incompatible with the source meaning	Source meaning backgrounded	Usual[a] (typically non-cancellable)
IV Convention-alization	The target meaning no longer needs to be supported by the context that gave rise to it; it may be used in new contexts	Target meaning only	—

[a] The term "usual" inference is adopted from Paul ([1880] 1920). Hopper and Traugott (2003: 80–1) use the term conventional implicature instead. We prefer not to refer to this inference as "conventional" since it is entirely dependent on the context and, hence, not conventionalized to the extent that it constitutes an independent grammatical meaning (see stage IV).

is, desemanticization; see below). Thus, semantic generalization (or bleaching) is an obligatory consequence of extension, and Haspelmath (1999a: 1062) goes so far as to argue that semantic generalization is in a sense the cause of the other processes of grammaticalization.

(14) English
    a. John died **in** London.
    b. John died **in** Iraq.
    c. John died **in** a car accident.

Invited inferencing is less general as a factor, even if it is involved in many instances of grammaticalization (see Traugott and Dasher 2002: 34–40).

It has the effect that, due to a specific context, a new meaning is foregrounded (stage II), even if it is still cancellable. But it may happen that the invited inference turns into a usual one in specific contexts—with the effect that the old meaning is backgrounded (stage III). Example (14) can illustrate this factor: In (14a), the item *in* serves its canonical function as a spatial preposition. This also applies to (14b), but on account of encyclopaedic (world) knowledge, in this case of contemporary history, (14b) may trigger an invited inference to the effect that the location expressed by *in* may have been causally responsible for the event described by the verb—hence a stage-II interpretation is possible. However, this interpretation is always cancellable, in that the speaker may insist that that particular location was not causally responsible for John's death. In a context like (14c), the causal inference is the usual one, in that an interpretation of *in* as a spatial or temporal preposition makes little sense—hence that interpretation is backgrounded and the only reasonable meaning of *in* is a causal one. While it is still theoretically possible to cancel the causal inference, this would be highly unusual.

It has been argued or implied that the main trigger of grammaticalization is frequency of use: The more often a given form or construction occurs, the more likely it is that it will reduce in structure and meaning and assume a grammatical function (Bybee 1985; Diessel 2005: 24; see especially the contributions in Bybee and Hopper 2001). In fact, extension to new (sets of) contexts implies a higher rate of occurrence of the items concerned, and more probable words are more likely to be reduced than less probable ones (cf. the Probabilistic Reduction Hypothesis of Jurafsky *et al.* 2001). Furthermore, when a grammatical item whose use is optional is used more frequently, its use may become obligatory. Nevertheless, we have found neither compelling evidence to support the hypothesis that frequency is immediately responsible for grammaticalization, nor that grammatical forms are generally used more frequently than their corresponding less grammaticalized cognates.

Two examples may illustrate this. A survey of a larger body of instances of grammaticalization suggests that, overall, non-grammaticalized items that serve as the source of grammaticalization do not necessarily belong to the most frequently used words of a language (see e.g. Heine, Claudi, and Hünnemeyer 1991: 38–9), nor are grammaticalized items necessarily used more frequently than their non-grammaticalized counterparts. The other example is of a different nature: As we saw in Section 1.1.3, the German verb *drohen* 'to threaten' has given rise to an aspectual-modal auxiliary meaning ("something undesirable is about to happen"). In Present-Day German, *drohen* is a main verb in 65.6 percent (173 instances) but an auxiliary only in 34.4 percent (96 instances) of its uses. In a similar fashion, the German verb *versprechen* 'to promise' has been grammaticalized to an aspectual-modal auxiliary ("something desirable is likely to happen soon") and, once again, the lexical uses (98.2 percent, 636 instances) clearly outnumber the auxiliary uses (1.8 percent, 12 instances) (Askedal 1997: 17; Heine and Miyashita 2006). To conclude, frequency of use appears to be an epi-phenomenal product of extension rather than being a trigger of it; neither do non-grammaticalized items that serve as the target of grammaticalization necessarily belong to the most frequently used words of a language (see e.g. Heine, Claudi, and Hünnemeyer 1991: 38–9), nor are grammaticalized items necessarily used more frequently than their non-grammaticalized counterparts.

1.2.1.3 Desemanticization

Desemanticization is an immediate consequence of extension: Use of a linguistic expression E in a new context C entails that E loses part of its meaning that is incompatible with C—in other words, the two are (as we observed in 1.2.1.2) Janusian sides of one and the same process.[27]

Desemanticization is frequently triggered by metaphoric processes (Lakoff and Johnson 1980; Lakoff 1987). For example, a paradigm case of grammaticalization involves a process whereby body part terms ('back', 'head', etc.) are reinterpreted as locative adpositions ('behind', 'on top of') in specific contexts, cf. English *in front of*. Via metaphorical transfer, concepts from the domain of physical objects (body parts) are used as vehicles to express concepts of the domain of spatial orientation, while desemanticization has the effect that the concrete meaning of the body parts is bleached out, giving way to some spatial schema. In a similar fashion, when an action verb (e.g. English *keep, use, go to*) is reinterpreted as a tense or aspect auxiliary (see Section 1.1.3), this can be understood to

---

[27] This view is at variance with that of Hopper and Traugott who argue that desemanticization ("weakening of meaning" in their terminology) is not involved in the beginnings of grammaticalization—that is, that it follows extension (or "pragmatic strengthening"): "There is no doubt that, over time, meanings tend to become weakened during the process of grammaticalization. Nevertheless, all the evidence for early stages is that initially there is a redistribution or shift, not a loss, of meaning" (Hopper and Traugott 2003: 94).

involve a metaphorical process whereby a concept of the domain of physical actions is transferred to the more abstract domain of temporal and aspectual relations.[28] Once again, this leads to the desemanticization of lexical meaning, namely that of the action verbs (Heine, Claudi, and Hünnemeyer 1991; Heine 1997b).

Rather than desemanticization, Hopper and Traugott (2003) prefer to describe the semantic development in terms of notions such as invited inferences, subjectification, or pragmatic strengthening. It would seem, however, that desemanticization is a fairly predictable component of the process while neither invited inferences, subjectification, nor pragmatic strengthening are: There are many grammaticalization processes that involve neither inferencing nor subjectification, nor strengthening and, conversely, there may be inferencing without there necessarily being desemanticization (Fritz Newmeyer, p.c.).

1.2.1.4 Decategorialization

Once a linguistic expression has been desemanticized, for example from a lexical to a grammatical meaning, it tends to lose morphological and syntactic properties characterizing its earlier use but being no longer relevant to its new use. In this way, a number of English verbs have been desemanticized in their gerundival form (-*ing*) and assumed prepositional functions, for example *barring, concerning, considering,* etc. (see Chapter 2). Consequently, they lost most of their verbal properties, such as to be inflected for tense and aspect, to take auxiliaries, etc. (see Kortmann and König 1992). Decategorialization entails in particular the changes listed in (15):

(15) Salient properties of decategorialization
    a. Loss of ability to be inflected.
    b. Loss of ability to take on derivational morphology.
    c. Loss of ability to take modifiers.
    d. Loss of independence as an autonomous form, increasing dependence on some other form.
    e. Loss of syntactic freedom, e.g. of the ability to be moved around in the sentence in ways that are characteristic of the non-grammaticalized source item.

---

[28] This is a simplified rendering of the process concerned; see our example of German *drohen* in Section 1.1.2, which gives a more detailed description of such a process from lexical verb to auxiliary.

f. Loss of ability to be referred to anaphorically.

g. Loss of members belonging to the same grammatical paradigm.

In accordance with this list, nouns undergoing decategorialization tend to lose morphological distinctions of number, gender, case, etc.; the ability to combine with adjectives, determiners, etc.; to be headed by adpositions; they lose the syntactic freedom of lexical nouns; and the ability to act as referential units of discourse.[29] In a similar fashion, when a demonstrative develops into a clause subordinator, as has happened in many languages of the world (see Chapter 5), it loses salient categorical properties. For example, the English demonstrative *that* is sensitive to number, having *those* as its plural form, it is fairly autonomous in that it can be used both as an attribute of a noun or as a pronoun, and it belongs to a morphological paradigm which also includes *this.* In its grammaticalized form as a relative clause marker, however, it is decategorialized, having lost this distinction (*The books that/*those I know*), it has lost the distinction between use as an attributive and a pronoun, and it has lost *this* as a co-member of the same paradigm.

Verbs undergoing decategorialization tend to lose their ability to be inflected for tense, aspect, negation, etc., to be morphologically derived, to be modified by adverbs, to take auxiliaries, to be moved around in the sentence like lexical verbs, to conjoin with other verbs, to function as predicates, and to be referred to (e.g. by pro-verbs). Finally, they lose most members of the grammatical paradigm to which they belong by changing from open-class items to closed-class items.

In more general terms, decategorialization tends to be accompanied by a gradual loss of morphological and syntactic independence of the linguistic item undergoing grammaticalization, typically proceeding along the scale described in (16) (see also Section 2.4).

(16) Free form > clitic > affix

But decategorialization is not restricted to open-class items such as nouns and verbs, it affects in the same way closed-class items, which are also likely to lose their categorial properties; as we just observed, demonstratives in many languages show distinctions in number, gender, and/or case, or between pronominal and attributive functions; such distinctions tend to disappear when decategorialization takes place.

---

[29] For an example involving English *while*, see Hopper and Traugott (2003: 107).

The generalizations just presented were hedged in the form of "tend-to" predications, for the following reason: The changes listed are not necessarily all present in a given case of decategorialization. For example, on their way from verb to auxiliary, English items such as *be going to* or *keep*, that we discussed in Section 1.1.2, have not lost their ability to combine with other markers of tense, aspect, and modality (e.g. *Mary would have kept coming*), and in the development from demonstrative to definite article *der/die/das* in German, distinctions in number, gender, and case were not lost either. There can be a number of different causes for decategorialization not taking place. One concerns language-internal factors: There may be specific structural constraints that prevent the loss of some categorical property. For example, verbal auxiliaries in many languages are the only grammatical category in the clause where distinctions of personal deixis, tense, aspect, or negation are encoded; thus, giving up these encodings might have dramatic consequences for the information structure of the clause. Another cause concerns the age of the grammaticalization process involved: Decategorialization does not happen overnight, that is, it takes some time to come in, and the younger a process is, the lower the amount of loss in categorical properties will be.

1.2.1.5 Erosion
Erosion means that as a result of undergoing grammaticalization, a linguistic expression loses phonetic substance. As we observed above, this parameter is usually the last to apply in grammaticalization processes, and it is not a requirement for grammaticalization to happen. For example, the German verb *haben* 'to have, possess' has been grammaticalized to a perfect aspect (and also a past tense) marker. But in spite of its history of nearly one millennium as a tense–aspect auxiliary it has not undergone erosion, being phonetically indistinguishable from the lexical verb; and much the same applies to its Latin equivalent *habēre* 'to hold, possess', which has also given rise to a tense–aspect auxiliary in the Romance languages, but its form as a verb of possession and as an auxiliary for example in French is the same[30] (Vincent 1995: 437).

---

[30] This does not apply to the development of Latin *habēre* as a future tense auxiliary, which has been affected by all parameters of grammaticalization, including erosion, in almost all of the Western Romance languages (Portuguese being an exception; anonymous referee).

Erosion can be of two kinds: First, it may involve entire morphological units. Thus, when the Old English phrase *þa hwile þe* 'that time that', or any of its variants,[31] was grammaticalized to the temporal and concessive subordinator *while* in Modern English, this meant that morphological segments were lost, and much the same happened in the case of its Old High German counterpart *al di wila daz* 'all the time that', which was grammaticalized to the causal subordinator *weil* 'because' in Present-Day German, which is also characterized by loss of morphological elements. We will refer to such cases as morphological erosion (see Heine and Reh 1984). More commonly, however, change is restricted to phonetic erosion, that is, to phonetic properties, in particular the ones listed in (17), or any combination thereof.

(17) Kinds of phonetic erosion
   a. Loss of phonetic segments, including loss of full syllables.
   b. Loss of suprasegmental properties, such as stress, tone, or intonation.
   c. Loss of phonetic autonomy and adaptation to adjacent phonetic units.
   d. Phonetic simplification.

The development in English from *because* to colloquial English *coz* is an instance of (17a): It entailed both loss of phonetic segments and reduction from a disyllabic to a monosyllabic unit. Since the eroded form *coz* does not occur in all varieties of English, there is reason to assume that erosion was a process that took place subsequent to desemanticization and decategorialization. The grammaticalization of the English adjective *full* to the derivational suffix *-ful* illustrates (17b), in that it entailed a loss of the ability to be stressed. (17c) can be illustrated with the following example. In the West African language Maninka, a locative construction [X be at Y], illustrated in (18a), has been grammaticalized to a progressive aspect (18b), as has happened in many other languages (Heine 1993; Bybee, Perkins, and Pagliuca 1994; Kuteva 2001). The result is that the copula *yé* (PM) and the locative postposition *ná* 'at' have been reinterpreted as a discontinuous progressive marker. Another result of this process is that *ná* has lost its phonetic autonomy: While it is invariable as a postposition, as a progressive marker it no longer is: It

---

[31] Such variants being *ðe hwile ðe, þa hwile þe, þa hwile þa, þa hwila þe, a hwile ðæ, ðe hwile ðæt*, etc.

assimilates to the preceding verb, taking the form *na* when the verb ends in a nasal but *la* elsewhere, and it has lost its high tone, adopting the tone of the preceding verb.

(18) Maninka (Mande, Niger-Congo; Friedländer 1992: 51, 67; quoted from Tröbs 1998: 37)
a. Seku yé bún ná.
   Seku PM house at
   'Seku is at home.'
b. à yé kàrán ná.
   3.SG PM learn at
   'He is learning, he learns.'

A paradigm case of (17d) concerns phonetically complex sound units that in the process of grammaticalization lose phonetic properties. For example, many West and central African languages have labial-velar consonants, such as *kp* and *gb*, which are phonetically more complex than the corresponding labial (*p, b*) or velar (*k, g*) consonants. Accordingly, in the Ewe language of Ghana and Togo, *gb* was "simplified" to a corresponding velar *g* in some words that underwent grammaticalization, for example when the noun *gbé* 'location, direction' gave rise to an ingressive aspect marker *-gé*, or the verb *gbɔ* 'to return' to the repetitive aspect prefix *-ga* (Heine 1993: 107).

In quite a number of cases, both morphological and phonetic erosion tend to be involved. For example, the grammaticalization of the phrase *by the side of* to the preposition *beside* in Modern English, or of *by cause of* to *because (of)* appear to have involved both morphological and phonetic erosion. Similarly, the development of the Latin phrase *(in) casa* 'in the house (of)' via Old French *(en) chies* 'at' to the Modern French locative preposition *chez* 'at' involved loss of both morphological and phonological substance.

In accordance with the definition provided above, we are using erosion as a technical term of grammaticalization theory, which needs to be distinguished from "phonetic reduction," for the following reasons: First, phonetic reduction does not necessarily involve grammaticalization; for example, the English copula verb *'s* in *She's a teacher* is a phonetically reduced version of *is*, but this is not due to grammaticalization. And second, erosion also differs from phonetic reduction in that it includes a loss of morphological elements, as we just saw.

1.2.1.6 Discussion

In the course of the past decades, a number of alternative approaches have been proposed to deal with grammaticalization phenomena. Some of these approaches highlight the pragmatic aspects of the process (Bybee and Hopper 2001), others focus on semantic aspects (Traugott and Dasher 2002), and still others seek to account for the process in terms of syntactic frameworks (Roberts and Roussou 2003; van Gelderen 2004; Kiparsky 2005). On the basis of the four parameters discussed in the preceding paragraphs, grammaticalization can be portrayed more generally as a process leading

(a) from concrete meanings to more abstract ones,[32]
(b) from fairly independent, referential meanings to less referential, schematic grammatical functions having to do with relations within the phrase, the clause, or among clauses,
(c) from open-class to closed-class items,
(d) from grammatical forms that may have internal morphological structure to invariable forms, and
(e) from longer grammatical forms to shorter ones.

Part of our reconstruction methodology is well known from what in historical linguistics is known as the method of internal reconstruction. When there is a development from category A to B, certain A-properties are likely to survive, while others will be replaced by B-properties. In specific cases, the presence of A-properties associated with a given B-category can be interpreted meaningfully only if there has been an earlier A. Such surviving A-properties can be used as evidence to reconstruct an earlier A.

But grammaticalization theory is not confined to internal reconstruction. We may illustrate this with the following example from English: The definite article freely combines with both singular nouns (*the tree*) and plural nouns (*the trees*) whereas the indefinite article combines with singular nouns (*a tree*) but not with plural nouns (**a trees*). We know that the indefinite article is historically derived from the numeral *one*, and it has retained some relics of its etymological source. One such relic is its

---

[32] Conceptual shift from concrete to abstract, as understood here, is anthropocentric in nature, in that it leads from meanings that are close to human experience and easy to describe, to meanings that are more difficult to understand and describe.

incompatibility with plural nouns,[33] another one can be seen in the fact that the indefinite article has retained the nasal consonant when preceding vowels (*an apple*). Accordingly, surviving properties such as the inability to qualify plural nouns and the presence of a nasal allow us to hypothesize that the indefinite article is historically derived from some element that had these properties. This is as far as internal reconstruction can go. Grammaticalization methodology can go one step further: We have a number of cases of attested developments of numerals for "one" developing into indefinite articles, so we can safely assume a parallel development in those unattested cases. At the same time, since no language has been found where an indefinite article has given rise to a numeral, we can strengthen the hypothesis by arguing that the English indefinite article must have had the numeral as its historical source (Fritz Newmeyer, p.c.).

But even if we had no diachronic knowledge of the fact that there is a universal process from numeral "one" to indefinite article, we would still be able to reconstruct further details of the process on the basis of the parameters described in the preceding sections. Whereas the numeral has a fairly concrete lexical meaning, the indefinite article has only a functional meaning, hence it is semantically "impoverished" vis-à-vis the numeral (desemanticization). Furthermore, the indefinite article is syntactically more restricted: It is a clitic that cannot occur without a following noun (**I have a.*), whereas the numeral "one" is a free word that can occur on its own (*I have one.*). Now, according to one of these parameters, decategorialization, free words tend to lose their independent status and to turn into clitics (sometimes further into affixes). Finally, the more a linguistic item is grammaticalized, the more it tends to lose in phonetic substance (erosion). In accordance with these three parameters, we will hypothesize that the indefinite article can be traced back in a principled way to an element that had a more concrete meaning, was a free form, and had more phonological substance than it now has.

1.2.2 Problems

Grammaticalization theory has been the subject of critical analysis, which centered mainly around two issues, namely its status as a theory and the

---

[33] As an anonymous referee of this book rightly points out, there are some exceptions, such as Spanish, Old French, and Bulgarian (see Heine 1997b).

significance of the unidirectionality principle. We will deal with each of these issues in turn.

That grammaticalization theory does not qualify as a theory has been claimed in particular by Newmeyer (1998a): He observes that this paradigm involves parameters (or mechanisms) relating to different components of language structure, that is, semantics (desemanticization), syntax (decategorialization), and phonetics (erosion), and that none of these parameters or mechanisms is restricted to grammaticalization; rather, each of them is also relevant to other kinds of language change. Accordingly, he concludes that grammaticalization is not a distinct process (Newmeyer 1998a; see also Campbell 2001).

While this observation is in fact true, the conclusion drawn by Newmeyer is not, for the following reason: Grammaticalization theory accounts for the development and structure of functional categories, and to this end it is concerned with the *interaction* of the three parameters (as well as of extension), rather than with each parameter individually. Therefore, to the extent that these parameters jointly provide a tool for describing and explaining the rise, development, and structure of functional categories through space and time, and to understand why they are structured the way they are, there is a distinct process that can be accounted for by means of a distinct theory. Note, however, that this is neither a theory of language nor of language change; its goal is restricted to the task just mentioned.

The second issue concerns the validity of the unidirectionality principle.[34] Grammaticalization theory relies on regularities in the evolution of linguistic forms and constructions, especially on this principle and the implications it has for the reconstruction of earlier language states. Still, this process is not without exceptions. A number of examples contradicting the unidirectionality principle have been proposed (see especially Joseph and Janda 1988; Campbell 1991; Ramat 1992; Frajzyngier 1996; and especially Newmeyer 1998a: 260ff.), usually referred to as instances of "degrammaticalization". A number of these examples, however, are controversial or have been shown by subsequent research not to contradict the unidirectional principle (see e.g. Haspelmath 2004b; Andersson 2005). While we lack reliable statistics, two authors have come up with more

---

[34] For an explanation of unidirectionality in terms of "rhetorical devaluation", see Haspelmath (1999a: 1059–61).

specific estimates on the relative frequency of grammaticalization vis-à-vis other processes. Newmeyer (1998a: 275–6, 278) argues that only about 90 percent of grammatical changes are in accordance with the unidirectionality principle (i.e. are „downgradings" in his terminology), and Haspelmath (1999a: 1046) says that about 99 percent of all shifts along the lexical/functional continuum are grammaticalizations. On a conservative estimate then, at most one tenth of all grammatical developments can be suspected to be counterexamples to the unidirectionality principle.

Three different stances have been taken to deal with "degrammaticalization". First, it has been argued that since there are some cases of "degrammaticalization", the unidirectionality hypothesis is false (Campbell 2001). Second, this hypothesis is largely though not entirely true; it takes care of a robust tendency of grammatical change (Haspelmath 1999a; Hopper and Traugott 1993, 2003). And third, the hypothesis is true and cases of presumed "degrammaticalization" can be accounted for by means of alternative principles. Principles that have been invoked are on the one hand morphosyntactic, like exemplar-based analogical change (Kiparsky 2005), and on the other hand cognitive and communicative forces, like lexicalization, euphemism, exaptation, and adaptation as proposed by Heine (2003).

Considering that no instances of „complete reversals of grammaticalization" have been discovered so far (Newmeyer 1998a: 263), we take the unidirectionality principle to provide a solid basis for linguistic reconstruction. In this respect, grammaticalization theory is comparable to the comparative method, which has also been used extensively for linguistic reconstruction on the basis of crosslinguistic regularities to be observed in sound change: Depending on what kind of evidence one decides to draw on, it seems safe to say that the quantitative magnitude of irregularities in sound change and sound correspondences is not drastically different from that to be found in grammaticalization.

But there is one difference between the two methodologies which is crucial for the purposes of the present book: While both draw on attested cases of language change that occurred roughly within the last five millennia, this exhausts the potential of the comparative method. As we hope to demonstrate in the following chapters, grammaticalization theory does not have this limitation: By extrapolating from findings on regular grammatical development to situations of unattested stages of language development, it is able to reconstruct earlier processes of grammatical development—that is, phases in the evolution of human languages that are not accessible to other methods of historical linguistics.

Newmeyer's reservations But the potential our framework has for reconstructing early language has not received full recognition in the specialized literature. To our knowledge, this framework, first presented in Heine and Kuteva (2002b), has not been challenged so far—with one exception: In his seminal papers on uniformitarianism in linguistic evolution, Newmeyer (2002: 366, 2003: 63) argues that there is a problem with this framework, which he describes thus:

The entire progression from full lexical category to affix can take fewer than 2,000 years to run its course. As they [= Heine and Kuteva 2002b; a.n.] note, if there were no processes creating new lexical categories, we would be in the untenable position of saying that languages remained constant from the birth of *Homo sapiens* until a couple of millennia ago at which point the unidirectional grammaticalization processes began. (Newmeyer 2002: 366)

There are a number of questions raised by this passage. First, to our knowledge, neither Heine and Kuteva (2002b) nor any other student of grammaticalization has ever proposed generalizations on the creation of new lexical categories, or argued that languages remained constant in any phase of their history. Second, there is no reason why unidirectional grammaticalization processes should have begun a couple of millennia ago. There is sufficient evidence, for example from Hittite, Ancient Egyptian, Akkadian, or Chinese (Deutscher 2000; Heine and Kuteva 2002a), to show that these processes occurred much earlier and were of the same kind as we observe them today. Third, there is also no reason to assume that the processes creating new lexical categories were any different two or five thousand years ago from what they are today—again, there is documented historical evidence. To conclude, on the basis of findings on grammaticalization we do not see any analytic or theoretical reason for dividing the evolution of human language into two phases: one that covers the last two thousand years, or any larger period for that matter, and another for the rest of the evolution.[35]

---

[35] To be sure, future research might reveal that there is justification to divide the evolution of grammar into two salient phases, say, an earlier phase characterized by some specific state of grammaticalization and a later one showing a different state of grammaticalization; for the time being, however, we see no reason to assume such a division.

Even if we were to adopt the position that prior to the last two millennia, or at some earlier point in time, there were no processes of creating new lexical categories, this would not mean that there was no grammaticalization or, if there was, that this would have had serious implications for the structure of the lexicon. Our concern here is not primarily with the lexicon; however, a few general remarks on the role of the lexicon are in order. When a verb is grammaticalized to, say, some tense marker then this need not, and frequently does not lead to the loss of that verb; cf. Hopper's (1991) principle of divergence. Thus, the grammaticalization of *keep* to an aspectual auxiliary did not have any implications for the lexical structure of English—*keep* is still available as a lexical item. In a similar fashion, the English possessive verb *have* (19a) has given rise to a number of grammaticalizations: It developed into a verbal aspect marker roughly a millennium ago (19b), and it also provided the conceptual template for a number of additional grammaticalizations, like that as a functional category of deontic modality (19c). Nevertheless, *have* is still available as a possessive verb in present-day English, that is, these grammaticalizations did not affect the English lexicon in any significant way.

(19) English
    a. They have no children.
    b. They have left.
    c. They have to pay.

To be sure, it may happen that with the grammaticalization of a given lexical item, that item will disappear from the lexicon—English *will* being a case in point: The verb of volition, from which it is historically derived, was—except for a few relics (e.g. *Do as you will*)—lost after it developed into a future tense marker; in fact, it is more likely that a lexical verb that has undergone grammaticalization is lost than one that has not.[36] But it does not seem possible to use this observation as a basis for generalization: Lexical loss also happens quite frequently without grammaticalization being involved; English has lost a wealth of verbs on the way from Old

---

[36] With reference to the development from verbs for 'say' to complementizers in Chadic languages, Frajzyngier maintains: "The development of verbs of saying into complementizers resulted in the bleaching of the lexical content of these verbs. As a result, new verbs of saying were introduced in many Chadic languages. This explains the large number of lexical innovations in the verbs of saying as compared with the large number of retentions for those verbs that have not served systematically as sources of grammatical morphemes" (Frajzyngier 1996: 470).

English to Modern English without there having been any kind of grammaticalization that could be held responsible for this fact. In other words, there is no significant relationship between the creation of new functional categories and that of new lexical categories.

More recently, Newmeyer (2006) has come up with further problems he has with applying grammaticalization theory to the reconstruction of early language. These problems are:

(a) Grammaticalization (Newmeyer argues) is a cyclical process, and why should one pick one point on the cycle, namely the point where lexical items are in place, as the starting point.
(b) There is no reason to assume that the earliest humans could not express concepts like "in" and "past time", and he wonders why categories such as prepositions and tense and aspect morphemes could not have existed at the outset of human language as independent categories.
(c) If a modern language like Riau Indonesian "can manifest grammaticalization as poorly as a language spoken 100,000+ years ago putatively did, then it follows that grammaticalization per se cannot tell us very much about the origin and evolution of language."
(d) If what is frequently expressed has changed over time, or if the balance of functional and counter-functional factors has not remained constant over time, then the process of grammaticalization might lack sufficient unidirectionality (or at least consistency).

One may wonder whether Newmeyer's worries are really justified. One worry relates to his assumptions about grammaticalization. With reference to (a), this concerns the notion "cyclical process". Grammaticalization may in fact be cyclic,[37] but it need not be. To take the example just looked at: As (19) shows, the erstwhile English verb *have* has been grammaticalized—among others—to a perfect aspect and a modality auxiliary. There is no evidence to assume that this was due to a cyclical process. What we know for sure, however, is that this process started out with a lexical verb of possession, hence there was an unambiguous starting point. And the same applies to the other cases of auxiliaries that we discussed in Section 1.1.3 (for a wealth of additional examples, see Chapter 2): Irrespective of

---

[37] This term refers to the fact that grammaticalization can be cyclical, leading from lexical to grammatical category and finally to the loss of the latter, and there may be renewal, in that the lost category is replaced by another lexical category (Givón 1979c; Heine and Reh 1984).

whether or not there was, or is, a cyclical process, the ultimate source of functional categories can be traced back in some way or other to lexical categories.

Problem (b) concerns the method of reconstruction. If Newmeyer believes that categories such as prepositions or tense and aspect markers could have existed at the outset of human language then one would like to know what the methodological basis of this belief is. We are not aware of any possible basis other than that provided by grammaticalization theory (see Section 1.2.1). And according to this theory, neither prepositions nor tense and aspect markers are diachronic "primitives" (see Section 2.2.3): Both are historically derived from lexical categories. For example, in Section 1.2.1.3 we mentioned the English prepositions *in front of* and *in back of*, which are both of nominal origin, and the tense and aspect auxiliaries *keep*, *used to*, and *be going to*, all of which are historically derived from verbs, and much the same observations can be made in many other languages. Accordingly, Newmeyer's belief must of necessity remain speculative as long as it is not supported by any method of reconstruction. To be sure, there are a number of functional items for which no reliable etymology exists; but we do not see any reason why they should behave diachronically any differently from those for which there is sufficient diachronic information.

With (c), Newmeyer raises a problem that is beyond the scope of grammaticalization theory. As we hope to show in this book, grammaticalization theory offers an account of grammatical evolution from early language to the modern languages; but it does not tell what happened thereafter in the history of modern languages. Riau Indonesian in fact manifests grammaticalization poorly: It is said to lack grammatical distinctions normally found in other languages (Gil 2001). But why this is so is open to question. That a language loses many of its grammatical distinctions in the course of its history is nothing unusual, French and English being cases in point: Both lost a large part of their inflectional and derivational morphology in the course of the last two millennia. And what led to the situation characterizing Riau Indonesian is also a question that concerns the particular history of this language rather than grammaticalization theory, which is concerned with regularities of grammatical change across languages. Assuming that a language such as Riau Indonesian, where it is not even possible to distinguish—linguistically—between nouns and verbs, let alone grammaticalized structures, represented an earlier stage of language evolution, this would in no way present a challenge for grammaticalization theory. On the contrary, such might be a language which can be used as living proof that the kind of system we arrive at, going back in time when we use grammaticalization theory as a reconstructing tool, is something not only plausible but also very possible, since it exists as part of our present linguistic diverse reality.

The nature of the last concern (d) would need further analysis; but irrespective of what the nature of the counter-functional, or counter-adaptive factors may be,[38] grammaticalization is overwhelmingly unidirectional and regular, as Newmeyer (1998a: 275–6, 278) himself admits (see above); hence, it provides an appropriate basis for linguistic reconstruction. To conclude, we do not see how Newmeyer's reservations could affect the application of grammaticalization theory to the reconstruction of early language.

1.3 The present volume

While previous work on language origin has been preoccupied in particular with the precursors and human-specific origins of language (Fritz Newmeyer, p.c.), our interest is primarily with subsequent development of language; the present book aims to reconstruct some major lines of grammatical evolution from its beginnings. Our concern will be with the questions listed in (1), and in Section 7.3 we will look for answers to these questions. More specifically, however, the kind of questions that we will try to answer in this book are the following: (a) How do functional categories arise? (b) What are the mechanisms to be held responsible for their rise? (c) What can all that tell us about the genesis and evolution of language? To this end, we have described in Section 1.2 the methodology to be used. An overview of pathways of grammaticalization is provided in Chapter 2, using a catalog of salient lexical and functional categories. The network of grammaticalization proposed in this chapter will provide the basis for our hypotheses on both language genesis and language evolution.

An important issue in any research on language evolution concerns the question of whether human communication can be linked to communication systems of non-human animals. To this end, we will try to determine

---

[38] Newmeyer cites Haspelmath (1999c) in support of this assertion.

in Chapter 3 whether there are any language-like abilities in animals and, if yes, what they can tell us about language genesis. Animal behavior has played an important role in discussions on early human language, but there is another phenomenon that has also frequently figured in such discussions, namely pidgin languages and other restricted linguistic systems. We have therefore decided to reserve Chapter 4 for these communication systems, which are widely held to provide some kind of analog to early language.

A crucial step in the evolution of language must have been the rise of clause subordination, which is the topic of Chapter 5. This topic is closely linked to the question of recursion, which we will look into in Chapter 6. And finally in Chapter 7 we will propose a general scenario of how human languages evolved, and we will relate our hypotheses to those volunteered by other students of language evolution. Finally, in Section 7.3 we will deal with the questions that we listed at the beginning of this chapter, providing answers wherever that is possible within the framework of grammaticalization theory.

We noted in the introduction that the scope of this book is a narrow one. We will not make any claims on theoretical issues such as whether there is, or when there arose, something like Universal Grammar, whether language is, or is part of, a distinct human faculty, is an organ, or a system, or whether language—however defined—is an entity that needs to be separated from some kind of cognitive faculty or capacity, nor with the question of whether or to what extent language is innately determined. Rather, we will simply be concerned with analyzing language structures as they manifest themselves in day-to-day interactions among their speakers, and as they are documented in grammatical descriptions. Accordingly, all we have to say about language genesis and evolution is based on typological generalizations on language structures. And even more narrowly, our goal is with applying typological observations made within the framework of grammaticalization theory to the reconstruction of earlier stages of linguistic evolution.

When working on specific reconstructions we found it tempting to adopt a cross-disciplinary perspective, immediately looking for evidence from other approaches or disciplines that might support our hypotheses, or discussing other views in relation to our findings. However, we decided not to do so but rather to confine ourselves strictly to our methodology, for the following reasons: First, we wish to avoid being biased by what other specialists have to say about the issues that we are dealing with, and second, we consider it important to keep the findings made by people working with different frameworks or theories separate in order to make sure that these findings can each be independently verified or falsified. This, however, in no way means that we are not interested in what other people do and what other theories have to offer; primatology, palaeoanthropology, evolutionary biology, neurosciences, psychology, psycholinguistics, computer sciences, philosophy, and other disciplines are all relevant for reconstructing language genesis. As we hope to demonstrate, our ultimate goal is to contribute to an interdisciplinary understanding of what language evolution is about.

Language structure is conveniently divided into various components, which are phonology, morphology, syntax, and semantics. We will be concerned with all of these components in some way or other—except for phonology. This omission may be justified by the fact that there are systems of linguistic communication, such as signed languages, that lack vocal phonology. However, this is not the reason for ignoring phonology in this book; rather, it is the fact that phonology is essentially not within the scope of grammaticalization theory; hence, we will have little to say about it.

Finally, there is another serious omission in our analysis of early language: Our approach is restricted to linguistic methodology, and we will not deal with the cognitive and neurological foundations of language evolution. The main reason for this is that we did not find any convincing evidence to establish that during the timespan we are concerned with, that is, the latest period of human evolution, possibly after the appearance of *Homo sapiens*, changes in the size and/or capacity of the brain affected linguistic communication in any significant way (see MacWhinney 2002: 250). We are aware that, given a more advanced stage of brain studies, this assumption might be in need of revision.

The linguistic materials presented in this book are frequently taken from African languages. This might seem appropriate considering that our concern is with the earliest forms of human language or languages, and Africa is commonly assumed to be the continent where both our species and our languages originated. As a matter of fact, however, this would not be an entirely convincing reason: We have found no evidence to suggest that African languages are in any way of special interest for the reconstruction of the prehistory of our species or languages, at least no more than any other languages in the world. The reason for drawing heavily on African languages is simply that these are the languages we are most familiar with. Even the fact that much of our field research has focused on languages of "archaic" hunter-gatherer societies such as Khoisan in southwestern Africa and Okiek (Dorobo) in eastern Africa does not turn out to be particularly rewarding since there is no convincing evidence to establish that the structure of these languages is in any way more closely linked to what may have characterized earlier forms of linguistic communication.

A word on our use of the term evolution may be in order. This term has received a wide range of different applications. Within linguistics, it has been employed variously for changes in the history of human language, the history of a given language or of languages, the history of a specific component of a language, or for changes towards a more "progressive" state of language or languages, for example an increase in structural complexity. Furthermore, there are more specific uses of the term, such as that by Nichols (1992: 276), who employs it to refer to the approximation to a standard profile in residual zones. And in works on grammaticalization, the term has been used for any crosslinguistic regularity in change as it can be observed in grammaticalization (see especially Bybee, Perkins, and Pagliuca 1994).

Our use of the term is based on the general definition proposed by Darwin ([1859] 1964), for whom evolution simply was "descent with modification". Accordingly, we will talk of "evolution" when any significant modification in the overall structure of human languages from early language to the modern languages is hypothesized to have taken place. This means that we will avoid the term for any of the many other uses it has been put to, including the one it has acquired in some grammaticalization studies. In doing so, we hope to avoid some of the misunderstandings that the term has given rise to in the past.

2 An outline of grammatical evolution

Among the questions that we raised in Chapter 1 there were the following: Which is older—the lexicon or grammar? What was the structure of human language like when it first evolved? How did language change from its genesis to now? Was language evolution abrupt or gradual? It is these questions that will be our concern in this chapter. After having provided a general introduction to grammaticalization theory in the preceding chapter we will now use this theory to reconstruct some major lines in the development of functional categories. The ultimate goal of this chapter is to trace grammar back to its beginnings in early language.

2.1 Introduction

The findings presented in this chapter are based on a wide range of data from over 500 languages across the world. In Heine and Kuteva (2002a), more than 400 common pathways of grammaticalization were identified; in the present chapter we will narrow down this range to a smaller number of more general grammatical developments. Most of the processes to be discussed have been documented to some extent in previous works on this issue (e.g. Lehmann 1982; Heine and Reh 1984; Heine, Claudi, and Hünnemeyer 1991; Hopper and Traugott 1993, 2003; Bybee, Perkins, and Pagliuca 1994; Heine and Kuteva 2002a; see also Dahl 2004), and the reader is referred to these and other works cited below for more details. But there are also a few pathways that have not been identified so far, and in such cases we will provide more detailed evidence to substantiate the hypotheses concerned. Since grammaticalization appears to be regular across different sensori-motor modalities, we will not be confined to spoken and written languages but will also include findings made on signed languages (e.g. Sexton 1999; Pfau and Steinbach 2005, 2006).

In our presentation we will begin with categories that previous research has established to be the least grammaticalized, that is, categories that cannot be derived historically from any other categories. Subsequently, we will proceed to reconstructing increasingly more strongly grammaticalized categories. Note that, in accordance with our methodology, we will not be able to analyze grammar as a whole but rather will be restricted to a range of morphosyntactic exponents of grammar.

A review of the grammatical descriptions available on the languages of the world yields a bewildering diversity of grammatical taxonomies, and reducing the taxa figuring in these descriptions to an uncontroversial and crosslinguistically stable set of categories is near to impossible. The categories figuring in our presentation therefore have to be treated with some caution. Selection was determined on one hand by what are particularly common patterns in the languages of the world—irrespective of whether these patterns are defined in terms of syntactic, morphological, or semantic criteria, or any combination of these. On the other hand, we are aiming to select the most inclusive categories available; to this end, we have chosen, for example, a more comprehensive category "pronoun" instead of less inclusive categories such as "personal pronoun", "indefinite pronoun".

2.2 Layers

In this section we present a skeleton of grammatical evolution. Our approach is reductionist in a number of ways, in that discussion is narrowed down to a range of notional categories and the major pathways of development linking these categories. Furthermore, space does not allow us to describe the cognitive and pragmatic foundations underlying these pathways, which have been the subject of many individual studies. Wherever possible, however, we will provide relevant information on the contextual frames that have contributed to these pathways.

We will describe linguistic evolution in terms of a set of grammatical categories that tend to be distinguished in the modern languages of the world. It is very likely, in the earlier development of human language, that these categories were not of the same kind as we find them today. For example, nouns in the modern languages usually have syntactic properties such as taking adjectives and demonstratives, or markers for number, gender, and/or case. As the following reconstructions suggest, such properties were presumably absent in the earliest stages of language evolution. The reader therefore has to be aware that the reconstructions proposed are based on the application of grammaticalization theory and can be accounted for with reference to this theory, but they are not necessarily of the same kind as those that may have characterized the structure of early language. In other words, if applying grammaticalization theory will allow us to reconstruct grammar back to an initial stage of nouns only, this does not mean that the first language(s) had nouns the way we know them from most languages nowadays, that is, as characterised by certain grammatical and distributional properties.[1]

We will describe grammatical evolution in terms of a set of "layers", that is, clusters of categories that show the same relative degree of grammaticalization vis-à-vis both the categories from which they are derived and which they develop into. For example, we observed in Section 1.1.3 that there is a regular development from lexical verbs to functional categories for tense and aspect. Accordingly, we will say that verbs belong to a different layer than categories of tense and aspect. Furthermore, as we will see in Section 2.2.3, aspect categories can further develop into tense categories, which allows us to argue that aspect and tense each represent a different layer; consequently, this example allows us to reconstruct three distinct layers.

2.2.1 *Nouns and verbs*

Nouns and verbs are the only items that are crosslinguistically fairly stable and clearly behave like open-class categories, even if there are languages that are claimed to lack verbs, or a categorial distinction between nouns and verbs (see, e.g., Gil 2001; Lüpke 2005: 93 ff.). With reference to their grammaticalization behavior, they can be called evolutionary primitives in that they are not derived productively from any other morphological or syntactic categories while they themselves commonly develop into other

---

[1] The reason for that is very simple: as Maggie Tallerman and Jim Hurford (p.c.) point out to us, one cannot talk of any distinct category until there is another category to contrast it with. Rather, it is reasonable to assume that the stage of *nouns only* reconstructed by means of grammaticalization theory corresponds to entities which served primarily the task of reference—rather than entities that served the task of predication—in the first language(s); we will return to this issue in Section 7.1.1.

categories. Nouns may behave like verbs (cf. English *head* in *He's heading our department*), but verbs may as well behave like nouns (English *in one go*); crosslinguistically there does not appear to be any general directionality regarding the grammaticalization of the two categories. However, as we will see in Section 2.2.6, the two are not of the same conceptual status.

2.2.1.1 The first layer: nouns

There are cases where nouns, or noun-like entities, are derived from other structural units. Such units can be whole sentences (e.g. *forget-me-not*), minor words (e.g. *the ifs and buts*), affixes (*Her ex is a monster*), or even "quasi"-forms (*I want an ade*). However, these tend to be idiosyncratic instances of lexicalization rather than being suggestive of regular processes leading towards grammatical categorization. In the present section we will outline the main pathways leading from nouns to other grammatical categories.

Noun > adjective Nouns typically denote tangible and/or visible things that refer, while adjectives denote qualities relating to such conceptual domains as dimension ('large', 'small'), age ('old', 'young'), color ('green'), or value ('good', 'bad'). In many languages a diachronic process can be observed whereby specific groups of nouns are grammaticalized to adjectives, such groups concerning nouns stereotypically associated with some specific quality. Groups commonly recruited include nouns denoting plants (or plant parts), specific animals, and metals. Thus, we find in English names of fruits such as *orange*, or metal names such as *bronze*, *brass*, or *silver* that have been grammaticalized to adjectives. This process involves on the one hand the parameter of desemanticization, whereby the nominal meaning is bleached out except for some salient property, referring to the color of the item concerned. On the other hand it involves decategorialization, in that nouns in such uses lose morphosyntactic properties characteristic of nouns, such as taking modifiers, determiners, and inflections and occurring in all the contexts commonly associated with nouns.

Another group of nouns widely grammaticalized to adjectives concerns sex-specific human items such as 'man' and 'woman' or 'father' and 'mother', which in many languages are recruited to express distinctions in sex. Thus, in the Swahili examples in (1), the nouns *mwana(m)ume* 'man' and *mwanamke* 'woman' are desemanticized and decategorialized in that their meaning is restricted to denoting the qualities "male" and "female", respectively, and they occur in the syntactic slot reserved for adjectives, namely after the noun they modify, and they agree in number with their head noun.

(1) Swahili (Bantu, Niger-Congo)
kijana   mwana(m)ume   'boy'
youth   man

kijana   mwanamke   'girl'
youth   woman

Adjectives tend to be closed-class categories, that is, unlike nouns and verbs, their number is severely limited in many languages. A number of languages lack adjectives altogether; however, such languages tend to have productive mechanisms to form what corresponds to adjectives in other languages. Not uncommonly, it is verbs of state ('be big', 'be bad') that are used for this purpose. The Ik language of northeastern Uganda is such a language: There are no adjectives whatsoever, even if state verbs introduced with the relative clause marker *na*, PL *ni* can be viewed as a weakly grammaticalized means of expressing recurrent distinctions of size, quality, etc.[2] But there is also a morphological distinction: Unlike other verbs of state, verbs expressing "adjectival" notions relating to dimension, weight, age, or color take an optional plural suffix -*ak* when used as nominal attributes, cf. (2b), while other verbs of state never use this suffix.

(2) Ik (Kuliak, Nilo-Saharan; König 2002)
   a. ɓisa   na   kúɗ
      spear.NOM   REL.SG   be.short
      'a short spear'

   b. ɓís-   ítíná   ni   kúɗa-   ak
      spear-   PL.NOM   REL.PL   be.short- PL
      'short spears'

To be sure, adjectives can be used in many languages as substitutes for nouns, typically by omitting the head noun in adjective–noun constructions. Such usage, however, does not normally lead to productive patterns of development from adjective to noun.

Noun > agreement marker Agreement markers may arise in a number of ways; the main pathways will be discussed below. One pathway, however,

---

[2] Only numerals and a few quantifiers do not use relative clause markers in Ik.

not to be analyzed further here, concerns the development from nouns to noun classifiers, overt noun class markers, and eventually to agreement markers (Aikhenvald 2000: 91). One possible development is discussed by Aikhenvald with reference to the noun class system of the Australian language Ngan'gityemerri, where the noun *gagu* 'animal' is used as a noun classifier in a pairing generic noun–specific noun, as in (3a), and the classifier use in this pairing may spread to the construction noun–modifier, where the generic noun is repeated on the modifier, as in (3b), paving the way for an obligatory agreement marker.

(3) Ngan'gityemerri (Australian; Aikhenvald 2000: 394)
  a. **gagu**   wamanggal kerre   ngeben-   da.
     animal   wallaby   big   1.SG.S.AUX- shoot
     'I shot a big wallaby.'

  b. **gagu**   wamanggal **gagu**   kerre ngeben-   da.
     animal   wallaby   animal   big   1.SG.S.AUX- shoot
     'I shot a big wallaby.'

Noun > adposition One of the main sources of adpositions, that is, prepositions and postpositions, is provided by nouns (or noun phrases), frequently but not necessarily in combination with some adposition and/or case inflection; we are not aware of any language which has not undergone such a process. This pathway of grammaticalization is well documented (Svorou 1986, 1994; Heine, Claudi, and Hünnemeyer 1991; Bowden 1992; Heine 1997b); we are therefore confined here to a few examples. The process almost invariably involves constructions of attributive possession [X of Y], where the head noun (phrase) X is grammaticalized to an adposition of the modifier noun (phrase) Y. New adpositions may be built on already existing adpositions or case markers, and in some languages, including English, this is the primary way in which new adpositions evolve, for example *in spite of, on behalf of, by means of.*

Desemanticization has the effect that the nominal meaning of X is reduced to some salient property, which in most cases is a spatial concept. Via decategorialization, X develops from a fully-fledged noun or noun phrase into an invariable functional category, that is, an adposition; accordingly, morphologically complex forms, such as *by means of*, turn into unanalyzable mono-morphemic functional markers. Finally, there may also be erosion, whereby the phonetic substance used for the expression is reduced in the process, as in the development of English *on gemang* >

among, *be sedan* > *beside(s)*, *in steede* > *instead* (König and Kortmann 1991: 110). The overall development of this pathway is described by Lehmann (1985: 304) thus:

(4) Relational > secondary > primary > agglutinative > fusional
    noun   adposition   adposition case affix   case affix

While relational nouns are in fact the major nominal source of adpositions, there may be other nouns as well; for example, the German causal preposition *wegen* 'because of' has the dative plural form of the non-relational noun *Wegen* 'ways' as its source. Note further that once a noun has entered the pathway from noun to adposition, it may (but need not) proceed further to developing into a case marker—as a matter of fact, in the vast majority of instances it does not; the grammaticalization process can be arrested at any stage in its development.

Perhaps the most salient group of nouns undergoing this grammaticalization concerns body parts, English *in back of* and *in front of* being typical examples of this process. Another group of nouns concerns what is technically known as environmental landmarks. Nouns for 'boundary' are instances of landmarks which not uncommonly give rise to delimitive adpositions ('until'). For example, the noun *tèka* 'boundary, end' of the Gur language Moré of Burkina Faso has developed into a temporal postposition *tèka* 'until', 'since', and in a similar fashion, the Swahili noun *m-paka* 'border, boundary' gave rise to a locative and temporal preposition *mpaka* 'up to, until' (Heine and Kuteva 2002a: 279–80).

Once the grammaticalization from noun to spatial adposition is concluded, that item may further develop into a temporal adposition. For example, the Icelandic body-part noun *bak* 'back' was grammaticalized to a locative preposition *bak(i)* 'behind', but also to a temporal preposition 'after', for example *bak jól-um* (after Christmas-DAT.PL) 'after Christmas' (Heine and Kuteva 2002a: 46–7).

In addition to body-part and landmark nouns there is a wide range of other nouns serving as sources for adpositions, acquiring adpositional functions other than locative and temporal ones, as can be seen in English prepositions such as *instead of, in spite of, by means of, on behalf of*, etc. One common group of nouns, expressing meanings such as 'matter', 'ground', 'cause', tends to give rise to cause or reason adpositions. An example is provided by the noun *mùqóá-sì* 'matter' (matter-F.SG) of the Khoisan language ‖Ani of Botswana, which developed into a cause postposition:

(5) ‖Ani (Central Khoisan)
tí   fìàâ-   tè   tsá   dì   **mùqóá-sì**   kà.
1.SG   come-   PRES   2.M.SG   POSS   reason- F.SG   LOC
'I come because of you.'

While constructions of attributive possession provide the most frequent source for adpositions, there may be alternative constructions as well. In English, for example, there are relational elements such as the prepositions *to* and *for* instead of the possessive particle *of*, as in *in addition to, contrary to, thanks to*, or *in exchange for*.

Noun > adverb When used in adverbial phrases, nouns or noun phrases may be grammaticalized to adverbs. One group of nouns undergoing this grammaticalization in many languages concerns concepts relating to spatial orientation, where desemanticization has the effect that the meaning of the noun is bleached out except for the property of denoting spatial orientation, and decategorialization means that most, if not all, nominal properties are lost and the item concerned turns into an invariable adverbial modifier of verbs or clauses. Thus, in the Gur language Moré of Burkina Faso, the noun *nyïngri* 'sky, firmament' has developed into an adverb *nyïngri* 'above, over, up', and so has the noun *yŏ p* 'sky, firmament' of the Cameroonian Bantu language Bulu, which has turned into an adverb 'above, up' (Heine and Kuteva 2002a: 279–80). This group also includes nouns meaning 'home(stead)', 'house', etc., which may develop into locative adverbs meaning '(at) home', as well as nouns for body parts which, on account of some salient spatial characteristic, may be grammaticalized to locative adverbs, English *back* (e.g. *Go back!*) being a case in point.

Another group of nouns that may develop into adverbs includes temporal items such as 'time', 'day', or 'hour' which are grammaticalized to temporal adverbs. Examples are Italian *ora* 'hour', a noun which gave rise to the temporal adverb *ora* 'now', or the Basque noun *ordu* 'hour', which is the base of the form *orduan* 'then' (with the locative case-ending *-an*; Heine and Kuteva 2002a: 176).

Decategorialization may have the additional effect that the relevant noun loses some or all of its morphological trappings in the process. For example, in East Nilotic languages such as Maasai and Teso, nouns are generally used with gender prefixes. But when grammaticalized to adverbs, they lose their prefixes, cf. (6).

(6) From noun to preposition in East Nilotic languages
Maasai (East Nilotic, ɛn-dʊkʊ́ya dʊkʊ́ya (Tucker and
Nilo-Saharan) 'in front', Mpaayei
'ahead' 1955: 248)
F-head (adverb)
'head' (noun)
Teso (East Nilotic, a-kuju kuju 'above, (Kitching
Nilo-Saharan) over, up' 1915: 74)
F-sky (adverb)
'sky, heaven'
(noun)

Not uncommonly, it is nouns headed by an adposition that grammaticalize into adverbs. For example, the Latin adverb *ergo* 'consequently, thus' is said to be historically derived from **e rego* (or **e rogo*) 'from the direction of' (Janson 1979: 99).

Noun > case marker Nouns (or noun phrases) provide one of the main sources for case markers, that is, grammatical forms, usually affixes or clitics, which have no function other than assigning a case property to the noun (phrase) they govern.[3] This process, which involves an intermediate stage where the noun (phrase) serves as an adposition (see Noun > adposition), entails most of all desemanticization, whereby the semantics of the noun is lost or, more precisely, is reduced to expressing a case property. Decategorialization has the effect that the noun loses its combinatorial potential and its internal morphological complexity and is reduced to an invariable grammatical form serving some specific syntactic function.

Possessive (genitival) case markers tend to occupy crosslinguistically a marginal position in case paradigms. One source for such markers can be seen in nouns meaning 'property', 'part', or 'thing'. The process underlying this development is based on the reinterpretation of a structure [X *property (of)* Y] as [X *of* Y], whereby the noun is desemanticized and decategorialized to a functional marker expressing a syntactic relation.

---

[3] We will not propose any rigid boundary between adpositions and case markers; this is an issue that is notoriously controversial in both typological works and grammatical descriptions. We will say that case markers are typically (though not necessarily) affixes whose function includes that of marking core participants (subject, object), while adpositions are typically free forms that serve the expression of a wider range of functional relations.

In the Aztecan language Pipil of El Salvador, the relational noun -*pal* 'possession' has turned into a preposition *pal*, and a case marker of attributive possession (Harris and Campbell 1995: 126–7). The Arabic noun *bita:ʕ* 'property' has provided the source for the genitive marker in the Arabic-based creole Nubi; in the process, the noun underwent erosion to *ta*. A similar process has taken place in the fellow Semitic language Maltese, where the noun *ta'* 'possession', 'property', has given rise to a case marker of a new pattern of attributive possession (Heine and Kuteva 2002a: 245–6; see there for more examples).

But there is also a more general process whereby nouns develop into adpositions, and some of the latter may turn into case clitics and case inflections. Most instances that have been documented involve nouns that turn into locative adpositions and finally into locative case inflections, although the use of the latter may be extended to also denote temporal, causal, and other case relations. But it is not only the spatial domain that is involved. The following examples illustrate an alternative pathway of grammaticalization. The Balto-Finnic noun *kansa 'people', 'society', 'comrade' developed into the comitative postposition *kanssa* in Finnish and *kaas* ('together with', 'in the company of') in Estonian, and eventually it turned into a comitative-instrumental marker -*ga*/-*ka* in Estonian. In a similar fashion, the comitative marker -*(gu)in* of the fellow Finnic language Sami appears to be etymologically derived from the Sami noun *guoibmi* 'comrade', 'fellow', 'mate' (Stolz 2001: 599–600; see Heine and Kuteva 2002a: 91–2 for additional evidence for this grammaticalization). A similar development appears to have happened in Basque, where the noun *kide* 'companion', 'fellow', 'mate', applied to both people and things, appears to be the source of the comitative case suffix -*ekin* (Heine and Kuteva 2002a: 91–2).

Rather than the meaning of the noun, it may be inflections on the noun that determine the function of the resulting case affix. The Hungarian case suffixes -*ben*/-*ban* 'in' (inessive) and -*ból*/-*bő̋l* 'away from' (elative) are both historically derived from a relational locative noun *bél* meaning 'interior'. The difference in case functions is due to the fact that the final segments *n* and *l* are themselves relics of the case suffixes (-*n* locative; -*öl* ablative) on the locative noun (Comrie 1981: 119; Lehmann 1982: 84–5; Hopper and Traugott 1993: 107–8).

Noun > complementizer Crosslinguistically, nouns form one of the main sources for complementizers, that is markers introducing complement clauses. It is general nouns such as 'thing', 'person', 'matter', and 'place' in particular which, when having argument status in the main clause, are reinterpreted as markers presenting complement clauses (see Chapter 5). Desemanticization has the effect that the nominal meaning disappears, giving way to the syntactic function of serving as the head of a subordinate clause. Decategorialization means that the erstwhile noun may no longer take modifiers or determiners and is restricted in its occurrence to the complementizer slot. Since we will deal in more detail with the process from noun to complementizer in Chapter 5, a few examples may suffice to illustrate it. The following example from a modifier–head language illustrates such a situation, where the noun !xái-sa 'matter', 'story' (oblique case) of the Khoisan language Nama has given rise to the complementizer !xái-sà 'that', 'whether':

(7) Nama (Central Khoisan; Heine and Kuteva 2002a: 211)
    tiíta   ke   kè   /'úú   'íí   !úū-   ts   ta
    1:SG   TOP   PAST not.know   PAST   go-   2.SG.M IMPFV
    !xái-   sà.
    COMP-   3.SG.M
    'I didn't know that you were going.'

Note that the complementizer exhibits an inflectional ending. The example is suggestive of an early stage of grammaticalization. At a more advanced stage, the complementizer is likely to be decategorialized to the extent that it loses its nominal properties and turns into an invariable grammatical marker. The following example from Japanese may illustrate this situation. The complementizer *koto* appears to be historically derived from a noun 'thing' (Lehmann 1982: 65). *koto* is no longer available as a noun, but it has survived as a complementizer (or nominalizer) that has lost essentially all its nominal properties, that is, it may no longer take case markers or any other elements associated with the use of nouns. Like Nama, Japanese has a rigid modifier–head word order; accordingly, the complementizer follows the subject complement clause in the following example:

(8) Japanese (Lehmann 1982: 65)
    Ano   hito   ga/no   hon   o   kai-   ta   **koto**
    that   person NOM/GEN   book   ACC   write- PART   NOMIN
    ga   yoku   sirarete   iru.
    NOM   well   known   is
    'That that person has written a book is well known.'

Noun > pronoun The process from noun to pronoun is a crosslinguistically widespread one, and it may take a number of different pathways. What all these pathways have in common is that they involve the same parameters of grammaticalization. This is on the one hand desemanticization, whereby the lexical semantics of the noun is lost; what remains is a schematic grammatical function. On the other hand it involves decategorialization, having the effect that the noun loses its categorial properties, such as taking nominal inflections and modifiers.

One of these pathways leads from nouns meaning 'man' to impersonal or indefinite animate pronouns. Much discussed examples concern the Latin noun *homo* 'man, person' which gave rise to the French impersonal pronoun *on*, or the German noun *Mann* 'man', which developed into the indefinite subject pronoun *man*. Another example is provided by Icelandic *maður* 'man, person', which assumed the function of an indefinite pronoun ('someone'; Heine and Kuteva 2002a: 208–9).

Other examples involve nouns meaning 'person' or 'people' that acquire uses of and may develop into indefinite pronouns (Heine and Kuteva 2002a: 232–3). This is a widespread process in the Chadic languages of the Afroasiatic family, where nouns for 'person' or 'people' have developed into indefinite subject pronouns, for example Margi *mjì* 'people', Kapsiki *mbelí* 'people', Guduf *údè* 'person', Gude *ənji* 'people' (Kim 2000). The following is an example from the Cameroonian language Baka, where the noun *wó* 'person, man' has been grammaticalized to an impersonal pronoun.

(9) Baka (Ubangi, Niger-Congo; Christa Kilian-Hatz, p.c.)
wó ndé a ye pòkì à mo- nda.
person without INF love honey LOC door- house
'One does not like the kind of honey that sticks at the house door.'

European examples include Albanian *njeri* 'person', which has uses of an indefinite pronoun ('somebody'), or Portuguese *pessoa* 'person', as the following example illustrates:

(10) Portuguese (Stolz 1991a: 13)
a **pessoa** não dev- e preocup- ar- se.
DET.F person.F NEG must- 3.SG.PRES worry- INF- REFL
'One should not worry.'

In a parallel fashion, nouns meaning 'thing' commonly grammaticalize into inanimate indefinite pronouns. For example, the Nahuatl noun *itlaa* 'thing' developed into an indefinite pronoun *tlaa* 'something', and the phrase *ohun kan* ('thing one') of the Niger-Congo language Yoruba of Nigeria turned into the indefinite pronoun *nkan* 'something'; for more examples, the reader is referred to Heine and Kuteva (2002a: 295–6; see also Lehmann 1982: 51–2; Heine and Reh 1984; Haspelmath 1997a: 182).

Evidence for a process from noun to indefinite pronoun also comes from signed languages, for example from German Sign Language (DGS) and Netherlands Sign Language (NGT), which both appear to have undergone a development from noun for 'person' to indefinite pronoun (Pfau and Steinbach 2006).

Another major pathway from noun to pronoun concerns the rise of personal pronouns. Nouns (or noun phrases) constitute one of the main sources for personal pronouns; we are confined here to illustrating the major lines of development. Perhaps the most salient line concerns general human nouns which develop into animate personal pronouns. Such pronouns have third person reference; in the Central Khoisan language ‖Ani of Botswana, the noun *khó(e)-mà* (person-M.SG) 'male person, man' has developed into a third person masculine singular pronoun *khó(e)-mà* or *khó-m̀* 'he'.

In a few African languages, both singular and plural forms of nouns have provided the source for third person pronouns. The following example concerns a language that has suppletive forms for human beings: In the Central Sudanic language Lendu there is a singular noun *ke* 'man' while the plural form is either *ndrú* or *kpà* 'people'. These exact forms appear to have given rise to markers of third person deixis: *ke* is also the third person singular pronoun, while *ndru* or *kpa* is the third person plural pronoun.

On the other hand, nouns for 'person' or 'people' can develop into first person plural pronouns. For discussion, see Heine and Kuteva (2002a: 233–4); two examples may suffice to illustrate this pathway. In the Central Sudanic language Ngiti of the Nilo-Saharan family, the noun *alɛ* 'person, people' appears to have given rise to a first person plural inclusive pronoun *àlɛ̀* 'we'. In the process, the erstwhile noun was adapted to the phonological paradigm of pronouns, taking the same tonal pattern as the other first- and second-person plural pronouns, and it underwent erosion, being shortened to *lɛ̀* or *l-* in fast speech (Kutsch Lojenga 1994: 195). A similar development is reported from the Spanish-based creole Palenquero of Colombia, where the noun *(ma) hende* 'people', ultimately derived from Spanish *gente* 'people', has given rise to a pronoun *(h)ende*, which serves both as an impersonal pronoun 'one' and as a first person plural pronoun 'we' (Schwegler 1993: 152–3).

Finally, mention should be made that nouns meaning 'body', 'head', etc. have given rise to reflexive pronouns; there are hundreds of languages that have been documented to have undergone this grammaticalization process (Schladt 2000; Heine 2000, 2005). As a rule, this process involves noun phrases consisting of a body or body-part noun plus a possessive modifier, as in example (11) from the Papuan language Yagaria of New Guinea; but there are also cases where the noun on its own turns into a reflexive marker, or where the possessive modifier is lost. Accordingly, there are a number of languages where the reflexive pronoun consists of the bare noun, as in (12):

(11) Yagaria (East Central Highlands; Renck 1975: 148)
d-   **ouva-**   di   begi-   d-   u-   e.
my-   body-   my   beat-   PAST- 1.SG- IND
'I hit myself.'

(12) Kabuverdiano (Portuguese-based creole; Holm 1988: 205)
ɛl   máta   **kabésa**.
(he   kill   head)
'He killed himself.'

To conclude, there is robust evidence to show that nouns, used either on their own or in combination with some modifier, commonly develop into markers of personal deixis. Most commonly, it is third person pronouns that arise, but given the right social and linguistic context, it may also be first- or second-person pronouns. Conversely, personal pronouns are unlikely to develop into nouns or noun phrases.

Noun > subordinator   Similar to the process for complementizers, general locative and temporal nouns provide a common source for subordinators of adverbial clauses. We will discuss more detailed evidence on this grammaticalization process in Chapter 5, suffice it here to illustrate the process with a couple of examples.

Like its etymological German counterpart *weil* 'because', English *while* is historically derived from a phrase having a temporal noun as its semantic nucleus, which was *al di wila daz* 'all the time that' in Old High German and *þa hwile þe* 'that time that' in Old English. The main difference between the two languages is that in German the process resulted in a cause/reason subordinator while in English it gave rise to a temporal and concessive subordinator.

To conclude, there is a ubiquitous process whereby nouns forming the semantic nucleus of noun or adverbial phrases are grammaticalized to adverbial clause subordinators; desemanticization leads to the loss of their lexical semantics, and decategorialization has the effect that the erstwhile noun loses most or all of its nominal properties, ending up as an invariable marker of clause subordination. In a number of cases, although not always, the process also involves erosion, in that part of the phonological and/or morphological substance of the phrase is lost; the German and English examples that we mentioned illustrate the effect of this loss.

2.2.1.2 The second layer: verbs

Like nouns, verbs provide a rich source for a range of functional categories. In the present section we give an outline of the main pathways leading straight from lexical verbs to items serving some grammatical function.

Verb > adposition   The primary source for prepositions and postpositions is provided by nouns (see above and also under Adverb > adposition). But verbs, as well, are a common source for adpositions, usually involving a process whereby a structure [verb + complement] is reinterpreted as a structure [adposition + noun phrase].

How this grammaticalization affects the structure of the verb undergoing the process has been shown in more detail by Matsumoto (1998) for Japanese. This language has a wide range of postpositions which are historically derived from verbs used in the participial *-te*-construction; Matsumoto (1998: 27–8) provides a list of 26 such postpositions. The following is a typical example of such a postposition. In (13a) the verb *tuku* (participial form: *tuite*) is exemplified, having the lexical meanings 'stick to, follow, take the post of', while (13b) illustrates the same item in combination with the dative marker *ni* as a functional category, namely as a postposition meaning 'about, concerning'. Since the verb follows its complement in Japanese, the resulting forms develop into postpositions rather than prepositions.

(13) Japanese (Matsumoto 1998: 27–9)
    a. Taroo wa [kare ni **tuite**] doko   made mo   itta.
       Taroo TOP he   DAT follow anywhere to   even go.PAST
       'Taro went anywhere, following him.'

b. Taroo   wa   [sono koto   **ni tuite**] setumee sita.
   Taroo   TOP   the   matter about   explain did
   'Taro explained about the matter.'

Desemanticization had the effect that the verbal semantics was reduced to a schematic function, in the case of *tuku* to a theme function. Decategorialization entailed the following changes in particular (see Matsumoto 1998: 29–33 for more properties): The verb lost the ability to be immediately preceded by particles such as *mo* 'too, even' or *dake* 'only', to be modified by manner adverbials, to take subjects and head clauses, or to take tense, negation, and other markers, and the verb can no longer be inflected for causativization, passivization, etc. The result is an invariable postposition *ni tuite*, where the erstwhile verb has merged with the dative postposition *ni*.

As a rule it is not finite verbs, inflected for person, tense, etc., that are grammaticalized but non-finite verb forms, such as infinitival or gerundival/participial verbs, as in our Japanese example. In English, prepositions are most commonly derived from present participle forms (*barring, concerning, considering, during,*[4] *excepting, failing, following, notwithstanding, pending, preceding, regarding*) or past participle verb forms (*given, granted*); in the Bantu language Swahili, it is generally the infinitive prefix *ku-* that turns verbs into prepositions, for example *ku-pitia* 'to pass by' > *kupitia* 'through, via', *ku-toka* 'to come from' > *kutoka* 'from', *ku-husu* 'to give a share to' > *kuhusu* 'concerning, about', for example:

(14) Swahili (Bantu, Niger-Congo)
   a-   me-   ni-   ita   **kuhusu**   ombi   langu.
   3.SG-   PERF- 1.SG-   call concerning   request   my
   'He's called me about my request.'

Depending on the particular contextual frame that is used for grammaticalization, however, a verb developing into an adposition may take other morphosyntactic forms. For example, this pathway may involve imperative singular verb forms, such as English *bar* or *come* (König and Kortmann 1991: 117), in which case the resulting adposition may consist of the bare verb stem only.

Paradigm cases of the verb-to-adposition pathway can be found in analytic-isolating languages of, for example, southeastern Asia and western Africa: In these languages, the verb is uninflected both in its lexical-verbal

---

[4] *During* is derived from an obsolete verb meaning 'to last, endure'.

and in its adpositional uses. The following example illustrates the process concerned. In the Ewe language of Ghana and Togo, the verb *tsó* 'to come from' has been grammaticalized to a source preposition 'from'. Example (15a) shows *tsó* in its lexical use as a motion verb, while (15b) exhibits its grammaticalized use as a functional category: Verbs in Ewe cannot occur without a formal subject (except for imperatives), nor can they appear sentence-initially; in its use as a preposition, however, the item *tsó* can, as (15b) shows.

(15) Ewe (Kwa, Niger-Congo; Hünnemeyer 1985: 76)
   a. é-   tsó   Lome.
       3.SG- come.from Lome
       'He came from Lome.'

   b. **tsó** éɟé ɖevíme   ke   m-   é-   té ŋú
       from his childhood PTC NEG- 3.SG- can
       kpɔ́- á   nú   o.
       see-   HAB thing   NEG
       'From his childhood, he cannot see.'

Evidence for this pathway can also be found abundantly in creole languages, such as the following example from Haitian, where the allative preposition *rivé* 'to' is historically derived from the French motion verb *arriver* 'arrive':

(16) Haitian (French-based creole; Sylvain 1936: 131)
     Li   broté   tut   pitit-   li   **rivé**   Pako.
     (s/he take   all   child- 3.SG to   Pako)
     'She moved all her children to Pakot.'

Verb > adverb One common way in which lexical verbs develop into adverbs is via the serialization of two verbs where one of them comes to assume a modifying function for the other and gradually turns into an adverbial modifier of the other verb. This grammaticalization is therefore particularly common in languages having serial verb constructions, even if it is not restricted to them. In the following example (17a), the Ewe item *dzó* 'to leave' is a lexical verb, while in (17b) it follows the verb *mli* 'to roll' and modifies the meaning of the latter. That *dzó* is no longer a verb in (17b) is suggested, for example, by the fact that it is not inflected for tense–aspect: It is invariable, exhibiting a morphosyntactic structure characteristic of adverbs.

(17) Ewe (Kwa, Niger-Congo; Hünnemeyer 1985: 103–5)
a. é-   **dzó**   le   xɔ   me.
   3.SG- leave   at   house   inside
   'He left the house.'
b. bɔlu   le   mi-   mli-   ḿ   **dzó**.
   ball   PRG   roll-   roll-   PROG   (leave)
   'The ball is rolling away.'

The verb developing into an adverb may take various morphosyntactic shapes; it can take a non-finite or a finite form. In Latin, a number of finite verb forms have given rise to adverbs; for example, the verb form *scilicet*, a compound of *scire* 'to know' and *licet* 'it is permitted', had the meaning 'one can know (that)' and took infinitival complements in early Latin. But from the earliest times, it was also used as an adverb meaning 'of course, naturally', and 'namely', mostly however by no means always used in the second position of the clause. And much the same applies to *videlicet* (composed of *videre* 'to see' and *licet*), which was used in the same construction with infinitival complements and developed into an adverb with much the same meaning as *scilicet* (Janson 1979: 97–8).

For more examples of this kind from verb to adverbial modifier from languages in other parts of the world, see Heine and Kuteva (2002a). What all these cases have in common is that the second verb in a verb–verb construction is desemanticized, whereby its lexical meaning is reduced to some specific schematic function (like 'away' in example (17b)), and decategorialized, in that that verb loses most or all of the verb properties characterizing its lexical use (like the ability of *dzó* to be used in the progressive aspect).

Verb > aspect Most languages of the world have undergone this process, whereby lexical verbs assume the function of an auxiliary of other verbs and turn into markers for aspect functions such as progressive, durative, habitual, completive, perfective, iterative, and the like. There is a rich literature on this grammaticalization process (e.g. Bybee, Perkins, and Pagliuca 1994; Kuteva 2001). We briefly discussed an example from English in Section 1.1.3 concerning the item *keep*, which is a lexical verb in most of its uses (e.g. *He kept the money*), but has also been grammaticalized to a durative aspect auxiliary when combined with main verbs (*He kept complaining*). In the following example from the Khoisan language Khwe (or Kxoé), the verb *xǔ* 'to leave, abandon, loosen' has been grammaticalized to a completive (terminative) aspect marker when serving as an auxiliary of another verb: It has lost its lexical meaning in favor of a schematic aspectual function, and it has been decategorialized to a verbal derivational suffix -*xu*:

(18) Khwe (Central Khoisan)
kx'ó:ró- xu 'è.
eat.meat- CPL IMP)
'Eat (the meat) up!'

But crosslinguistically verbs meaning 'finish' that give rise to completive, perfect, and perfective aspect markers are more common (Bybee, Perkins, and Pagliuca 1994; Heine and Kuteva 2002a). This is also a process that has happened in a number of signed languages, such as American Sign language (ASL),[5] Israeli Sign Language, and Italian Sign Language (Janzen 1995; Sexton 1999; Morford 2002: 331; Pfau and Steinbach 2006), for example:

(19) American Sign language (ASL; Fant 1994: 245, quoted from Sexton 1999: 115–16)
TOUCH FINISH JAPAN YOU
'Have you ever been to Japan?'

Suffice it to add an example of a specific but widespread pathway of grammaticalization whereby verbs meaning 'to return, come back' develop into iterative (or repetitive) markers when combined with other verbs. In the Central American Chibchan language Sanuma, the lexical verb *kō* 'return' has given rise to a repetitive marker when in combination with other verbs:

(20) Sanuma (Yanomam, Chibchan; Borgman 1990: 180–1)
ĩ   hamö sa   pili- a-   mö   ku- a   kō-   ki
REL LOC 1:SG live- DUR- PURP be- DUR return- FOC
pia   salo.
intend RESULT
'I intend to live in that place again.'

Verb > case marker Verbs are a common source of adpositions (see above), and in some cases they may further develop into case markers. Givón (2006: 24) observes that the Uto-Aztecan language Ute derives all locative

---

[5] Janzen (1995) observes that FINISH in ASL has acquired a wide range of tense–aspect meanings, namely completive, perfect (or anterior), perfective, past, as well as a conjunctive function.

case markers from historically-still-traceable precursor verbs which have turned into noun suffixes, no longer carrying any discernible residue of verbal properties; a few examples are provided in (21).

(21) Ute (Uto-Aztecan; Givón 2006)
Case suffix	Verbal source
-*caw* 'toward'	-*cawi* 'come to'
-*kwa* 'to'	-*kwa* 'go to' (defective verb)
-*naagh* 'in'	-*naagha* 'enter'
-*ruk* 'under'	-*rukwa* 'descend'
-*tarux* 'on (top)'	-*tarugwa* 'climb'

Since case affixes derive almost invariably from adpositions and, next to nouns, verbs provide a major source for adpositions (see Verb > adposition), this grammaticalization appears to be part of the more general pathway (22).

(22) verb > adposition > case marker

Verb > complementizer Next to demonstratives (see below), lexical verbs provide one of the most important sources for the development of markers introducing complement clauses. Since we will be dealing with this pathway of grammaticalization in more detail in Section 5.3.2.3, we are confined here to one example. Speech act verbs meaning 'say' or similative verbs meaning 'be like', 'be equal', or 'resemble' are the ones that commonly give rise to complementizers (see Heine and Kuteva 2002a: 257–8; 261–5; 273–5). The following example from the West African language Hausa illustrates the use of a 'say'-verb as a complementizer: The verb *cê* 'say' is used in its nominalized form *cêwā* to present complement clauses (though not when the main clause verb is *cê*):

(23) Hausa (Chadic, Afroasiatic; Newman 2000: 97)
yā musà **cêwā** shī ɓàrāwò nē.
he   deny   that   he   thief   be
'He denied that he was a thief.'

Verb > demonstrative This is a pathway, which appears mainly to concern verbs meaning 'to go',[6] to a lesser extent also verbs for 'see' (Heine and Kuteva 2002a: 172–3, 159, 294–5), for which Frajzyngier (1995: 197)

---

[6] Frajzyngier (1995: 1991–2) argues that in Chadic languages there is evidence for verbs meaning 'say' to form the source of demonstratives, but the data provided are not entirely satisfactory to strengthen this hypothesis.

proposes the following pathway: 'go' > remote deictic > demonstrative > pronoun.

Archaic Chinese provides one example: The motion verb *ZHI* 'to go' has developed into a proximal demonstrative 'this' (Heine and Kuteva 2002a: 159). But clearly cases where the resulting demonstrative function is distal rather than proximal are more common. For example, in the Chadic language Mupun (or Mopun), the verb *'di* 'go' appears to have given rise to a distal demonstrative (Frajzyngier 1987), and similar examples can be found in the Khoisan family. In the E3 dialect of !Xun (Ju |'hoan) of the North Khoisan family, the verb *tò'à* 'go' seems to have provided the source for the distal demonstrative *tòàh* 'that', cf. (24a). That this hypothesis is correct is supported by the fact that there is another verb for 'go', *'úú*, which is also used as a demonstrative marker, added to the distal demonstrative to form another demonstrative denoting extreme distance ('that, far away'), cf. (24b).

(24) !Xun (Ju|'hoan, E3 dialect, North Khoisan; Köhler 1973b: 48)
    a. dzháú-   s-   à   tòàh
       woman- PL- T DI
       'the women there/ those women'
    b. dzháú-   à   'úú-   tòàh
       woman- T go- DI
       'the woman over there (far away)'

Verb > negation The development of negation in French, where the noun *pas* 'step' (as well as a set of other nouns) was introduced as an optional pseudo-object reinforcing the inherited pre-verbal negative particle *ne*, later turning into the primary means of expressing verbal negation (Schwegler 1988; Hopper and Traugott 1993: 58–9), is usually cited as a paradigm example of how negation markers arise. Crosslinguistically, however, this pathway, leading from noun to negative particle is not really common, even if it is widespread in European languages (Ramat and Bernini 1990); perhaps the main way in which new negation markers evolve is via the grammaticalization of verbs. In many cases, the process is confined to modally marked contexts, especially to prohibitive or negative imperative constructions, where verbs meaning 'stop' are reinterpreted as negation markers. The following example from Welsh illustrates this case of restricted grammaticalization: The verb *peidio* 'cease, stop' has acquired the function of a prohibitive auxiliary, as in (25). Similar cases abound in the Kru languages of West Africa (Marchese 1986; Heine and Kuteva 2002a: 283–4); example (26) is representative of this situation, where (26a) shows the lexical and (26b) the functional use of the verb.

(25) Welsh (Wiliam 1960: 78)
Paid â mynd!
(stop.IMP.2.SG and go.VN)
'Don't go!'

(26) Sapo (Kru, Niger-Congo; Marchese 1986: 191)
a. ɔ bɔ kò dī- ε̄.
he stop rice eat- NOMIN
'He stopped eating rice.'
b. b- ɔ bɔ́ kò dī- ε̄.
that- he stop rice eat- NOMIN
'He mustn't eat rice.'

Negative imperatives and prohibitives have a number of other verbal sources in addition, in particular verbs of negated volition, that is, negative verbs of desire, such as Latin *nōlī*, the imperative of 'not want, be unwilling' which introduces negative imperatives. In the following example from the Philippine language Tagalog, the item *huwag* expresses negative desire in a declarative sentence such as (27a), but a negative imperative in (27b) (see Croft 1991: 14–16 for more details).

(27) Tagalog (Austronesian; Croft 1991: 15)
a. **Huwag** [nga- ng] wala- ng pera si Ben.
NEG.want POL- LINK not.have- LINK money PNT Ben
'I don't want Ben to be without money.'
b. **Huwag** kayo- ng magsayaw ng pandanggo.
NEG.IMP 2.PL- LINK dance NT fandango
'Don't dance a fandango.'

This grammaticalization process does not involve any dramatic degree of desemanticization; what appears to happen in most cases is that a negative verb of desire is extended from a modally unmarked to a modally marked context and in this process loses its volition meaning. Not uncommonly, such verbs are historically derived from a combination of a negation marker plus a verb of volition.

Otherwise, verbs giving rise to negation markers are in particular 'lack', 'miss', 'leave', and 'fail' (Heine and Kuteva 2002a: 188, 192–3). For example, the verbs WU and WANG of Archaic Chinese, both meaning 'lack', developed into negative markers (Alain Peyraube, p.c.). Once again, there are a number of Kru languages which underwent this grammaticalization (Marchese 1986: 183), and in the Khoisan language !Xun (W2 dialect), the negation marker in modally marked sentences is *n‖à* (or *n‖àn*), which is likely to be derived from the verb *n‖àn* 'leave, abandon', for example:

(28) !Xun (W2 dialect; North Khoisan; field notes, König in prep.)
 n‖à̀ n!!ú- hŋ̄- ā hà̀ kē tūú.
 NEG CAU- see- T N1 TR book
 'Don't show him the book!'

Worldwide perhaps the most common manifestation of the Verb > negation pathway is provided by negative existential verbs ('not to exist') gradually developing into markers of verbal negation (for a detailed description of this process, see Croft 1991). Extension has the effect that the negative existential spreads from nominal complements to verbal complements; desemanticization leads to the bleaching out of the predicate function and the 'existence' meaning, and decategorialization means that the negative existential loses all the verbal properties it may have had and be restricted in its occurrence to the position next to the other verb.

Frequently, these negative existentials have only a limited number of verbal properties, or they may even be invariable particles; historically it is not unusual for them to be combinations of a positive existential plus a negation marker. At the initial stage, the negative existential may simply serve as an emphasizing device for an already existing negative construction. The final stage is reached when the negative existential replaces the earlier negation marker, or becomes the main exponent of negation; we will discuss an example in Chapter 4 (Section 4.3).

In a number of languages, however, the new negation marker is not generalized to all contexts but remains restricted to certain tenses or aspects. The negative existential *bâ* 'there is not' of the Saharan language Kanuri of Nigeria, for example, seems to have developed into a negation marker of the imperfective aspect, while in the perfective there is a different negation marker (*-nyi*). In Chinese, the "normal" declarative negation marker is *bu*, illustrated in (29a), while the negative existential *méi* [*yǒu*] (cf. (29b)), consisting of the special negator *méi* and the existential/possessive verb *yǒu*, appears to have been grammaticalized to a negation marker of completed actions, cf. (29c).

(29) Chinese (Croft 1991: 11; cited from Li and Thompson 1981)
a. tā   bu   sǐ.
   3.SG NEG   die
   'S/he refuses to die/won't die.'
b. méi [yǒu]   rén   zài wìmian.
   NEG.EXIST   person at   outside
   'There's no one outside.'
c. tā   méi [yǒu]   sǐ.
   3.SG NEG.EXIST   die
   'S/he hasn't died', or 'She didn't die.'

Verb > passive Passives are the result of a wide range of different conceptual sources and constructions (Haspelmath 1990; Givón 2004; see Pronoun > passive for one of these sources). In a number of these constructions, there is a verb that is grammaticalized to a passive marker, either in combination with some other morphological exponent (cf. English *be* plus participle, or German *werden* 'become' plus participle), or on its own. In a number of East Asian languages, such as Thai, Vietnamese, Mandarin Chinese, or Korean, there is a verb meaning 'suffer' that has assumed the function of a passive marker (see Heine and Kuteva 2002a: 284). For example, the Chinese verb *bei* 'to receive', 'to suffer', 'to be affected'[7] of the Warring States period turned into a passive marker *bei* in Early Medieval Chinese (second–sixth century AD), for example:

(30) Early Medieval Chinese (*Shi shuo xin yu:* fang zheng; quoted from Peyraube 1996: 176)
Liangzi **bei** Su Jun hai.
Liangzi PASS   Su Jun kill
'Liangzi was killed by Sun Jun.'

Verbs meaning 'get, acquire', including English *get*, crosslinguistically provide a common source for passive markers. For example, in the French-based Indian Ocean creoles, the French verb *gagner* 'to gain' has turned into a passive marker *gaŷ*:

---

[7] Originally, *bei* was a noun meaning 'blanket', later it turned into a verb meaning 'to cover', 'to wear', before it acquired the meanings 'to receive', 'to suffer', 'to be affected' (Peyraube 1996: 176).

(31) Seychellois (French-based creole; Corne 1977: 161)
zot   pa   ti   **gaŷ**   êvite   dâ   sa   festê.
3:PL   NEG   PAST   get   invited   in   that   party
'They did not get invited to that party.'

Verb > subordinator Subordinators, that is, markers introducing adverbial clauses, can be derived from a wide range of verbs depending on their particular function (see Heine and Kuteva 2002a for examples). Since we will deal with this issue in more detail in Chapter 5 (Section 5.3.3.3), one example may suffice to illustrate this pathway. In this process, the parameter of extension has the effect that the use of the verb is extended from nominal complements to adverbial clauses, the lexical meaning of the verb is desemanticized, what remains is the syntactic ability of the verb to take a complement, and via decategorialization the verb loses most, if not all, verbal properties. The following example illustrates the grammaticalization of the verb *bang* 'go' to *-bang*, a subordinating conjunction of goal or purpose in the Central American language Rama.

(32) Rama (Chibchan; Craig 1991: 457)
tiiskama   ni-   sung-   **bang**   taak-   i.
baby   1.SG-   see-   SUB   go-   TNS
'I am going in order to see/look at the baby.'

Verb > tense In the same way as they give rise to markers of verbal aspect, lexical verbs combine with other verbs, assume the role of auxiliaries and turn into tense markers. In this way, English acquired two future tense categories, *will* and *be going to*, by grammaticalizing a verb of volition (*will-an*) and of motion (*go to*) to tense markers. Similar processes have occurred in a wide range of languages all over the world, whereby verbs meaning 'want', 'come to', or 'go to' turned into future markers. The following example from the South African Zulu language is a case in point: The lexical verb *-ya* 'go' (33a) has developed into a remote future prefix (33b).

(33) Zulu (Bantu, Niger-Congo; Mkhatshwa 1991: 97)
   a. Ba-   **ya**   e-   Goli.
      (3:PL-   go   LOC-   Johannesburg)
      'They are going to Johannesburg (eGoli).'
   b. Ba-   **ya-**   ku-   fika.
      (3:PL-   FUT-   INF-   arrive)
      'They will arrive.'

The reader is referred in particular to Bybee, Perkins, and Pagliuca (1994) for ample illustration of the ways in which verbs give rise to tense categories, either directly or via an intermediate stage of an aspect category (see Verb > aspect). Evidence for this pathway is also found in signed languages (see Pfau and Steinbach 2005, 2006).

And in much the same way as lexical verbs commonly develop into tense and aspect markers, they also give rise to markers of modality—both in spoken languages (Bybee, Perkins, and Pagliuca 1994) and in signed languages (Pfau and Steinbach 2006).

2.2.2 The third layer: adjectives and adverbs

We saw in Section 2.2.1 how nouns and verbs give rise to minor categories, including categories that are widely treated as lexical categories, such as adjectives and adverbs. In the present section we are concerned with the latter two, that is, with lexical modifiers of nouns (adjectives) and of verbs or clauses (adverbs).

Adjectives may develop into all kinds of functional markers, into clitics and affixes. The English adjective *full* has acquired some productivity as a nominal suffix *-ful*, as in *joyful, shameful, beautiful*, etc. But more so than in English, adjectives in German have provided the source for a number of derivational suffixes, such as *-frei* 'free from' (< *frei* 'free'), *-leer* 'devoid of' (< *leer* 'empty'), *-voll* 'full of' (< *voll* 'full'), etc.

That adjectives may also give rise to adpositions has been mentioned in a number of studies. König and Kortmann (1991: 109), for example, observe: "That adjectives may develop into prepositions is shown by German examples such as *fern, unweit, längs, entlang, frei*, and English examples such as *along, near, worth, subsequent to, precious to*, etc." Quantifying adjectives such as 'some' or 'all', in particular, have experienced various grammaticalizations. For example, words for 'some' have occasionally given rise to indefinite articles, as has happened with *some* in earlier forms of English. Words for 'all' have developed into number markers in some languages, as in the French-based creole Tayo of New Caledonia, where French *tous les* 'all the' provided the source for the nominal plural proclitic or prefix *tule* (frequently reduced to *tle* or *te*) (Heine and Kuteva 2002a: 36).

Furthermore, in a number of languages, words for 'all' have given rise to comparative markers of superlative, as for example in the Finnic language Estonian, where the form *koīk* 'all' serves as a superlative marker ('of all'), and in the Baltic language Latvian, the quantifier *viss* 'all' has developed into the superlative prefix *vis-* (Stolz 1991b: 50–4).

Adjectives may give rise to adverbs. When this happens, adjectives are decategorialized in losing the ability to be inflected for number, gender, and/or case and turn into invariable forms that modify verbs or clauses rather than nouns. Latin adjectives are inflected for all these functions, but in those cases where they gave rise to adverbs, they were frozen in one of the possible inflectional forms. In this way, the adjective *ver-us* 'true' provided the source for the adverb *vero* 'truly, in fact', and *prim-us* 'first' for the adverb *primō* 'in the first place', and the degree adverbs *multum* 'much' and *quantum* 'how much' are accusative case forms of adjectives that have been decategorialized to invariable adverbs (Janson 1979: 101; Blake 1994: 182). In the Bantu language Swahili, adjectives are inflected for noun class gender; but those adjectives that developed into adverbs became unanalyzable morphemes, having the frozen noun class markers such as *ki-* (class 7) or *vi-* (class 8) on them, for example *-dogo* 'small' (adjective) > *kidogo* 'a little' (adverb), *-zuri* 'nice' (adjective) > *vizuri* 'nicely, well'. On the basis of such data one might argue that there is a pathway adjective > adverb. However, the overall evidence is not entirely conclusive and we have therefore decided not to postulate such a pathway, which is in need of further investigation.

So far, no crosslinguistically regular grammaticalization patterns leading from adjectives to minor functional categories have been identified; their potential for grammaticalization appears to be limited. Our focus in this section therefore is on adverbs.

Adverb > adposition Next to nouns and verbs, adverbs form the third major source of adpositions. Adverbs belong to those categories that show a particularly free word order behavior; with their grammaticalization to adpositions they are decategorialized in that they lose that freedom and become tied to the position next to a noun (phrase). Normally it is not the adverb on its own that develops into an adposition; rather, it tends to require some linking element, such as a case inflection or an adposition. In Swahili, for example, it is the comitative preposition *na* 'with' or the gender-sensitive possessive/genitive particle *-a* which link the adverb with the noun. Thus, the adverb *karibu* 'nearby', illustrated in (34a), requires *na* to form an adposition (34b), while the adverb *nje* 'outside' in (35a) requires the genitive particle *-a* for the same purpose (35b).

(34) Swahili (Bantu, Niger-Congo)
    a. Rafiki yangu a-   na-   kaa **karibu.**
       friend my   3.SG- PRES- live near
       'My friend lives nearby.'
    b. Rafiki yangu a-   na-   kaa **karibu na** shule.
       friend my   3.SG- PRES- live near   school
       'My friend lives near the school.'

(35) Swahili (Bantu, Niger-Congo)
    a. Tu-   me-   lala **nje.**
       1.PL- PERF- lie   outside
       'We have slept outside.'
    b. Tu-   me-   lala **nje y-a** nyumba.
       1.PL- PERF- lie   outside   house
       'We have slept outside the house.'

Adverb > demonstrative It has been argued that demonstratives are diachronically basic or „semantic primitives", in that they may give rise to other kinds of grammatical categories, while they themselves cannot be historically derived from other entities such as lexical items (see Plank 1979; Diessel 1999b: 150 ff.). This view is in need of revision: There is massive evidence to show that demonstrative determiners have locative adverbs as their primary diachronic source; in addition, verbs such as 'go' or 'see' may also form the source of demonstratives (see Heine and Kuteva 2002a: 172–3, 159, 294–5; see also Section 2.2.1.2).

The way that the process from adverb to demonstrative determiner is likely to happen is that ad-verbial modifiers typically denoting proximal ('here') and distal ('there') location are added appositionally to nouns (e.g. 'the house here/there') and grammaticalize to nominal determiners ('this/that house'). This process usually does not involve desemanticization, in that the markers concerned retain their locative semantics, but decategorialization has the effect that the markers become restricted in their occurrence to the position next to the noun; they are no longer part of the paradigm of adverbs, and tend to change from free words to nominal clitics, sometimes even to affixes.

In a number of languages, the locative adverb is added while the already existing determiner is retained, thereby giving rise to double marking. In French, the locative adverbs *ici* 'here' and *là* 'there' have been added to the noun. The earlier demonstratives *ce(t)* (masculine), *cette* (feminine), and *ces* (plural) have not been lost, but the semantic distinction proximal vs. distal is expressed by the erstwhile adverbs, for example *cet homme-ci* 'this man' vs. *cet homme-là* 'that man'. In Afrikaans, the locative adverbs *hier* 'here' and *daar* 'there' precede the determiner to form demonstratives, for example *hier-die wa* (here-the car) 'this car' vs. *daar-die wa* (there-the car) 'that car'.

As a rule, the adverb developing into a demonstrative is simply juxtaposed to the noun; there are however also languages where the adverb requires some linking morphology to assume the role of a demonstrative. In the Chadic language Mupun of Cameroon, a relative clause structure is used: The adverbs (called deictics by Frajzyngier 1995: 190) *sə̀* 'here' and *sə́* 'there', cf. (36a), require the relative clause marker *ɗə́* to form demonstratives, cf. (36b).

(36) Mupun (Chadic, Afroasiatic; Frajzyngier 1995: 190–1)
    a. wu   wa   sə́.
       3.M   come there
       'He came from there.'

    b. nləər   ɗə́sə̀   ret.
       shirt   DEM   nice
       'This shirt is nice.'

Creole languages offer a wide range of examples of this grammaticalization. For example, the Portuguese-based creole Angolar has developed a second series of demonstrative attributes derived from the locative adverbs *aki* (< Port. *aqui* 'here') 'here', *nha* 'there', *nhala* 'there', and *nge* 'here', which are added in apposition after a noun or adjective, for example:

(37) Angolar (Portuguese-based creole; Maurer 1995: 44)
    mbata aki *or* mbata nge   'this side'
    mbata nha   'that side'

In a similar fashion, Kouwenberg and Murray (1994: 49) report that in the Spanish-based creole Papiamentu, the adverbs *aki* 'here', *ei* 'there', and *aya* 'yonder' can take the final position in a definite noun phrase to mark demonstrative deixis. Note that the adverb-derived demonstrative is not necessarily restricted to nouns but may in some languages also be added to adjectives or determiners, as in the English-based creoles Jamaican and Guyanese:

(38) Jamaican (English-based creole; Mufwene 1986: 168)
   a. Da   buk-   de   a   fi   mi.
      that book- there   is   for   me
      'That book is mine.'

or

   b. Dat-   de   buk   a   fi   mi.
      that- there book   is   for   me
      'That book is mine.'

There are, however, a few counterexamples to the directionality adverb > demonstrative. In Bantu languages, demonstratives usually consist of a combination of noun class marker, agreeing with its head noun in gender, plus demonstrative stem. Now, the demonstrative can also occur without its head noun, and in some cases—when used with the agreement markers of the locative noun classes—such demonstratives can be reinterpreted as locative adverbs. For example, the locative adverb *hapa* 'here' of Swahili is historically a demonstrative consisting of the demonstrative stem *ha-* plus the agreement marker *-pa* of the locative noun class 16.

Adverb > subordinator It is mostly locative and temporal adverbs that provide the source for adverbial clause subordinators. We will deal with this pathway in Chapter 5 (Section 5.3.3.5), where we are concerned with patterns of clause combining. By way of illustration, one historically attested case may be mentioned. In early Latin there was a temporal adverb *dum* 'now' which was grammaticalized to the temporal clause subordinator *dum* 'while, as long as'. While the adverb was lost, except for a few relics, the subordinator survived in classical Latin (Janson 1979: 104–5).

Adverb > tense Clearly the most widespread source of tense markers is provided by verbs turning into auxiliaries of other verbs (see Verb > tense). While being less common than verbs, temporal adverbs nevertheless constitute a crosslinguistically significant source for tense markers. Desemanticization has the effect that the adverb loses its specific temporal significance, its temporal function as a tense marker being defined with reference to the verb to which it is attached. For example, in some languages, adverbs meaning 'tomorrow' have given rise to future tense markers, whereby their specific meaning ('one day later than now') gave way to the more general meaning 'later than now'. This process can be observed, for example, in a number of West African Kru languages, for example in Neyo, where the adverb *kɛɛlɛ* 'tomorrow' developed into the future marker *lɛ* (Marchese 1984: 206–7, 1986: 257). As this example shows, the process from adverb to tense marker may trigger erosion, whereby the phonetic substance of the form is reduced, even if not all instances of this process involve erosion.

Perhaps the most common instance of this pathway concerns adverbs meaning 'then' which are grammaticalized to future tense markers. In the East Nilotic language Bari of the Sudan, the adverb *(e)dé* 'then, afterwards' appears to have given rise to the future tense marker *dé*, and in the Bantu language Lingala, the adverb *ndé* 'then' turned into the future tense prefix *ndé-*. One may also mention that in the English-based pidgin/creole Tok Pisin the adverb *baimbai* 'afterwards, later' (historically derived from English *by-and-by*) developed into a future tense marker (see Heine and Kuteva 2002a: 293–4, 299).

Adverbs are fairly uncommon as sources for verbal aspect markers. However, there are occasionally grammaticalizations of adverbs as aspect markers. In American Sign Language (ASL), an adverb preceded by the (optionally) modulated verb can be used to express habituality:

(39) American Sign Language (ASL; Sexton 1999: 134)
 I GO$^{([+ \text{ habitual modulation}])}$ REGULAR
 'I go regularly.'

2.2.3 *The fourth layer: demonstratives, adpositions, aspects, and negation*

It is hard to find a common semantic denominator for categories of the fourth layer of grammaticalization. What they have in common is their grammaticalization behavior, that is, the fact that they tend to be derived from categories of the first three layers and that they provide sources for categories of subsequent layers. In fact, three of the four categories of this layer, demonstratives, adpositions, and verbal aspects, exhibit rich grammaticalization behavior, while negation does not: Negative markers do not normally grammaticalize to other functional categories.

Demonstrative > pronoun Demonstrative pronouns provide perhaps the most common source for third person pronouns. Desemanticization has the effect that the spatial deixis of the demonstrative is bleached out, giving way to contextual or co-textual deixis, while decategorialization means that the inflectional structure that the demonstrative may have is reduced, the result being, as a rule, that an invariable personal pronoun emerges, which is no longer part of the paradigm of demonstrative categories but becomes part of the paradigm of personal pronouns (Heine and Kuteva 2002a: 112–13).

A paradigm example is provided by the Latin distal demonstrative *ille* (masculine), *illa* (feminine) 'that', which provided the source for third person pronouns in the modern Romance languages (see Vincent 1995: 442), for example French *il* 'he', *elle* 'she'. Erosion had the effect that the disyllabic Latin source items were reduced to monosyllabic personal pronouns in the Romance languages. A similar development appears to have taken place in Ancient Egyptian: The proximal demonstrative pronoun *pw* 'this' was also used as a general third person pronoun ('he', 'she', 'it', 'they'; Gardiner 1957: 85 f., 103). Lezgian offers another example, where the distal demonstrative *a* 'that' in combination with the absolutive case serves as a third person singular pronoun *am* 'he', 'she', 'it' (Haspelmath 1993: 190, 401).

Demonstrative > definite marker   Crosslinguistically this appears to be the most common way in which definite articles arise (Greenberg 1978, 1991; Diessel 1999b; Heine and Kuteva 2002a: 109–11). The transition from demonstrative attribute to definite article appears typically to lead from exophoric to endophoric demonstrative and finally to definite marker (Diessel 1998: 7–8), and it is both proximal and distal demonstrative attributes that can be recruited.

In this process, desemanticization has the effect that the demonstrative loses its deictic (locative) content, such as the ability to express relative distance (e.g. proximal vs. distal). Via decategorialization, the demonstrative loses its independent status, becoming an appendage of its head noun, it changes from free word to clitic, and eventually it may turn into an affix. Finally, erosion may enter the process as well, in that the erstwhile demonstrative tends to be shortened and loses the ability to carry stress. A canonical instance of such a process can be observed in Chinook Jargon: the general demonstrative *úkuk* 'this, that' (deictic pronoun) was grammaticalized to a definite article *uk-* in Grand Ronde Chinook Jargon. This process involved three of the parameters distinguished above: The demonstrative lost its deictic semantics (desemanticization), its status as an independent word, becoming an NP-prefix (decategorialization), and it also lost in phonetic substance (erosion), in that *úkuk* was reduced to *uk-*.[8]

---

[8] The definite article of Grand Ronde Chinook Jargon is used in broadly the same contexts as the definite article in English, which suggests that contact-induced replication played a role in the process (see Heine and Kuteva 2005).

(40) Grand Ronde Chinook Jargon (Grant 1996: 234)
   uk-   háya-   haws
   (the- big-   house)
   'the big house'

In the development of the French definite article, erosion meant that the Latin distal demonstrative *ille* (masculine), *illa* (feminine) 'that' was reduced to *le* and *la*, respectively (see e.g. Vincent 1995: 442), and the Old Norse postposed demonstrative **hinn* 'that' (e.g. **úlfr hinn* 'that wolf') ended up in Danish as a definite singular marker for common gender *en*, as in *dreng-en* 'the boy' (Hopper and Traugott 1993: 9).

Observations made in a number of creoles suggest that this is a ubiquitous process. For example, Bruyn observes that the demonstrative *da* 'that' (presumably derived from English *that*) of the English-based creole Sranan assumed roughly the function of a definite article: "The 18th century determiner *da* could convey demonstrative force. Over the course of time, it increased in frequency, lost some of its deictic value, and has been reduced in form, via *na* to *a*" (Bruyn 1996: 39; see also Boretzky 1983: 97).

Demonstrative > relative clause marker   Demonstrative pronouns are presumably the most frequent source of markers introducing (restrictive) relative clauses, English *that* being a paradigm example (O'Neil 1977; Hopper and Traugott 1993: 190 ff.; see also Section 5.3.1.1). In contradistinction to the pathway from demonstrative to definite article, the present pathway concerns demonstratives used as pronouns rather than as nominal determiners, but as in the former case, both proximal and distal demonstratives may be used. Since we will return to this pathway in more detail in Chapter 5, a few observations may be enough to illustrate this grammaticalization process. The process involves on the one hand desemanticization, in that the spatial deixis of the demonstrative is bleached out, and on the other hand decategorialization, in that the demonstrative pronoun loses its freedom to occur on its own as an argument of the clause and is restricted to the function of presenting relative clauses.

The following example from the Central African language Baka illustrates the grammaticalization process. In example (41a), the proximal demonstrative *kɛ̀* 'this' serves as a nominal determiner, which like all other modifiers follows its head noun. In (41b), the same item serves as a pronoun that is used to present the following clause. The resulting relative clause is bracketed in that in addition to the clause-initial relativizer there is a clause-final marker *nè*, diachronically an adverb meaning 'here'.

(41) Baka (Ubangi, Niger-Congo; Kilian-Hatz 1995: 30)
a. wósɛ̀   kɛ̀
    woman   this
    'this woman'
b. bo   kɛ̀   ʔá   jao   wósɛ̀   kɛ̀   **nè**
    person   REL   s/he.PERF   marry.PAST   woman   this   REL
    ʔé   gɔ̀ɛ   a   gba   a   ngɛ́.
    s/he   go.PAST   ALL   village   POSS   3.SG.POSS
    'The man who married this woman went to his village.'

Demonstrative > complementizer We will deal in some detail with this pathway of grammaticalization in Section 5.3.2.4; suffice it to cite an example from English, where *that* was originally a distal demonstrative pronoun (*þæt*) but was—already by the stage of Old English—grammaticalized to an object clause complementizer and, beginning with the fourteenth century, also to a subject clause complementizer (see Hopper and Traugott 1993: 185–9). Roughly the same process happened in German, where the singular neuter form of the demonstrative *thaʒ* of Old High German developed into the complementizer *dass* (distinguished orthographically, though not phonetically, from the Modern High German neuter article *das*, the successor of the demonstrative *thaʒ*). The process from demonstrative pronoun to complementizer is universally attested; it is most commonly found in Africa, where several hundred or more languages appear to have undergone this development.

Aspect > tense All the available crosslinguistic data suggest that the structuring of events is directional in that there appears to be a unidirectional process whereby linguistic expressions for linguistic aspect, such as completive or perfect markers, highlighting the boundary behavior of events, may gradually develop into tense markers, that is, expressions for deictic time such as past tense, while we are not aware of any convincing evidence to show that a past tense marker turned into an aspect marker such as a perfect (see Bybee, Perkins, and Pagliuca 1994). The evidence that is available essentially concerns two pathways of development, namely that from perfect (anterior) to past tense markers, and from progressive to present tense markers.

A number of European languages are experiencing a process from perfect expressing current relevance (= present anteriors) to past tense markers, and this development can be described in terms of the following four-stage model (see Heine and Kuteva 2006 for details).

Stage 0: This stage is characterized by the presence of a perfect, that is, a present anterior form expressing current relevance but having no aorist/preterite function.

Stage 1: The perfect spreads into the domain of the past, assuming functions of an aorist/preterite, and it competes with the already existing markers expressing past time reference. At a more advanced stage, aorist/preterite is marginalized.

Stage 2: Aorist/preterite is completely lost, there is now only one form for the semantic fields of current relevance, perfective past and imperfective past. The erstwhile perfect can no longer be combined with past time markers to form pluperfects.

Stage 3: The erstwhile perfect marker starts to take over other typical features of past time markers. It can no longer be combined with future markers to form future perfects, and it acquires modal uses.

Adposition > case marker When discussing the development from noun to case marker we observed that there is an intermediate stage where the relevant form serves as an adposition (see also Noun > adposition); accordingly, there is a more extensive development noun > adposition > case marker (see Lehmann 1982: 79–86, 1985: 304). However, since adpositions are not only derived from nouns but also from other categories (see Verb > adposition, Adverb > adposition), the present pathway is not restricted to case markers of nominal origin. When turning into case markers, adpositions lose most or all of the semantic content they may have had, such as expressing spatial deixis (desemanticization), they become restricted in their occurrence to contexts where they express a syntactic function, they develop into noun phrase clitics or even affixes (decategorialization), and they tend to lose in phonetic substance (erosion). The following general remarks on western European languages also apply in some way or other to many languages in other parts of the world:

Thus the prepositions Engl. *of,* French *de,* German *von* all had a fuller ablative meaning, but are now largely devoid of it and mostly used as attributors. The fate of English *to*, Romance *a* is similar: they have been grammaticalized from directional prepositions to case markers of the dative and, in Spanish, even the accusative. (Lehmann 1982: 82)

In many languages, like the ones just cited, the resulting case markers do not develop into case affixes but rather remain clitics, or clitics bordering affixation, and rather than cliticizing on the noun, they can also merge with a determiner, as has happened in a number of languages that use both prepositions and pre-posed determiners, such as French (*à le* > *au* 'to the') or German (*zu dem* > *zum* 'to the').

There is massive evidence that this development is essentially unidirectional. For example, in its preliterary period before 1200 AD, Hungarian had postpositions; but as from the beginning of the literary tradition, the postpositions appear as case suffixes (Lehmann 1982: 85).

Adposition > complementizer As we will see in more detail in Section 5.2, prepositions and postpositions, being heads of noun phrases, commonly develop into markers of complement clauses, that is complementizers. For example, *for* was a preposition of location and purpose in Old English, as in (42a), but came to be used as a complementizer by early Middle English, cf. (42b) (van Gelderen 2004: 30).

(42) English (van Gelderen 2004: 30)
    a. Old English (*Beowulf*, 358)
        þæt   he   **for**   eaxlum   gestod
        that   he   before shoulders stepped
        'that he stood in front of [. . .].'
    b. Early Middle English (Layamon, Otho, 5523)
        moche   he   lofde   echn(e)   cniht.   þat   lofde
        much   he   loved   every   knight that loved
        for   to   segg(e)   riht.
        for   to   say   truth
        'Much he loved every knight who loved for to say the truth.'

Adposition > subordinator Adpositions becoming case markers may develop further into markers of adverbial clause subordination (see Case marker > subordinator), but they may as well move straight to marking adverbial clauses without an intervening stage of case marker. This will be discussed in Section 5.2. The following example illustrates this pathway: The ergative/instrumental marker *-na* of the Tibeto-Burman language

Newari, described as a postposition of nouns (43a), is hypothesized to have given rise to the temporal clause subordinator -*na* suffixed to verbs (43b).

(43) Dolakhari Newari (Tibeto-Burman; Genetti 1991: 227)
a. cotan-   **na**   pol-   ju.
   spoon-   INSTR   strike-   3.SG.PAST
   'He ate with a spoon.'
b. chē-   ku   yer-   **na**   wā   ām-   e   naku
   house- LOC   come-   when   EMPH   he-   GEN cheek
   moŋ-   an   coŋ-   gu.
   swell- PART stay-   3.SG.PAST.HAB
   'When he came to the house his cheek was swollen.'

2.2.4 The fifth layer

The preceding sections have demonstrated how linguistic expressions having fairly transparent lexical meanings are grammaticalized to forms that still preserve properties of their lexical sources but at the same time form crosslinguistically stable functional categories: Most languages distinguish demonstratives, adpositions, or verbal aspects as salient concepts of linguistic categorization. The present section concerns a range of grammatical functions that are commonly derived historically from such categories. It is hard to find a common semantic or syntactic denominator characterizing this group of functional taxa other than that they are the product of the grammaticalization of other functional taxa of layer IV or any other preceding layer. Furthermore, it is also virtually impossible to narrow down the range of functional taxa that could reasonably be allocated to this layer—considering that there is a bewildering variety of grammatical categories in the languages of the world that in some way or other belong to this layer. To discuss all the pathways of grammaticalization that emanate from the layers examined in the preceding sections would clearly be beyond the scope of the present treatment, quite apart from the fact that appropriate crosslinguistic typological information is absent for most of them; we will be confined to a small range of such pathways. The categories concerned are devoid of any lexical content, their primary, or their only function being to establish relations among participants of linguistic discourse, such as linking constituents of discourse (agreement), manipulating the argument structure (passives), and marking relations between different levels of syntactic structure (subordination).

As elsewhere in this section, we are confined to pathways that are crosslinguistically salient, ignoring further pathways that are only locally found. For example, in the Cariban language Panare of South America, a development from demonstrative pronoun to tense marker has been observed (Gildea 1993), which is ignored here since we are not aware of similar developments in other languages.

Case marker > subordinator There is a wide range of conceptual sources leading to the growth of adverbial clause subordinators (see Section 5.3.3), such as Noun > subordinator, Verb > subordinator, Adposition > subordinator, Complementizer > subordinator, but these do not exhaust the range of possibilities; mention should be made, for example, of the fact that relative markers may give rise to other forms of subordination including adverbial clause subordination (Heine and Kuteva 2002a: 254–5; see also Section 5.3.4). Since we are confined here to the crosslinguistically most salient developments, we have excluded such pathways. But one development that is salient concerns that from nominal to clausal syntax. Examples demonstrating that the use of markers expressing distinctions of case on nouns or noun phrases is also extended to marking subordinate clauses can be found across the world in languages that have productive paradigms of case inflections. Most of these instances of extension from nominal case marker to clause subordinator, however, involve non-finite subordinate clauses, that is, verbs in such clauses are not inflected for person, number, tense–aspect, etc. (see Section 5.2). However, there are sufficient examples to show that this extension may also lead to case-marked finite subordinate clauses.

We will deal with this process in more detail in Section 2.2.6; the following example illustrates the process. In the Ik language of northeastern Uganda, which has an elaborate paradigm of case inflections, one of these inflections, the dative suffix -k^e, has been extended to use as a clause subordinator (Heine 1990; König 2002). Example (44a) illustrates the use of the suffix as a nominal case inflection, while (44b) shows the very case marker being used as a subordinator. That this was a unidirectional process from nominal to clausal inflection is suggested by the fact that the former, but not the latter, can be reconstructed back to Proto-Kuliak, the hypothetical ancestor of the family to which Ik belongs.

(44) Ik (Kuliak, Nilo-Saharan; Heine 1990)
    a. …kʲe-   esá   ntsa   awá- k^e.
        go-   FUT   he   home-   DAT
     '…and he will go home.'

b. ńtá   k'ó- í- í   ma-   í-   í-   kᵉ.
   NEG   go-   I- NEG   be.sick-   I-   OPT- DAT
   'I cannot leave because I am sick.'

Complementizer, relativizer > subordinator Complementizers and adverbial clause subordinators are both means of clause subordination, but the evidence available suggests that they do not belong to the same level of grammatical development. We will discuss this evidence in Chapter 5 (Section 5.3.4; see also 6.4.1.2), hence we are confined here to summarizing the main lines of development leading from markers of complement clauses to markers of adverbial clause subordination. The development of complementizers includes two major pathways of grammaticalization, one involving demonstrative pronouns (e.g. English *that*) and the other involving a verb meaning 'say' (e.g. Ewe *bé* 'say'). Once these two kinds of forms have been established as markers introducing clausal complements, they may undergo extension in that their use is extended to present subordinate clauses that are not part of the valence of the main (or matrix) verb.

Note, however, that the pathways discussed here concern only a small range of subordinators; in the vast majority of cases where new categories of subordinators evolve there is no intermediate stage of complementizers. An example from relative clause marker to subordinator will be discussed in Section 6.4.1.1.

Pronoun > agreement Morphological forms described as agreement markers are, as a rule, the result of processes whereby functional categories having a clear-cut lexical or grammatical function are desemanticized to the effect that they have no more meaning other than signalling syntactic relations across words and phrases, and decategorialized in losing their independent status and becoming clitics or affixes. A paradigm instance is provided by personal subject pronouns which become obligatory clitics or affixes on verbs (see Donohue 2003 for examples in the Skou language of New Guinea). In the Palu'e language of Flores, Indonesia, there is no agreement, and none of the closely related languages has agreement. But Palu'e does have one agreement clitic, the proclitic *ak-*, related to *aku* 'first person singular' (45a), which can only mark a nominative argument (45b). It is clearly part of the same phonological word as the verb it attaches to, although it has not become an obligatory part of the verb, being exclusory of any free pronoun, cf. (45c).

(45) Palu'e (Austronesian; Greville Corbett, Linguist List, 2005; data from Mark Donohue, p.c.)
a. **Aku** pana.
   1.SG go
   'I went.'
b. **Ak-** pana.
   1.SG.NOM- go
   'I went'
c. *Aku ak- pana.
   1.SG 1.SG go

Existing personal pronouns can be replaced by a set of new personal pronouns but survive as verbal clitics or affixes, although they have no more distinctive function since distinctions of personal deixis are now expressed by the new set of personal pronouns. The only function that the old pronouns may have is to express coreference with the pronominal or nominal subject, that is, to mark agreement between the subject and the verb.

For example, the personal suffixes of Latin verbs functioned as personal pronouns since they expressed distinctions of personal deixis, for example *am-at* 's/he loves', *am-amus* 'we love'. In the development from Latin to French, distinctions of personal deixis came to be encoded by either absolute pronouns or nouns, with the result that the old personal pronouns lost their function as deictic markers and survived as predictable markers of agreement between verb and subject, for example *il aime* 'he loves', *nous aim-ons* 'we love'. But French appears to be undergoing a second development from personal pronoun to agreement marker: The new French personal pronoun *il* 'he' (derived from the Latin distal demonstrative *ille* (masculine) 'that (one)') has become an agreement marker in non-standard French, that is, a predictable form bound to the verb, no longer distinguishing number or gender, cf. (46). Accordingly, French *il* is suggestive of an extended pathway Demonstrative > pronoun > agreement (see Demonstrative > pronoun).

(46) Non-standard French (Hopper and Traugott 1993:17; Lambrecht 1981: 40)
Ma femme il est venu.
my.F wife AGR is come
'My wife has come.'

The final stage of grammaticalization is in fact reached when the erstwhile personal pronoun is no longer functionally distinctive, thereby becoming a "meaningless" appendage of the verb. For example, the English personal pronoun *he* turned in English-based Melanesian pidgins such as Tok Pisin of Papua New Guinea or Bislama of Vanuatu into a redundant proclitic *i* placed before the verb, initially only after third person subjects and later on via extension also to first and second person subjects, as in the following example:

(47) Bislama (English-based pidgin; Meyerhoff 2001: 254)
Afta   mi   stap   wokbaot   i   go.
after   1.SG   PROG   walk   AGR   go
'And I started to walk off.'

While less common, roughly the same process may also lead to object agreement, whereby personal pronouns in object function become obligatory markers on the verb (see Lehmann 1982: 42).

Evidence for this pathway also comes from signed languages, where the signing space, in combination with pointing signs (or eye gaze), is recruited as the primary means for defining distinctions of personal deixis and agreement. Thus, Pfau and Steinbach (2006: 34) reconstruct the following grammaticalization chain for agreement markers: Pointing gesture > locative > demonstrative pronoun > personal pronoun > agreement marker.

Pronoun > passive There is a wide variety of forms and structures leading to the emergence of passive markers and constructions (Haspelmath 1990; Givón 2004; see also Verb > passive). Here we are concerned with only one pathway, whereby personal pronouns are reinterpreted as markers of passive constructions. The way this happens is that third-person plural subject pronouns of transitive sentences undergo desemanticization, losing their meaning and turning into markers whose only function it is to signal a syntactic configuration (see Haspelmath 1990; Heine and Kuteva 2002a: 236–7 for examples; Givón 2004). Many languages exhibit the initial stage of this process, even if this is usually not appreciated by grammarians; the following is an example from Basque:

(48) Basque (Heine and Kuteva 2002a: 236)
Hil   z-   u-   te-   n.
kill[PFV]   PAST-   AUX-   3.PL.ERG-   PAST
'They killed him' = 'He was killed.'

In a number of languages, this process has given rise to fully-fledged passive constructions, where an adjunct noun phrase has been introduced to present the agent which was formerly expressed by the third-person plural subject pronoun (see Givón 1979b: 188, 211 for a canonical example and discussion; see also Givón 2004). In the following example from the Congolese Bantu language Luba, the pronoun *ba-* 'they' of the human noun class 2 (C2) has been reinterpreted as a passive prefix, the patient is structurally a direct object, and the new agent is presented in the form of an existential phrase:

(49) Luba (Bantu, Niger-Congo; Heine and Reh 1984: 99)
 bà- sùm- ìne mu- âna kù- dì nyòka.
 C2/they- bite- PERF C1- child there.where- is snake
 'The child has been bitten by a snake.'

The Maasai dialect of the East Nilotic language Maa shows the final stage of this process, where the third-person plural pronoun *ki* 'they' has lost its function as a personal pronoun and is now exclusively a passive suffix (*-ki* or *-i*) (Greenberg 1959; Heine and Claudi 1986: 79–84).

An equally widespread pathway from pronoun to passive marker concerns reflexive pronouns. Unlike the process just sketched, in this pathway it is not pronouns as subject arguments but rather pronouns serving as complements, typically as clausal objects, that undergo grammaticalization. Once again, this process entails desemanticization, in that the function of the reflexive marker to express reference identity is lost and that marker is reduced to signaling a syntactic relation. However, this process does not lead straight to the passive but involves one or more intermediate stages, where the reflexive first assumes an intransitivizing, deobjective, and/or anticausative function before becoming a passive marker (Kemmer 1993; Haspelmath 1990; Heine & Miyashita 2004); we will return to this process in Section 2.4.

2.2.5 The final stages

The developments sketched in the previous section by no means constitute the endpoint of grammaticalization; rather, as has been demonstrated independently in many works, the process may continue until both the meaning and the phonetic substance of the item undergoing grammaticalization are lost. For example, when definite articles arise out of demonstrative attributes in accordance with the Demonstrative > definite article pathway, they may further undergo extension, spreading from definite to all specific nouns (= "Stage-II articles"), and finally their use can be generalized to occur with virtually all nouns, except, for example, for some generic uses, becoming an obligatory appendage of nouns, thereby acquiring the status of signaling nominality (= "Stage-III articles"; Greenberg 1978, 1991). In a similar fashion, when a tense marker arises, this need not be the end of grammaticalization. Tense markers may develop further into markers of epistemic modality and/or they may assume subordinating functions (see, e.g., Section 1.1.2 (2d)).

The following example, being a special instance of the Verb > complementizer pathway, illustrates what may happen in the final stages of the process. The more a verb in combination with another verb loses in meaning content, the more it tends to merge with the other verb, and in the end the two may lexicalize into a new verb. In the Saharan languages of the Nilo-Saharan family, a verb for 'say' has been added productively to other verbs, adjectives, and nouns serving as its complements, one of its functions being to derive verbs from other word categories.[9] In the course of this process, the 'say'-verb has merged with its complement to the extent that it has lost its meaning entirely, the result being a new verb. The following example from the Nigerian Kanuri illustrates the various stages of development. Sentence (50a) shows the lexical use of the verb, (50b) that of 'say' as an element deriving (inchoative) verbs from adjectives (in this case, kúrà 'big'), and (50c) illustrates the use of 'say' as a semantically empty inflectional appendage of a verb, that is, its erstwhile complement.

(50) Kanuri (Saharan, Nilo-Saharan; Crass *et al.* 2001: 132–8)

    a. fátòrò   lèné   **wònò**.
       house.to go.IMP say.PAST.3.SG
       'He said: 'Go home!'

    b. kùrà-   **jin.**
       big.ADJ- say.IMPFV.3.SG
       'He's becoming big.'

    c. gùl-   **wónò**
       say-   say.PAST.3.SG
       'He said'

---

[9] It is largely unclear what its function was when combined with other verbs.

The erstwhile verb has undergone massive erosion, surviving in most cases solely as an alveolar nasal consonant, and in specific contexts even this consonant disappears, leaving no segmental trace of the erstwhile verb. The result of this process is that in Kanuri and other Chadic languages, a new class of verbs has evolved, where the verb 'say' has been reduced to a phonological relic of the verb stem. This class constitutes one of the three verb classes commonly distinguished in Saharan languages; it is actually the largest and most productive of the three classes, accounting for up to 90 percent of all verbs.

Thus, one possible final stage of grammaticalization is that the item undergoing the process loses its function and ends up as a phonological appendage of another form, serving no purpose, or no purpose other than identifying these forms as belonging to the same lexical or morphological paradigm.[10] In this way, they may become markers identifying noun classes or nouns (Greenberg 1978), or verb classes or verbs. Alternatively, and perhaps more commonly, the item may be lost entirely without leaving any semantic or formal traces.

2.2.6 Treating events like objects

We observed in Section 2.2.1 that nouns may behave like verbs and verbs like nouns and that so far it has not been possible to establish any significant directionality regarding the grammaticalization of the two kinds of categories. Nevertheless, the relationship between the two is far from symmetric; rather, there appears to be a unidirectional development whereby expressions reserved for nouns, or nominal concepts, are exploited for encoding actions or events, that is, concepts that are typically expressed by verbs; conversely, there is no evidence to suggest that verbs are regularly grammaticalized to express nominal concepts. Crosslinguistic data suggest that there are cognitive-communicative strategies such as the following:

(a) verbal predicates tend to be structured in terms of nominal morphology (subordination);
(b) verbs tend to be presented like nouns (auxiliation);
(c) verbal complements are treated like nominal complements (negation);
(d) actions and events tend to be treated pronominally like nouns.

---

[10] It happens occasionally that such function-less items are re-grammaticalized, or "exapted" (Lass 1990, 1997).

We will deal with each of these strategies in turn; for a more detailed treatment of some of the issues discussed here, see Chapter 5.

(a) Treat subordinate clauses like nouns (subordination) As we will see in more detail in Chapter 5, this strategy can most readily be described by looking at case languages, that is, languages that express arguments of the clause by means of case clitics or affixes. In most case languages that we are familiar with, the use of some case marker or markers is extended productively from nouns to verbal predicates—with the effect that a subordinate clause structure arises. When a nominal structure is reinterpreted as a clausal structure, the result may be that a genitive case marker, marking the nominal modifier, is reinterpreted as an ergative or agent case marker (see e.g. Dixon 1994). Conversely, when in a given language there is an ergative case marker which is the same as, or is historically derived from a genitive marker, this is likely to reflect a historical process leading from nominal to clausal structure.

One possible effect of this grammaticalization from nominal case to verbal subordination marker is that the resulting subordinate clause exhibits nominal properties whereas the main clause does not; we will deal with this effect in more detail in Section 5.2. Dixon (1994: 192) observes that in the Carib languages of South America, subordinate clauses generally have the status of nominalizations and show an ergative pattern. In the following example from the Nilo-Saharan language Ik of Uganda, the (non-finite) complement verb 'to eat' in (51a) appears in the nominative case (NOM)[11] and the "object" (patient) of the complement clause in the genitive case (GEN). What characterizes this pathway is that complement clauses are structured on the model of nouns. In (51a), the verb bɛd- 'want' has a nominal complement in the nominative, and this structure is similar to that of the complement clause of (51b), exhibiting morphosyntactic properties characteristic of the nominal structure. Thus, the literal meaning of (51b) is something like 'I want the eating of food'.

(51) Ik (Kuliak, Nilo-Saharan; König 2002)
    a. bɛd-   ía   mes-   ᵃ.
        want- 1.SG beer- NOM
        'I want beer.'

---

[11] Ik has a general rule according to which clausal objects are encoded in the nominative (NOM) when the subject has first- or second-person reference, as in the present example, but in the accusative (ACC) when the subject has third-person reference.

b. bɛɗ-   ía   atsʲ-   ésa   ŋƙáƙá-   é.
   want-   1.SG   eat-   INF.NOM   food-   GEN
   'I want to eat food (or meat).' (Lit.: 'I want the eating of food'.)

In other languages, the extension from nominal case suffix to verbal predicate does not affect the structure of the subordinate clause, which takes the form of a finite-verb clause. Thus, in the following example from Imbabura Quechua the complement clause (in square brackets) has the finite structure of a main clause but receives the case marking that would be used if there were a nominal complement:

(52) Imbabura Quechua (Cole 1982: 43)
    Pedro   ya-   n   [ñuka   Agatu-   pi   kawsa-   ni]   -ta.
    Pedro   think-   3   I   Agato-   in   live-   1-   -ACC
    'Pedro thinks that I live in Agato.'

Languages of this type tend to use this kind of construction only for complement clauses with a limited spectrum of main clause (matrix) verbs, most of all speech-act, cognition, and/or verbs of volition as matrix verbs. Furthermore, many of these languages restrict this structure to complement clauses where the subject of the complement clause is co-referential with that of the main clause subject.

A widespread strategy concerns the extension of allative/goal or benefactive/dative case markers from nouns to verbs in order to introduce purpose clauses. Thus, the benefactive/dative case suffix *-gu* or *-wu* (DAT) of the Jaminjung language of Northern Australia serves as a nominal case suffix in (53a) but as a marker of non-finite purpose clauses in (53b).

(53) Jaminjung (Jaminjungan; Schultze-Berndt 2000: 56, 111)
    a. [...] janju   jalig-   **gu**
              DEM   child-   DAT
       '[...] for that child'
    b. burr-   irriga   jawaya-   **wu.**
       3.PL.3.SG-   COOK.PAST   eating-   DAT
       'They cooked it for eating.'

In the South Caucasian language Laz, spoken in Turkey, it is possible to have a dative marker cliticized to a finite verb form, thereby turning a main clause, as in (54a), into a subordinate one (54b):

(54) Laz (South Caucasian; Nino Amiridze; Funknet, April 2005)
    a. ali   oxori-   sha   mo-   xt-   u.
       Ali   house-   in   PREVERB-   come-   S3.SG.AOR
       'Ali came home.'

b. ali   oxori-   sha   mo-   xt-   u-   **shi**.
   Ali   house-   in   PREVERB-   come-   S.3.SG.AOR-   DAT
   'When Ali came home (...).'

Essentially the same process appears to have happened in the Ik language of northeastern Uganda: The dative suffix -k^e, illustrated in (55a), has been added to the finite verb form, which is constructed in the optative mood (using the suffix -i), and in this usage has been generalized both as a clause subordinator and as a complementizer, as in (55b) (see Heine 1990 for details).

(55) Ik (Kuliak, Nilo-Saharan; Heine 1990; König 2002)
   a. ia   ŋóká   ńci- k^e.
      be   dog-NOM   I-   DAT
      'I have a dog.' (Lit.: 'A dog is to me.')
   b. bɛɗá   yakw-   á   ńci-   a   wet-   í-
      want   man-   NOM   1.SG-   ACC   drink-   OPT-
      í- k^e.
      1.SG-   DAT
      'The man wants that I drink.'

In a similar fashion, nominal case markers have been extended in Omotic languages of Ethiopia to serve as clause subordinators. For example, the dative (benefactive) case suffix -m preceded by the absolutive case suffix -o, illustrated in (56a), appears to have developed into the purpose clause subordinator -$óm$ (PURP) in the Maale language, cf. (56b). Decategorialization had the effect that the two nominal case suffixes merged into one form.

(56) Maale (Ometo, Omotic, Afroasiatic; Amha 2001: 58–9, 186–7)
   a. ʔíiní   ʔaʃk-   ó   ʔas-   ó-   m   pák'-
      3.M.SG.NOM   meat-   ABS   people-   ABS-   DAT   divide-
      é-   ne.
      PFV-   A.DCL
      'He divided the meat among the people.'
   b. ba-   at-   á   késk-   **óm**   karr-   ó   búll-
      cattle- PL-   NOM   go.out-   PURP   door-   ABS   open-
      é-   ne.
      PFV-   A.DCL
      '(Someone) opened the door so that the cattle go out.'

Massive evidence showing that nominal case markers are grammaticalized to clause subordinators comes from the Tibeto-Burman Bodic languages. There is a wealth of data describing the move from prepositions or postpositions, being heads of noun phrases, to adverbial clause markers, that is, heads of adverbial clauses (see Genetti 1986 for examples from Bodic languages; we will return to this situation in Section 5.2).

Obviously, the treat-subordinate-clauses-like-nouns strategy is found to be at work most of all in what Givón (2006) calls nominalizing languages, such as Tibeto-Burman, Turkic, Carib, Quechuan, Northern Uto-Aztecan, Omotic, Finno-Ugric languages, or the Gorokan languages of the Papuan Highlands. But the very same strategy, whereby adpositions, case affixes, or other noun-specific morphologies are grafted onto verbal structures, can also be observed in many other languages, even if the implications are less pervasive. English has a number of instances of it, involving prepositions such as *after*, *before*, *for*, *until*, etc.; consider the example in (57), where the use of the prepositional form *in terms of* appears to have been extended to nominal to adverbial clause complements—in accordance with the general conceptual and syntactic process that we sketched earlier.

(57) English (Hopper and Traugott 2003: 185)
They're a general nuisance **in terms of** they harass people trying to enjoy the park.

What distinguishes this English example from the ones discussed earlier is the fact that it represents only an incipient stage of grammaticalization: *in terms of* has not, or not yet, developed into a regular marker of adverbial clause subordination; rather, it appears to be a discourse option that is available in some registers of spoken English but has not been generalized across the speech community.

(b) Treat verbs like nominal complements (auxiliation) There is a universally attested diachronic process of grammaticalization leading from structure (58a) to (58b).

(58) A morphosyntactic model of auxiliarization[12]
    a. Verb   nominal complement
    b. Auxiliary main verb

---

[12] The constituent ordering in (58) is the one to be expected in verb-initial (VSO) and verb-medial (SVO) languages; in verb-final (SOV) languages, the reverse order would be expected.

Since this is a gradual process, it involves many intermediate stages; we are restricted here to the most salient stages of the process. Examples that we discussed in Section 1.1.3 of Chapter 1 illustrate this process. We may add two further examples to show that the process is crosslinguistically regular. The first example comes from the West African language Maninka, where verbs in the progressive aspect (59a) are treated like locative nouns (59b).

(59) Maninka (Mande, Niger-Congo; Friedländer 1992: 51, 67; quoted from Tröbs 1998: 37)
    a. Seku   yé   bún   **ná**.
       Seku   PM   house   at
       'Seku is at home.'
    b. à   yé   kàrán   **ná**.
       3.SG   PM   learn   at
       'He is learning, he learns.' (Diachronically: 'He is at learning.')

The second example is taken from the Ik language, where the use of the verb *cɛm-ɛ́s* 'to fight' was extended from nominal complements, as in (60a), to non-finite verbs, and finally to verbs that were reinterpreted as new main verbs—with the effect that *cɛm-ɛ́s* was "downgraded" to an auxiliary, grammaticalized to a progressive marker, as in (60b), where its lexical semantics is completely bleached out. That *cɛm-ɛ́s* is fully grammaticalized, that is, has reached the stage of conventionalization, and thus is a fully-fledged functional category, can be seen in the fact that the auxiliary can co-occur with its lexical source (60b). And just as the auxiliary requires the main verb to take the ablative suffix in (60b), so does the main verb with its complement in (60a).

(60) Ik (Kuliak, Nilo-Saharan; König 2002)
    a. cɛm-   í-   á   ɛ́ɓ-   ^o.
       fight-   1.SG-   *a*   gun-   ABL
       'I fight with a gun.'
    b. cɛm-   í-   á   cɛ́m-   on-   ^o.
       PROG-   1.SG-   *a*   fight-   INF-   ABL
       'I am fighting.'

(c) Treat verbal complements like nominal complements   Another manifestation of the extension from nominal to verbal structures can be observed in the rise of negation markers that we mentioned above in Section 2.2.1.2: In some languages the use of negative existential verbs has been extended

from nominal (N) complements ('there is no N') to verbal (V) complements ('there is no V' > 'V does not happen'), and in the latter case the existential verb may assume the function of a negation marker.

We saw one example of this process earlier (Section 2.2.1.2): In Chinese, the negative existential *méi* [*yŏu*] takes nominal complements, as in (29b), but its use appears to have been extended to verbal complements, as in (29c), with the result that there now is a new negation marker of completed actions. For more examples and details of this process, see Croft (1991).

(d) Treat verbal interrogatives like nouns That there is a unidirectional development whereby the use of nominal structures is extended to verbal structures can be demonstrated with other kinds of evidence. One piece of evidence concerns pronominalization. For example, a typological survey of question pronouns suggests that there is a widespread process whereby interrogative pronouns referring to inanimate objects ('what?') are extended to also refer to actions and events (Heine, Claudi, and Hünnemeyer 1991: 56 ff.). Evidence for this directionality comes in particular from languages where the interrogative pronoun is etymologically transparent: In such languages the pronoun is not infrequently derived from a phrase 'which thing?' For example, in the Ewe language of Togo and Ghana, the pronoun *nú-ka* 'what?' means historically 'thing-which?', but is used in the same way for nominal and for verbal referents, as in (61), and the interrogative pronoun m̃tcí 'what?' of the !Xun language of southwestern Africa, which is historically composed of the interrogative element *m̃ [13] and the noun *tcí* 'thing', is not restricted to nominal referents but is used in much the same way to refer to actions and events, cf. (62).

(61) Standard Ewe (Kwa; Niger-Congo)
nú-ka   wo-   ḿ   ne-   le?
what   o   PROG   2.SG-   PROG
'What are you doing?'

(62) !Xun (W2 dialect, North Khoisan; König in prep.)
m̃tcí   á   hà-   è   ò?
Q.thing Q   N1   REL   do
'What does he do?'

Further data is found in pidgins and creoles, where not uncommonly the question word referring to actions ('what?') is transparently derived from

---

[13] *m̃ is no longer a productive morpheme in !Xun.

the phrase 'which thing?', as in the following example from the Spanish-based creole Papiamentu:

(63) Papiamentu (Holm 1988: 180)
    Ta **kiko** Wan ta hasi?
    (is what.thing John TAM do)
    'What is John doing?'

Essentially the same kind of evidence comes from affirmative pronouns ('that'): It is demonstrative pronouns, that is, items typically reserved for nominal referents, that tend to be extended to also refer to actions and events, for example English *Don't do **that** again.*

Evidence for directionality To conclude, the evidence presented in this section suggests that there is a unidirectional relationship between nominal and verbal concepts, whereby the former tend to be recruited as templates for expressing the latter—a relationship that appears to be based on a common human strategy to conceptualize and describe non-time-stable, dynamic phenomena (actions and events) in terms of time-stable, thing-like phenomena (nouns). This relationship is of a different kind than the one we were dealing with in all other cases discussed elsewhere in this chapter (hence, it is marked with a dotted line in Figure 2.1)—however it is one that shows the same kind of regularity.

The evidence for this regularity is taken from diachronic observations: There are historically attested examples of change from nominal case marker to clause subordinator while we are unaware of any examples where clause subordinators developed into case suffixes[14] (Heine and Kuteva 2002a). And much the same applies to the auxiliation process: There are attested cases where, in accordance with (58), a structure [verb–nominal complement] was grammaticalized to another structure [auxiliary–main verb]. For example, in the history of English it was possible to say something like *John is going to town* before it was possible to also say *John is going to come*, that is, English *to go* could take nominal complements long before its use was extended to verbal complements in a future tense construction. And once again, there is essentially no evidence for a process in the opposite direction, that is, where a tense or aspect construction developed into one of nominal complementation.

---

[14] Even if such examples should become available they would clearly be exceptional.

And finally, there is also evidence to show that nouns meaning 'thing' were recruited to form interrogative pronouns ('which thing') that refer not only to nominal participants but also actions and events. In Swahili, the noun *kitu* 'thing' is unambiguously a noun denoting thing-like entities. As we will see in Chapter 4, in Kenya Pidgin Swahili this noun has been grammaticalized in combination with the attributive interrogative *gani* 'which?' to a question word (*kitu gani* 'what?') referring not only to thing-like entities but in the same way to actions and events, for example:

(64) Kenya Pidgin Swahili
Kamau   na-   taka   fanya   **kitu**   **gani**?   Na-   taka
Kamau   NF-   want   do   thing   which   NF-   want
kimbia   tu.
run.away   only
'What does Kamau want to do? He just wants to run away.'

2.3 Evidence from signed languages

As we concluded in the preceding section, the first layers of grammatical evolution were restricted to lexical categories, which are the earliest forms of grammar that are accessible via grammaticalization in spoken (and written) languages. This raises the question of whether it may not be possible to push reconstruction back to linguistic forms that may have preceded lexical categories. There is no lack of works dealing with this question, nor of hypotheses that have been voiced; nevertheless, we will not deal with these hypotheses here because they are generally based on premises that are empirically not satisfactory in every respect.

But there is one possible exception and it concerns signed languages. As has been demonstrated in recent work, signed languages show roughly the same kinds of grammaticalizations that we observe in spoken languages (see e.g. Janzen 1995, 1998, 1999; Janzen and Shaffer 2002; Morford 2002; Sexton 1999; Pfau 2004; Pfau and Steinbach 2005, 2006; Shaffer 2000; Wilbur 1999). We saw above that a number of the pathways outlined for spoken languages have analogs in signed languages (see the seminal work of Pfau and Steinbach 2006 for a general treatment).

But there are also what appear to be modality-specific grammaticalizations in signed languages which are undocumented in spoken languages. For example, in German Sign Language (DGS), a grammaticalization from noun to auxiliary has been observed (Pfau and Steinbach 2006), and—to our knowledge—the development of the adjective WRONG of American Sign Language into an adverbial marker for unexpected events (Morford 2002: 331) also has no equivalent in spoken languages. However conceptually plausible these processes may be, we are not aware of any parallel of them in spoken or written languages (for an exceptionally interesting development of the noun 'thing' → 'long object used as an instrument' → purpose marker, see Epps 2007).

Perhaps more importantly, research on signed languages may also shed light on the question posed above, namely on whether it may not be possible to push reconstruction back to linguistic forms that preceded lexical forms. There is little doubt that at least some lexical items of signed languages that were grammaticalized to functional categories have their source in gestures, that is, in communicative expressions using the hands and face. Thus, Janzen and Shaffer (2002) propose the reconstructions in (65); note further that Wilcox and Wilcox (1995) argue that there are gestural origins for markers of evidentiality (see also Pfau and Steinbach 2006).

(65) American Sign Language (ASL; Janzen and Shaffer 2002)
    a. Gesture 'to leave, go' > lexical verb 'to go' > future tense marker
    b. Gesture '(be) strong' > lexical 'strong' > 'can' > epistemic modal 'can'
    c. Gesture 'to owe' > lexical verb 'to owe' > 'must, should' > epistemic modal 'should.'

But work on signed languages provides yet another perspective that is immediately relevant for the reconstruction of language evolution. Our discussion in Section 2.2 suggested that functional categories can be traced back ultimately to lexical categories. As it surfaces from recent research (especially Janzen 1999; Janzen and Shaffer 2002; Pfau and Steinbach 2005, 2006), however, there is an alternative pathway for functional categories, namely one that bypasses the lexicon, leading straight from manual or non-manual gesture to functional marker, such as the ones summarized in (66). In (66a) for example, the gesture proposed as the origin of the polar (= yes–no) question marker in both American Sign Language and British Sign Language (Woll 1981) is an eyebrow raise extending over the entire proposition being questioned, which became the obligatory sign for polar questions (usually along with a forward head tilt), that is, a marker for a grammatical function. Subsequently, this marker was further grammaticalized to a topic marker, first within the pragmatic domain, subsequently within the syntactic, and finally the textual domain.

(66) From gesture to functional category (Janzen and Shaffer 2002; Wilcox 2004a; Pfau and Steinbach 2006)
 a. Questioning gesture > polar question sign > topic sign
 b. Gesture > sentence-final word question particle (Indopakistani Sign Language)
 c. Gesture > classifier
 d. Negative head shake > negation marker
 e. Gesture 'wait a second' > negative completive marker (Jordanian Sign Language)

On the basis of such observations one might conclude that ultimately lexical material might not be the only source for functional categories. Conceivably, such a hypothesis could be strengthened by more detailed analysis of suprasegmental grammatical forms as they show up, for example in tone or intonation contours of polar question marking. Unfortunately, the evidence available does not allow for any generalizations on this issue. But this raises more generally the question of what role gesture and prosody may have played in the evolution of spoken languages. Since we are restricted here to segmental speech, we have little to say about this issue except that it requires more research.

2.4 A scenario of evolution

The developments sketched in the preceding sections can be conflated in the form of an evolutionary network as presented in Figure 2.1, where six layers of evolution are proposed. As we observed in Section 2.2, the term "layer" refers to clusters of categories that show the same relative degree of grammaticalization vis-à-vis both the categories from which they are derived and which they develop into.

As a rule, individual instances of grammaticalization processes extend over only two or three layers; there are however a few cases where a grammatical item in a given language may cover a larger range of layers. We mentioned in Section 2.2.4 that there is a fairly widespread grammaticalization path leading from reflexive pronoun to passive marker. Now, one of the most common lexical sources for reflexive pronouns is nouns meaning 'body' or 'head', and there are some languages exhibiting the whole pathway from noun via reflexive to passive marker. The !Xun language of

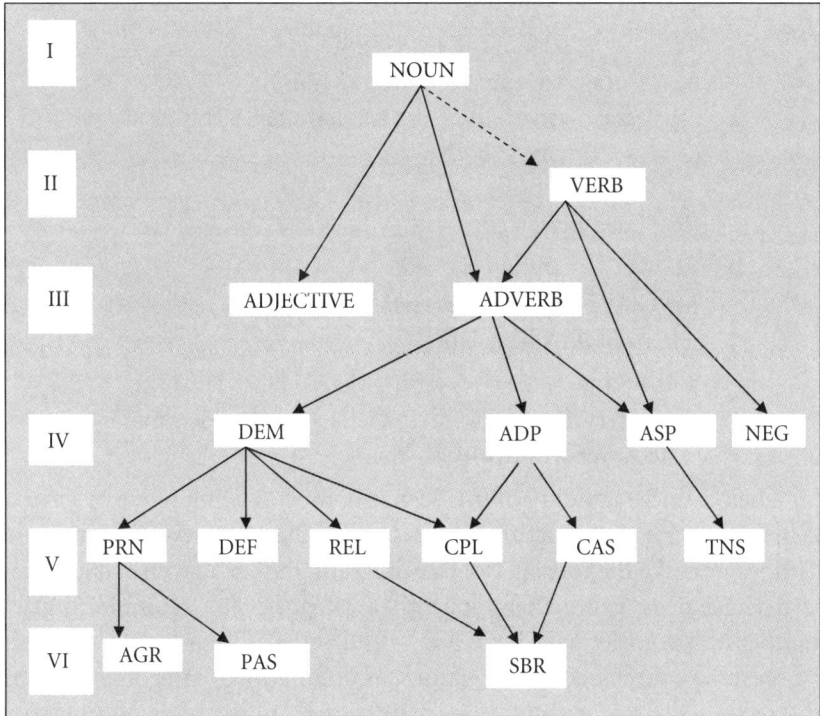

Fig. 2.1. Layers of grammatical development

Abbreviations: I, II, etc. = layers; AGR = agreement marker; ADP = adposition; ASP = (verbal) aspect; CAS = case marker; CPL = complementizer; DEF = marker of definiteness ("definite article"); DEM = demonstrative; NEG = negation marker; PASS = passive; PRN = pronoun; REL = relative clause marker; SBR = subordinating marker of adverbial clauses; TNS = tense marker.

southwestern Africa is such a language: There is a noun /é 'body', cf. (67a), but it can also be interpreted as a reflexive marker. While the lexical meaning is still available in the eastern E2 dialect, the northern N1 dialect of !Xun has grammaticalized it into a reflexive and eventually also into a passive marker. Thus, example (67b) exhibits the reflexive construction, and (67c) the passive construction, where there is an inanimate subject; note that we are dealing with a fully-fledged passive that can take an agent introduced by the multi-purpose preposition kē having a transitivizing function (TR). However, the passive construction still bears witness to its nominal origin since the passive marker /é behaves morphosyntactically like a head noun obligatorily requiring a possessive modifier. Accordingly, cases like this one cover the whole spectrum from layer I (noun) via layer V (reflexive pronoun) to layer VI (passive marker).

(67) !Xun, N1 dialect (North Khoisan, Khoisan)
    a   mi   she   mi |'é.   (E2 dialect)
        I   see   my   body
       (i) 'I see my body.'
       (ii) 'I see myself.'
    b   yà   kē   !hún   yà |'é.
       he   PAST kill   his   self
       'He has killed himself.'
    c   g‖ú   má   kē   tchù   ká'ŋ |'é   kē   mí.
       water TOP PAST drink its   self TR me
       'The water was drunk by me.'

For a better understanding of this scenario, the following remarks seem to be in order. First, the structure it represents is *non*-transitive, that is, Figure 2.1 does not take the format of a tree diagram, in that a given category can be derived from more than one other category. For example, quite a number of categories have both nouns and verbs as their lexical sources.

Second, we discussed the development of functional categories in terms of a restricted set of grammatical distinctions. On the basis of our cross-linguistic observations, these distinctions are typologically salient; however, there are many conceivable other distinctions that could have been considered. For example, we discussed a category of definite markers but not of indefinite markers, and our network includes the categories tense and aspect but not that of modality.

Third, the processes depicted in Figure 2.1 are salient ones but they are not the only ones that have been identified so far. For example, there is no one source for passives, relative clause markers, subordinators, and other categories, that is, there are alternative sources in addition (see e.g. Lehmann 1984; Haspelmath 1990; Heine and Kuteva 2002a). The main reason for not including further pathways is either that they do not appear to be cross-linguistically stable on the basis of our present knowledge, or that they are genetically or regionally more restricted in their occurrence. For example, interrogatives constitute a salient source of grammaticalization, nearly comparable to that of demonstratives, giving rise in particular to relative clause markers (e.g. English *who*) and subordinators in European languages (e.g. English *when*), and they are not restricted to European languages; still, this pathway is not really widespread outside Indo-European languages, hence it does not figure in the evolutionary schema of Figure 2.1. And fourth, discussion in this chapter was confined to the evolution of what we called the most inclusive types of linguistic categorization. In doing so, we ignored a number of alternative aspects of grammaticalization.

In more general terms, one may say that these developments lead

- from concrete meanings to more abstract ones;[15]
- from open-class to closed-class items;
- from fairly independent, referential meanings to less referential, schematic grammatical functions having to do with relations within the clause or between clauses.

One of the most pertinent distinctions within the parameter of decategorialization concerns the relative degree of morphological (and phonological) independence of linguistic items, commonly described with reference to the following scale (see Section 1.2.1.4):

(68) free form > clitic > affix > zero

In some way, this scale underlies many pathways of grammaticalization, including the ones sketched above: There is a positive correlation to the effect that, the higher up a category is located in the network of Figure 2.1, the more likely it is to be expressed by a free form, while categories at the lower end of the network are more likely to be encoded as affixes. Another line of grammaticalization that is ignored in this chapter concerns the following crosslinguistically stable morphological development:

(69) compounding > derivation

When two kinds of lexemes are compounded, one of them may undergo desemanticization, losing salient components of its meaning, becoming semantically dependent on the meaning of the other part of the compound, and decategorialization, losing most or all of its lexical morphosyntax. For example, one of two compounded nouns may be used more frequently and, being associated with a wider range of contexts, may acquire a more general meaning, gradually assuming a derivational function, and eventually turning into an affix of the other noun. This pathway, as well as many others, are ignored in this chapter for the reasons outlined above.

---

[15] Conceptual shift from concrete to abstract, as understood here, is anthropocentric in nature, in that it leads from meanings that are close to human experience and easy to describe, to meanings that are more difficult to understand and describe.

114 *The genesis of grammar*

Fifth, the network is highly abstract in that it conflates a number of, to some extent, disparate pathways. There are two kinds of pathways that make up the network, namely direct and indirect ones. Whereas the latter involve one or more intermediate stages, the former do not. For example, tense markers may be the result of the direct pathway, as is the case with the English *will*-future, which arose directly from a verb of volition in accordance with the Verb > tense pathway (see Section 2.2.1.2). But they may as well result from the indirect pathway Verb > aspect > tense. In a similar fashion, subordinators can arise directly from the Noun > subordinator or the Verb > subordinator pathway, but they can also be the result of a chain of pathways, for example Noun > adverb > adposition > case marker > subordinator. As we saw above, nouns and verbs each provide the source for a number of direct pathways, most of all the following:

Noun > adjective, adverb, adposition, pronoun, complementizer, case marker, agreement marker, subordinator

Verb > adverb, adposition, aspect, negation, tense, complementizer, subordinator.

Functional categories resulting from indirect pathways are likely to differ from those resulting from direct pathways in that they are relatively older, having gone through more than one process of grammaticalization. Since they involve parameters such as desemanticization and decategorialization two or more times, they are likely to be more strongly grammaticalized, that is, we hypothesize that they have lost more of the properties of their lexical sources than categories resulting from direct pathways; but more research is required on this issue.

2.5 Conclusions

The evolutionary pathways sketched above may have conveyed the impression that human languages are becoming increasingly more grammaticalized, that is, grammatically more complex.[16] However, on the basis of the materials we have worked with we are not justified to say that this is

---

[16] An anonymous referee of this work draws attention to the fact that the perspective adopted in the present chapter is a semasiological one, while increasing grammatical complexity would be an onomasiological problem.

really the case; Modern English is demonstrably no more morphologically complex than its predecessor Old English. There are a number of reasons for this fact. First, grammaticalization need not, and frequently does not, run through the entire range of pathways such as the ones sketched in Figure 2.1; it can be arrested at any point in time. Second, functional categories may fall into disuse at any stage of their development. And third, and most importantly, once functional categories have reached the final stages of development, that is, layer VI and beyond, they have lost most of their meaning (desemanticization) and phonetic substance (erosion), and they may subsequently be given up entirely, in accordance with the scale of morphological development sketched in (68). This is what happened to the system of case inflections on the way from Old English to Modern English. To be sure, loss can be made up for by new grammaticalization processes, for example by adpositions taking over the functions of lost case inflections, thereby giving the impression of a cyclic development (see Section 1.1.4); but this is by no means always the case. To conclude, the evolutionary network described in this chapter does not allow us to tell whether a given language, or human language in general, is heading towards a higher or a lower degree of grammatical complexity.

The main goal of the present chapter was to apply the methodology of grammaticalization theory to the reconstruction of salient pathways of grammatical change. In doing so, we had to reduce the complex—and to some extent still poorly understood—structure of conceptual relations underlying grammaticalization processes to a few salient pathways. The outcome was a scenario of grammatical evolution described in Figure 2.1 in the form of a network.[17] There are two kinds of implications that this scenario has for our understanding of language.

The first kind is synchronic in nature. It allows us, on the one hand, to understand why there are a number of homophonous structures in the languages of the world that, on a synchronic analysis, appear to be unmotivated. Thus, it would seem possible to account for the fact that the English item *that* has a number of contrasting uses, including those of a demonstrative, a relative clause marker, and a complementizer. On the basis of the developments sketched above we argue that the three are

---

[17] Such a network was outlined in Heine and Kuteva (2002a) but no evidence was provided there to substantiate the hypotheses on which the network rests.

interrelated in a principled way: the latter two are derived from the former via the pathways Demonstrative > relative clause marker and Demonstrative > complementizer; we will return to this example in Chapter 5. In a similar fashion, we are able to hypothesize that the different uses of the English item *after* as an adverb, an adposition, and an (adverbial clause) subordinator is consistent with the pathways Adverb > adposition and Adverb > subordinator. In other words, the different developments summarized in Figure 2.1 may surface in the synchronic structure of a given language in the form of polysemous (or homophonous) forms belonging to different functional categories.

On the other hand, the findings made also allow for some generalizations on the structure of linguistic categories as we find them in modern languages. A number of different taxonomies of linguistic categories have been proposed, and with the present discussion we do not wish to question the validity of any of these taxonomies. However, we argue that with reference to their evolutionary behavior, categories can be related to one another in a principled way, and the resulting structure of relationship does not necessarily match that proposed by some authors. For example, in some schools of linguistics, the following four taxa are treated as major lexical categories: nouns, verbs, adjectives, and adpositions (= "prepositions"). While there is solid syntactic evidence to support this taxonomy, such a treatment is not without problems. One problem relates to the fact that there are quite a number of languages that lack adjectives. Another problem—one that relates generally to our discussion in this chapter—concerns the fact that there are some properties distinguishing these categories. According to the observations made above, adjectives and, more conspicuously, adpositions do not belong to the same layer of grammaticalization as nouns and verbs. While the basis of our framework is strictly diachronic, grammaticalization theory predicts that this difference is also reflected in the synchronic structure of the taxa concerned. Ignoring the notoriously controversial status of adjectives, we will expect that on account of their relative degree of grammaticalization, adpositions differ drastically from nouns and verbs.[18] In accordance with the various parameters defined in Chapter 1, we will expect that crosslinguistically the following generalizations hold true:

---

[18] As Fritz Newmeyer (p.c.) rightly observes, this does not necessarily apply to adpositional phrases vis-à-vis other phrasal categories.

(a) Nouns and verbs are semantically more complex, while adpositions are likely to have some schematic meaning (desemanticization).
(b) Nouns and verbs have a much larger syntactic and discourse-pragmatic potential of use than adpositions, which are confined to occurrence next to a noun or noun phrase (decategorialization).
(c) Nouns and verbs can be the target of WH-questions (e.g. *What is this?*, *What does he do?*), while this is usually not possible in the case of adpositions (*Where is the book? *In.*)
(d) Nouns and verbs take a maximal range of inflectional and derivational morphology, while the range of inflectional and derivational morphology that can be used with adpositions is severely limited, typically being zero (decategorialization).
(e) Adpositions are on average shorter than nouns and verbs (erosion).

But our concern in this chapter was not with the synchronic implications of grammaticalization; rather, we were concerned with reconstruction. Using grammaticalization theory as a framework, we arrived at a network of grammaticalization, summarized in Figure 2.1. On the basis of this network we hypothesize that certain functional categories are more grammaticalized than others; for example, adpositions are more grammaticalized than adverbs and tense markers are more grammaticalized than aspect markers.

Research that was carried out in the course of the last decades on the typology of grammatical categories in both their diachronic and synchronic manifestations has brought about a number of observations that have a bearing on how human language evolved. On the basis of these observations, Givón proposes the following hypotheses:[19]

(70) Hypotheses on language evolution (Talmy Givón, Funknet, August 8, 2005)
    a. All other things being equal, typological features that are more widely attested across languages may have appeared earlier in evolution.
    b. Typological features that are more frequent in live communication may appear earlier in evolution.
    c. In grammaticalization chains, earlier stages, those that tend to be "source constructions", may also have evolved earlier.

---

[19] Hypotheses proposed by Talmy Givón in the e-mail network Funknet, August 8, 2005.

d. Likewise in morphemic development, concrete lexical senses have most likely evolved before more abstract, metaphoric and/or grammatical, uses.
e. Communicative behavior (function) most likely has preceded grammaticalization (structure) in evolution.

Two of these hypotheses, (70c) and (70d), are fully in accordance with the findings presented above. By using the methodology applied in this chapter it is possible to reconstruct present-day functional categories back to earlier stages of language evolution where such categories must have been absent—that is, there is reason to argue that functional categories located on earlier stages of grammaticalization chains and/or expressing concrete rather than abstract grammatical meanings can be hypothesized to have appeared earlier in the evolution of human language. And it would also seem plausible that, in accordance with hypothesis (70e), communicative behavior preceded grammaticalization in language evolution; we will return to this issue in Chapter 7 after having looked at a larger range of phenomena.

But two of the hypotheses proposed in (70) are not fully supported by our reconstructions. (70a) is in fact an attractive hypothesis: Properties that are found in all languages of the world must be old, therefore they are suggestive of having been part of early language, or at least of some earlier stage of language development; conversely, properties that are found in a few languages only are more likely to be of more recent origin. In fact, hypothesis (70a) receives partial support from the observations we made in the preceding section with reference to Figure 2.1: Most languages of the world have some equivalent of the categories of the less grammaticalized layers I through IV, while some categories figuring in the more grammaticalized layers V and VI are less commonly distinguished in the languages of the world.

Nevertheless, there are also problems with (70a): For example, crosslinguistic observations suggest that functional categories such as adjectives and verbal aspect, located at layers III and IV respectively, are not distinguished in all languages, while pronouns, located at layer V, do exist in all languages known to us. On the basis of such observations there is reason to question the contribution that an approach that is restricted to synchronic typological generalizations can make to the study of early language. Accordingly, we follow Givón (2002a; 2005), who himself insists that any linguistic work aimed at studying earlier states of language evolution that does not take diachronic evidence into account is likely to miss certain insights that are crucial for reconstruction. And (70b) is also not uncontroversial. While this is an appealing hypothesis that requires further analysis, it is not entirely supported by the findings made in this chapter, as can be shown with the following example. Among the twenty most frequently occurring words in written English there is none that clearly belongs to any of the layers I, II, or III, that is, which is distinctly lexical in nature; rather, it is function words of layers IV and V that exhibit the highest frequency, such as articles (*the, a*), adpositions (*in, to, of*, etc.), personal pronouns (*I, he*), or connectives (*and, but*).

We are now in a position to return to the questions raised in the introduction to this chapter (see also Chapter 1, (1)). The first of these questions was: Which is older, the lexicon or grammar? As our scenario in Figure 2.1 suggests, it was the lexicon that must have preceded grammar in evolution: Lexical categories such as nouns and verbs are a prerequisite for other categories to arise. And there is also a clear answer to the question of what the structure of human language was like when it first evolved: All evidence from grammaticalization leads to the same hypothesis, namely that the earliest structure of human language was lexical in nature, first consisting only of noun-like utterances before verbal utterances appeared, thereby making it possible to form propositional constructions.

This leaves us with the remaining two questions, namely: How did language change from its genesis to now? Was language evolution abrupt or gradual? The grammaticalization scenario that we proposed allows for no interpretation other than that the evolution of grammar proceeded gradually from fairly rudimentary, lexically based structures to increasingly differentiated grammatical structures. Each new layer contributed new grammatical distinctions, leading incrementally to increasing morphological and syntactic complexity, culminating in layer VI, where there now was a fully-fledged grammatical architecture akin to that found in modern languages. To conclude, on the basis of our reconstruction there is little support for any hypothesis to the effect that human language evolved abruptly in a discrete leap from non-language to modern languages. However, the possibility that the evolution from layer I to VI proceeded relatively fast—thus giving the impression of a sudden change—cannot be entirely ruled out; we will return to this issue in Chapter 7.

There were a number of other questions in addition that we raised in Chapter 1, but they cannot be answered on the basis of the data analyzed in this chapter. These questions will be the subject of subsequent chapters, such as the significance of animal communication for studying language evolution (Chapter 3), the contribution that the study of restricted linguistic systems like pidgins can make in reconstructing language genesis (Chapter 4), the rise of clause subordination (Chapter 5), or the question of how modern languages acquired their recursive structures (Chapter 6).

3 Some cognitive abilities of animals

One of the questions raised in the introduction to this book was: Can language genesis be related to the behavior of non-human animals? This question forms the main subject of the present chapter. There is also fairly wide agreement that apes and other non-human animals lack language and that the emergence of language was a hallmark in human evolution. In a number of works it has been argued that early language can be traced back to animal communication, but this view is not uncontroversial. The problem that we are confronted with in this chapter is not whether animals "have language" but rather whether they show cognitive and/or communicative abilities that are a requirement for developing something corresponding in any significant way to human language abilities.

3.1 Introduction

Opinion on what typical human language abilities are differs greatly depending on which theoretical stance one wishes to adopt. Work on generative grammar, for example, has focused on specific syntactic features that are taken to be characteristic of human languages, such as long-distance dependencies, binding, movement/displacement, or recursion, while other linguistic schools highlight semantic or discourse-pragmatic properties to be observed in languages. We will use "language" in a more general sense—one that also incorporates restricted linguistic systems arising in special situations. Accordingly, our interest will be with features that can be expected to be present in any kind of linguistic system and we will ignore features that characterize only specific ("fully-fledged") varieties of human languages.

Our discussion has to be taken with care, for a number of reasons. First, research on animal cognition and communication is an area of much ongoing research; we now know a lot more than we did ten years ago.

Quite conceivably, our understanding of this issue might change again in the years to come on account of new research findings. Second, the question of how observations on animal behavior are to be interpreted differs greatly from one author to another. Differences are found depending on whether authors report about their own findings or whether other authors comment these findings. They are also found between authors analyzing the same animals. After having studied Washoe and other chimpanzees (*Pan troglodytes*), the Gardners (e.g. Gardner and Gardner 1969, 1975, 1978) and Fouts and Mills (1997) concluded that these apes had acquired a wide range of signs, including signs for objects, actions, names, locations, qualities, etc., and patterns of combinations of signs, such as [agent + object], [agent + action], and [attribute + object], and that they exhibited a range of different communicative motivations. The same animals were later analyzed by Rivas (2005) on the basis of videotaped material,[1] and he found only few of the achievements in these animals that earlier authors had described. For example, the chimpanzee Loulis had been reported to have acquired at least fifty-one signs from the other chimpanzees (e.g. Fouts and Mills 1997) while Rivas found this chimpanzee to produce no more than four signs (see also Terrace 1979, 1985; Seidenberg 1986).

Third, the findings that are available are the result of a variety of different research designs. At one end there are findings gained by observing animals in their natural environment, for example in the rainforest of the Ivory Coast or the savannahs of East Africa; at the other end there are findings gained in laboratory situations involving intense training and social interaction between human caretakers and the animals. In accordance with the way they were gathered, these findings may carry quite different weight. For example, the chimpanzee Washoe studied by the Gardners (Gardner and Gardner 1969) exhibited only minimal syntactic abilities compared to the chimpanzees studied by Premack (1971), and McNeill (1974) suggests that this can be accounted for by the fact that the authors concerned used contrasting techniques of training and interaction with the animals. The study by Savage-Rumbaugh *et al.* (1980: 922–4) also reveals that the extent to which non-human primates can develop

---

[1] This material consisted of recordings of unstructured, relaxed, and naturalistic daily interactions between humans and five chimpanzees, containing altogether 3,448 analyzable utterances (Rivas 2005).

concepts is contingent upon how they are trained. The training of their chimpanzees Sherman and Austin emphasized the pragmatic and semantic functions of form–meaning pairings, and Savage-Rumbaugh and colleagues suggest that Sherman's and Austin's training led them to link the use of an object and the label of an object together. For the chimpanzee Lana on the other hand, who had initially been trained mainly specific paired-associative responses, the two skills appeared to remain separate, and Lana did not acquire the concepts concerned.

That the extent to which animals acquire communicative and linguistic skills is contingent on the kind of training method employed has also been demonstrated by Miles (1983). For the present purposes, however, we will ignore such differences, our interest being exclusively with what abilities animals exhibit, irrespective of where and how they acquired these abilities. We are also not interested in the learning abilities of animals; rather, we want to find out the communicative capacities that these animals can be assumed to have, irrespective of how they may have been acquired.

Another issue that we are also not able to do justice to is what kind of intelligence or knowledge systems are available to animals: Do animals discriminate on the basis of stimulus generalization (from specific exemplars and their features) or of concept formation? Are they restricted to "implicit intelligence", that is, to practical, sensorimotor, or situational knowledge, or do they have, or can they acquire "explicit intelligence", that is, conceptual, representative, or discursive knowledge? While we will discuss a number of findings that have some bearing on this issue, such issues are generally beyond the scope of this chapter.

A number of other problems will also be largely ignored, although they must have some bearing on the findings reported. One such issue concerns the difference between comprehension and production: To what extent are there correlations between the two? Which precedes which in the acquisition process? The research findings that are available show that the two need not go together, and which comes first depends in particular on the training techniques applied and the kind of animals studied (Herman 1987; Pepperberg 1999b: 126). For example, work on marine animals such as dolphins and sea lions has focused on comprehension, while that on primate behavior has emphasized production.

Another problem concerns the question of how the training of animals compares to the learning process of children. To what extent is the fact that animals can be trained to acquire characteristics of word order comparable to the way children learn word order via observational learning? What effects does training have on the cognitive abilities of animals? Kako (1999: 12) observes that Alex and the dolphins had to be prodded and cajoled into syntactic competence, and, although the bonobo Kanzi acquired both *Yerkish* (named after the Yerkes Laboratories of Primate Biology at Orange Park, Florida) and English without intensive training, his competence in both systems never exceeded that of an average 2-year-old child. In comparing linguistic performance of animals with that of children, the researcher is confronted in particular with the problem that young children learn to understand the communicative intentions of other persons, while this does not appear to be the case with apes or other non-human animals (but see Tomasello, Call, and Gluckman 1997: 1067). Since our concern is with achievements rather than with the way these achievements are arrived at, we will have little to say on this issue; and more generally, we will not attempt to compare the behavior of animals with that of children in the process of first language acquisition—in spite of the fact that this has been a major theme in previous works on animal communication.

The animals that have been the target of detailed study based on intense training are of various kinds. They include non-human primates, such as Washoe and other chimpanzees (Gardner and Gardner 1969, 1978), Sarah and three other African-born chimpanzees studied by Premack (e.g. 1976), Kanzi, the bonobo (*Pan paniscus*) studied by Savage-Rumbaugh and her associates (e.g. Savage-Rumbaugh and Lewin 1994), Koko, the lowland gorilla of Patterson (1978a, 1978b), or Chantek, the orangutan (*Pongo pygmaeus*) studied by Miles (1978, 1983); there are also non-primates, such as bottle-nosed dolphins (*Tursiops truncatus*) (e.g. Herman 1987), Californian sea lions (*Zalophus californianus*) (e.g. Schusterman and Gisiner 1988), or Alex, the African grey parrot (*Psittacus erithacus*) (e.g. Pepperberg 1999b).

And the techniques to establish communication with the animals also differ greatly from one researcher to another. For example, the Gardners (1969) taught the chimpanzee Washoe American Sign Language (ASL), raising her in an environment like that of a human child, stuffing her living quarters with furniture, a kitchen, bedroom and bathroom, toys, tools, etc., and Washoe's days were made up of meals, naps, baths, play, schooling, and outings, while the fellow chimpanzee Nim studied by Terrace (1979, 1980, 1983) was exposed to a clearly less luxurious physical and social environment. Premack's chimpanzees were trained with pieces of plastic, backed with metal adhering to a magnetized slate, each standing for some English word. Kanzi, the pygmy chimpanzee (bonobo; *Pan paniscus*), on the other hand was exposed without explicit training to the artificial lexigram system *Yerkish*, and he received a portable, folding posterboard containing printed lexigram symbols (Greenfield and Savage-Rumbaugh 1990).

The communication systems used were frequently designed in accordance with what the researchers concerned considered most appropriate given the general situation and the task facing them. In studying the orangutan Chantek, Miles (1983: 48) decided on a "pidgin" Sign English based on gestural signs of ASL with English word order but with little grammatical morphology; Terrace (1983: 22) and associates also used a kind of "pidgin" sign language with the chimpanzee Nim. An artificial gestural language was used in the study of the two Californian sea lions Rocky and Bucky (e.g. Schusterman and Krieger 1984), and the two female bottle-nosed dolphins studied by Herman (e.g. 1987, 1989) were also taught artificial languages: Phoenix learned an acoustic language using computer-generated sounds broadcast into the tank through an underwater speaker, while Ake was taught a "dolphinized" version of a gestural language, where the gestures consisted of movements of a trainer's arms and hands.

Other animals were studied under their natural behavioral and environmental conditions. A paradigm example is Cheney and Seyfarth (1992), who studied communication among vervet monkeys (*Cercopithecus aethiops*) over a continuous period of eleven years in Kenya's Amboseli National Park, and other *Cercopithecus* species were studied by Zuberbühler (2002; Zuberbühler, Cheney, and Seyfarth 1999: 40) in the rainforest of the Ivory Coast.

3.2 What linguistic abilities do animals have?

The scope of the following survey is limited. We will simply be concerned with the question of which traits in animal behavior may relate to language-related cognitive abilities; accordingly, we will ignore capacities of animals that have no demonstrable analog in human cognition or languages. For example, the fact that some bird songs, the grooming patterns of rats, etc., exhibit structural properties that have been called syntactic is not taken into account since such phenomena cannot demonstrably be related to the syntactic combining of meaningful elements in human languages. Accordingly, our survey will be anthropocentric in a double sense—first, because most substantial findings on language-like abilities in animals have been made not in their natural environment but instead in environments designed by humans for specific research purposes, and second because we will have to describe animal behavior relative to human abilities and with reference to notions of linguistic categorization, which are taken as heuristic devices to evaluate animal performance. It is quite possible that monkeys and other non-human animals have what Cheney and Seyfarth (1992: 146) call "a kind of laser beam intelligence," which allows them, for example, to solve problems of social interaction but less so problems that they are confronted with when dealing with a human researcher looking for language-related skills (see McNeill 1974). It goes without saying that we are therefore not able to do justice to the specific nature of animal communication in all its manifestations.

3.2.1 *Communicative intentions*

Non-human animal communication has been claimed to be restricted to what has been called imperatives; these animals are claimed to be incapable of declaratives, that is, of communicating something for its own sake, simply trying to affect the listener's mind (see e.g. Baron-Cohen 1992: 149). But what is the communicative intention, for example, of the leopard, eagle, and snake alarm calls of vervet monkeys?

Cheney and Seyfarth (1990, 1992) is a seminal study of the communicative system of vervet monkeys and an important contribution to our understanding of intentionality[2] in human communication vs. animal communication. On the basis of this study we know that the vervet monkeys have a repertoire of four distinct calls:

(a) *Wrr*—usually given when a neighboring group has first been spotted;
(b) snake call—alarm call for snakes;
(c) leopard call—alarm call for leopards;
(d) eagle call—alarm call for eagles.

---

[2] Intentional phenomena are about some other thing, be it a physical stimulus or another mental state. Whenever an individual thinks, believes, wants, likes, or fears something, s/he is said to be in an intentional state.

These calls are phonologically distinct and arbitrary, representational signals which stand for an object even when that object cannot be seen. And they evoke clear reactions: When hearing the leopard call, other vervets on the ground run into trees, on the eagle call they look up or run into bushes, and on the snake call they stand bipedally and peer into the grass around them.

What about the level of intentionality observed in the behavior of vervet monkeys? Cheney and Seyfarth (1990: 139–43) distinguish between the following levels of intentionality and types of intentional systems: (i) zero-order, where there are no beliefs or desires at all; (ii) first-order, where there are beliefs and desires but no beliefs about beliefs; and (iii) second , third-, or higher-order, where there is some conception about both one's own and other individuals' states of mind. Cheney and Seyfarth conclude that the calls of the vervet monkeys are suggestive of first-order intentionality because the monkeys producing these calls want others to run into trees, bushes, etc., but not necessarily of second-order intentionality, in which case they would want others to think that there is a leopard, eagle, or python nearby. Cheney and Seyfarth (1992: 174) furthermore observe that there is no evidence that vervet alarm calls are declarative rather than imperative.

Studies carried out on animals in captivity suggest that by far the most important motivation of animal communication is manipulative rather than declarative—a motivation which Terrace (1985) calls acquisitive: The communicative behavior of the animals is geared primarily at expressing requests. 96 percent of the lexigram utterances made by the bonobo Kanzi were interpreted as requests (Greenfield and Savage-Rumbaugh 1990), in Rivas's (2005) sample on five chimpanzees, 86 percent of all utterances were requests (65 percent for objects and 18 percent for actions), and Terrace (1985) also describes chimpanzee signing as being predominantly acquisitive in nature.

But animal communication is not exclusively acquisitive, and some authors (e.g. Gardner and Gardner 1978; Fouts and Mills 1997) even reject the claim that the communicative intentions of their chimpanzees are primarily request-oriented. Greenfield and Savage-Rumbaugh (1990) classified 4 percent of Kanzi's utterances as being indicatives or statements, and the chimpanzees studied by Rivas (2005) made 4 percent of their utterances to answer questions posed by humans, and 2 percent to name or label objects and pictures, while for 8 percent of their utterances the communicative intention was not evident. For the chimpanzee Ally, action requests and naming were the largest categories of speech acts, but almost one-fourth of his communications consisted of other kinds of sign acts (Miles 1978: 113). And signing by the gorilla Koko was not exclusively acquisitive either; it also included comments about the state of the environment, for example when she signed LOOK BIRD to draw attention to a picture of a crane in a stereo viewer, or LISTEN QUIET when an alarm clock stopped ringing in the next room (Patterson 1978a: 87–8).

That the motivations of apes cannot be reduced to being manipulative is also suggested by the observation that chimpanzees were found signing when they were alone, for example looking at pictures in magazines, and the gorilla Koko engaged in imaginative play using signing, or talked to herself by signing what she saw (Patterson 1978b: 191, 195). While the purpose of this signing has not been reconstructed, it is unlikely to have been request-oriented.

3.2.2 Concepts

Recent work suggests that categorical perception is not due to unique human adaptation but rather can also be found in chinchillas, macaques, and birds, suggesting that the basis for categorical perception is "a primitive vertebrate characteristic" (Hauser, Chomsky, and Fitch 2002: 1572), and these authors go on to observe: "Studies using classical training approaches as well as methods that tap spontaneous abilities reveal that animals acquire and use a wide range of abstract concepts, including tool, color, geometric relationships, food, and number" (Hauser, Chomsky, and Fitch 2002: 1575). In fact, non-human animals have been found to have a range of conceptual abilities. In their study of four wildborn squirrel monkeys, Burdyn and Thomas (1984: 411) claim that these animals have "natural concepts" and are capable of using working memory with conceptual information, and one of their subjects was capable of remembering conceptual information for at least 16 seconds in order to make a conceptual choice. Burdyn and Thomas also draw attention to the work of Herrnstein and his associates (Herrnstein, Loveland, and Cable 1976) who show that pigeons in the laboratory respond reliably and discriminatively to exemplars of water (pictures of puddles, streams, lakes, etc.), trees, people, leaves, and fish even when new pictures were presented. Diana monkeys studied in the wild show referential abilities: Female Diana monkeys react with their eagle alarm call not only on eagle shrieks but also on male Diana monkeys' eagle alarm calls (Zuberbühler, Cheney, and Seyfarth 1999: 40).

So, do non-human animals have, or are they able to acquire concepts, rather than simply perceiving associations or generalizing stimuli? Concepts have been defined in a wide range of different ways. For the purposes of the present discussion we will say that in order to understand a concept, the following abilities have to be in place:

(a) to understand that different referents are instantiations of one and the same entity;
(b) to understand that such instantiations include referents that are outside the here and now of a given situation;
(c) to use the learned entity in new contexts;
(d) to reconstruct the presence of a conceptual entity even if only parts of that entity are perceptually accessible;
(e) to produce novel instances of instances of that entity;
(f) to relate different conceptual entities to one another on the basis of size, shape, color, etc.

It would seem that these abilities can be found, in some way or other, in at least some animal species, and not just primates.

The grey parrot Alex "identified objects that differed somewhat from items used in his training; for example, although all paper initially consisted of fairly similar pieces of white, unlined index cards, he could identify pieces that varied considerably in size and shape and, without training, transferred them to items such as computer and notebook paper. Hide, wood, cork, and clothes pins ("peg wood") were consistently identified correctly even if these items were chewed and barely resembled their original forms (Pepperberg 1999b: 45). Alex could not only respond "green" in the presence of a green rather than a blue object, but also knew to respond "green" rather than "three-corner" based on the type of question; "[h]e could comprehend a vocal question, extract the relevant category from the question and from an object that could be classified in multiple ways, and respond with the label for the one, correct instance of this category" (Pepperberg 1999b: 61).

Thus, there is evidence of the fact that animals might have a concept of an object that is not present in a given situation of interaction. A number of animals, including the grey parrot Alex "do not use symbols only for objects and actions in the here and now. They, like humans, will request an absent object, or an action not currently being performed, and accept that object or action and no other.... They generally demonstrate label comprehension as well as production" (Pepperberg 1999b: 43). Alex consistently showed object permanence in his behavior, in that objects continued to exist when no longer in view for him, and they were unaffected by simple movement (Pepperberg 1999b: 168–85). All apes raised in human culture can refer to things not present (Miles and Harper 1994), and the gorilla Koko was able to remember events that had happened one or more days earlier and comment on them (Patterson 1978b: 197).

Extension   Conceptual reasoning may also be seen in the ability of animals to "over-generalize", that is, to extend the use of a form–meaning pairing to referents beyond the ones canonically associated with that unit. The following are a few examples illustrating the cognitive mechanism that appears to underlie the strategy employed, which has no analog in the behavior of the human counterparts concerned:

- The orangutan Chantek used the sign LYN not only for his caregiver Lyn Miles but for all caregivers, but never for strangers, and the sign for 'dirty' was used more generally to refer to bad things, until he learned the sign for 'bad' (Miles 1990: 525–8).
- The chimpanzee Washoe used the signs she had been taught in ASL not only for specific objects or events but rather for classes of referents, including absent ones; for example, the sign for 'dog' was used to refer not only to live dogs and pictures of dogs of many breeds, sizes, and colors, but also for the sound of barking made by an unseen dog (Gardner and Gardner 1978).
- The lowland gorilla Koko extended the sign STRAW, learned initially with reference to drinking straws, to label plastic tubing, cigarettes, and a car radio antenna. She learned the sign NUT as a name for packaged nuts, but later extended it not only to roasted soybeans and sunflower seeds but also to pictures of nuts in magazines and peanut butter sandwiches. And TREE was acquired for acacia branches and celery, but was over-generalized to asparagus, green onions, and other thin objects presented vertically (Patterson 1978a: 83).
- Bottle-nosed dolphins were taught the signal HOOP in reference to a particular, large, octagonal floating plastic hoop, but they generalized immediately to hoops of different sizes, shapes, thicknesses, and colors, as well as to hoops that sink to the bottom of the tank instead of floating. The dolphin Ake was taught the sign WATER to refer to a

thin stream flowing from an ordinary garden hose; when she was moved to another tank, she immediately responded to a waterfall entering the tank when WATER was used in a sentence for the first time at that new location (Herman *et al.* 1984; Herman 1987, 1989).
- The sign OPEN was initially used by the gorilla Koko with reference to locked doors, but was generalized to boxes, covered cans, drawers, and cupboards (Patterson 1978a: 82).

However, the possibility that these over-generalizations are due to associative rather than conceptual logic cannot be ruled out on the basis of the data that are available. That the dolphins had conceptual knowledge may be suggested by the fact that they could generalize from parts to the whole: The signal PERSON, originally taught to dolphins with reference to a particular trainer who held her arm in the water, was generalized immediately to a leg in the water, an elbow, or the whole person floating, as well as to other trainers. Note also that when one dolphin "says" something in reaction to a shown object, another dolphin is able to pick up another piece of the same object without knowing what the first individual saw. And these animals could also respond to an object in the absence of that object (Herman *et al.* 1984; Herman 1987, 1989).

That animals differ in their conceptual abilities can be seen in a comparison of the chimpanzees Sherman and Austin with the bonobo Kanzi: Whereas the former showed an ability to refer to absent objects only after specific training and no earlier than the age of 5, Kanzi was able to do so without formal training (Herman 1987: 12).

Same vs. different   That there is conceptual knowledge at least in some animals is also suggested by their treatment of the comparative distinction between same and different. Squirrel monkeys (*Saimiri sciureus*) can associate triangularity with choosing same and heptagularity with choosing different. Burdyn and Thomas (1984) therefore conclude that for these subjects, triangularity and heptagularity were abstract symbols for sameness and difference, respectively, and the monkeys were able to represent them over intervals in which neither symbol nor referent was present. And dolphins and sea lions respond not only to novel combinations of attribute and object labels but also to novel combinations of actions and object labels (Pepperberg 1992: 301).

Pepperberg (1987a, 1992: 301, 1999b: 144) demonstrates that her trained grey parrot was able to compare objects with one another on the basis of relational concepts of 'same' and 'different', and he could discriminate between objects on the basis of color, shape, and material, for example when given the question: "What color is (item designated by shape-X and material-Y)?". He made these distinctions not only on known objects but also on exemplars that were novel to him. In another series of experiments, Pepperberg was able to establish that Alex distinguished fairly consistently between relative differences in the size of objects: He transposed size relationships to objects not involved in training and transferred his knowledge to items of novel colors, shapes, and sizes, and he was in a position to indicate when items did not differ in size (Pepperberg 1999b: 166–7).

One problem associated with some of the research reported above is that the animals analyzed were usually exposed to only a limited range of contexts and to tasks whose solution was predictable to some extent on the basis of the training they had received. It is unclear whether or how this affects the findings obtained.

Symbols? The term symbol is used in a wide range of works on early language as a key notion to describe or account for the specific nature of human language, but it has been defined in a number of different ways. Most of all, it has been used for:

(a) signs that, unlike indexical and iconic signs, exhibit an arbitrary relation between a meaning and a form used for its expression;
(b) objects whose reference is context-independent, including objects displaced in space and/or time (i.e. outside the here-and-now);
(c) a convention, or shared cultural understanding, whereby different symbol users interpret the symbol the same way;
(d) signs that are intentional, or at least "functionally referential";
(e) signs that are connected with other signs of the same kind in a network of internal relations, that is, not through relations between respective referents.

Do animals have the ability to understand and/or use entities in the sense of (a)? The answer is clearly in the affirmative:[3] Not only some great apes but also the grey parrot Alex are able to produce arbitrary form–meaning relationships (Pepperberg 1999b: 43). Savage-Rumbaugh and Lewin

---

[3] Not everybody, however, agrees with this conclusion. Li and Hombert (2002: 177), for example, argue that non-human primate communicative signals have functions but not meanings—hence they are not symbolic. Note, however, that our concern here is not with animals in the wild but rather with the abilities shown by trained animals.

(1994: 160) maintain that from a very early age, the bonobo Kanzi demonstrated an understanding of a one-to-one relationship between a symbol and an object or action. Of three trained chimpanzees analyzed by Savage-Rumbaugh *et al.* (1980), two had acquired concepts of 'food' (= edibles) and 'tool' (= inedibles) that were functionally based, generalizable, and—as these authors argue—symbolically encoded.

Premack (1976: 165) argues that, after intensive training, the chimpanzee Sarah was able to comprehend the relationship between a "word" (which in his studies was a piece of plastic) and the corresponding object, for example the word for 'apple' and the fruit apple: "For instance, she was asked 'Apple ? object apple' (What is the relation between the word 'apple' and the object apple?); as well as the negative version of the same question, viz., 'Apple ? object banana?' (What is the relation between the word 'apple' and the object banana?)." The subject was able to make the correct choice between "name of" and "not-name of" at a significant rate (Premack 1976: 163–4), and Premack concludes: "The fact that we could teach Sarah a property by the name of an object no less than by the actual object was, of course, highly encouraging: it was the first unqualified suggestion that the pieces of plastic had the referential function of words" (Premack 1971: 813).

To conclude, trained animals have the ability to acquire arbitrary mappings between signals and concepts in accordance with (a), they can learn that signs are understood as surrogates for things. And the same applies to a number of untrained animals: The calls of vervet monkeys, macaques, Diana monkeys, meerkats, prairie dogs, and chickens are suggestive of "arbitrary" form–meaning combinations, as Hauser, Chomsky, and Fitch suggest:

First, individuals produce acoustically distinctive calls in response to functionally important contexts, including the detection of predators and the discovery of food. Second, the acoustic morphology of the signal, although arbitrary in terms of its association with a particular context, is sufficient to enable listeners to respond appropriately without requiring any other cotextual information. (Hauser, Chomsky, and Fitch 2002: 1576)

Are these calls also in accordance with (b)? Here the answer is in the negative since the calls just mentioned do not appear to be independent of the contexts in which they are used. But the situation is different in the case of trained animals if we assume that context-independence manifests itself, for example, in the fact that a signal can be used even in the absence of the object to which it refers. That Premack's chimpanzee Sarah was able to understand the relation between a signal and an object that was not present is suggested by examples such as the following: She could select the correct color for a food item such as an apple or a banana with only the signal for that food present and the signal not being the color of the food (Premack 1976). And much the same applies to the chimpanzees Sherman and Austin, who could indicate through use of their lexigram symbols on a keyboard which one of a variety of specific foods they desired, even though no food was immediately present (Savage-Rumbaugh *et al.* 1980). But (b) does not only apply to primates: The grey parrot Alex was also able to produce arbitrary form–meaning relations and use them, not only for objects and actions in the here and now (Pepperberg 1999b: 43).

An interesting observation on displaced reference has been made in the case of the orangutan Chantek: He placed the object before the verb (object-GIVE) when the item referred to was present, but after the verb (GIVE-object) when the object was absent. When Chantek was eight years old, 38 percent of his signings showed displacement of reference (Miles 1990: 519, 528).

And these trained animals also comply with (c) if one assumes that they and their human caretakers are different "symbol users", and with (d): The findings of the researchers referred to above show clearly that there was a significant overlap in the interpretation of signals between the animals and their human counterparts, and that the signs used by trained animals are at least functionally referential, if not intentional (Hauser 1997; Johansson 2001).

But evidence that animals in captivity are able to conform to (e) is scanty. As we will see below, animals, including non-primate species such as grey parrots and bottle-nosed dolphins, are able to distinguish a range of different colors, perhaps even a term for 'color', but whether such terms are in fact suggestive of networks of internal relations does not become entirely clear. Perhaps the most remarkable observation that could be interpreted in favor of (e) can be seen in the taxonomic abilities exhibited by some primates, such as the ability to comprehend inclusion relations (see Section 3.2.9); but one might wish to have more evidence of this sort to establish that these animals really conform to (e).

Accordingly, whether such mappings indeed qualify as "symbols" or words is an issue that we do not wish to decide here, considering all the discussions that have been led on this term (see especially Deacon 1997,

2003; Johansson 2001, 2002; Li and Hombert 2002; Bickerton 2003: 82–3); suffice it to cite Hurford (2003: 48), who concludes that "[a]n ape can make a mental link between an abstract symbol and some object or action, but the circumstances of wild life never nurture this ability, and it remains underdeveloped."

There are in fact abilities that are not found in animals. Savage-Rumbaugh (1986: 13) observes that language-trained chimpanzees used pointing to request, for example, desired foods from another chimpanzee, but that there was no evidence to suggest that they comprehended the indexical or referring function of pointing when it was displayed to them by human experimenters. And Hauser, Chomsky, and Fitch (2002: 1576) find "that many of the elementary properties of words—including those that enter into referentiality—have only weak analogs or homologs in natural animal communication systems, with only slightly better evidence from the training studies with apes and dolphins," and they conclude that additional evidence is required before the signals that have been described can be considered as precursors for, or homologs of, human words. This is largely in accordance with Pinker and Jackendoff's (2005: 215) view according to which words are a distinctive language-specific part of human knowledge.

3.2.3 "*Lexicon*"

Which types of form–meaning pairings were distinguished by animals was to a large measure contingent on what they had been taught. The artificial languages used to train bottle-nosed dolphins, for example, included objects, actions, and properties, so these were the units that the animals comprehended and reacted to (Herman *et al.* 1984; Herman 1987, 1989). But different researchers did not always arrive at the same conclusions on lexical classification. Analyzing videotaped human–chimpanzee interaction, where the animals had been taught ASL, Rivas (2005) found only the following classes of signs: objects, actions, request markers (GIMME, HURRY), the deictic sign THAT/THERE/YOU, and the chimpanzee's own name sign, although the relevant animals had been taught a larger range of unit types by the Gardners, their original caretakers.

One domain where explicit training of animals yielded particularly notable results is that of the "lexicon", or at least of what corresponds to lexical categories in human languages. The orangutan Chantek had acquired 26 signs at age 2;2, he could use 56 signs when he was roughly three and a half years old and eventually acquired 150 signs (Miles 1983: 49–50; 1990; Miles and Harper 1994), and the gorilla Koko is said to have acquired a "vocabulary" of 100 signs in American Sign Language (ASL) within 30 months, and 243 signs in 50 months of training (Patterson 1978a: 72, 1978b: 173). The chimpanzee Nim had acquired 40 signs at age 2;2, learned to use a vocabulary of 125 signs at age 4, and eventually knew 140 signs (Terrace 1983: 23). The grey parrot Alex acquired more than 90 form–meaning pairings (or vocalizations) with explicit training, including labels for foods and locations (Pepperberg 1994). Teaching the chimpanzee Washoe started when she was about 11 months old, and 51 months later she had a vocabulary of 132 signs of ASL, using space, eye gaze, facial expression, and repetition (Gardner and Gardner 1978). As we noted in Section 3.2.2, these authors claim that Washoe used her signs not for specific objects or events but rather for classes of referents; for example, the sign for 'dog' was used to refer not only to live dogs and pictures of dogs of many breeds, sizes, and colors, but also for the sound of barking made by an unseen dog.

Overall, trained apes have been found to acquire between 120 and 200 different form–meaning pairings, with the gorilla Koko topping the list, having a reported inventory of 243 learned signs. These units include most of all concrete objects and physical actions, while there is a conspicuous lack of items for non-tangible objects and abstract concepts.

Most of these form–meaning pairings were acquired via transmission from human caretaker to animal. However, some of the animals showed the ability to create signs on their own; thus, the chimpanzee Washoe and the orangutan Chantek created five, and the gorilla Koko four new signs (Miles and Harper 1994). Koko was taught a sign TICKLE by Patterson (1978a: 84), but she instead created a more iconic gesture, done by drawing her index finger across her underarm. Furthermore, after three months of training Koko appears to have created a sign to express polar (sentence) questions, by using gestural intonation. That new creations are likely to be iconically motivated is also suggested by the sign for 'stethoscope' that Koko created by placing an index finger to each ear (Patterson 1978a: 79, 1978b: 193).

While apes cannot vocalize speech, Savage-Rumbaugh and Lewin (1994: 150) argue that they do understand it. These authors performed three testing sessions with the bonobo Kanzi in which their requests alternated between spoken English and lexigrams. The tests included thirty-five different items, used in 180 trials in English and 180 with lexigrams. Kanzi scored 95 percent correct on the lexigram trials and 93 percent on the English trials; he understood 150 spoken form–meaning pairings at the end of the seventeen-month period.

There is, however, a caveat to some of these findings. We noted above that the chimpanzee Washoe had acquired a vocabulary of 132 signs of ASL. But roughly two decades later, Rivas (2005) found on the basis of 612 videotaped utterances that Washoe distinguished no more than 43 different signs, and it remains largely unclear how this difference is to be accounted for.

In spite of such observations, we assume that most of the findings summarized above are empirically sound. That Washoe and the other chimpanzees were really capable of comprehending lexical distinctions can be demonstrated perhaps more convincingly by the errors they made than by their correct responses. Most of the semantic errors made by the chimpanzees studied by the Gardners (1969, 1978) involved cases where a sign was used incorrectly for a similar meaning, for example COMB for BRUSH, while the form errors concerned mostly pairs of contrasting meanings but signs that were similar in their physical form, like MEAT and OIL. This behavior would seem to suggest that these animals had some understanding of the nature of the sign language they were taught (Fouts 1987: 64).

Numbers   A much discussed question in the cognitive sciences is whether animals can represent numerosity: (a) Can they identify a property of the stimulus that is defined by the number of discriminable elements it contains, (b) can they count, and (c) can they use numerical representations recursively?

There are findings that indicate that question (a) can be answered in the affirmative. Many animal taxa can discriminate stimuli differing in number, such as pigeons, parrots, rats, dolphins, monkeys, and chimpanzees (see Brannon and Terrace 1998). Ravens (*Corvus corax*) and jackdaws (*Corvus monedula*) succeeded on numerical match-to-sample tests on quantities up to 8, and tested on sets up to 4, the chimpanzee Sheba demonstrated ordinality and labeled, with a card depicting an Arabic numeral, the sum of two arrays separated in time and space (Koehler 1943, 1950; Pepperberg 1987a; Boysen and Berntson 1989). The trained grey parrot Alex replied correctly to the question 'How many?' (Pepperberg

1999b: 131) and learned to produce vocal numerical labels for sets of two to six objects and showed remarkable abilities in handling numerical quantities. For example, when presented with two pieces of cork or five pieces of wood, the responses of Alex would be *two cork* and *five wood* (Pepperberg 1987a, 1987c: 42, 1994).

But it seems that questions (b) and (c) have to be answered in the negative: Neither do non-human animals show a concept of counting, nor do they appear to have the capacity to create open-ended generative systems—that is, numbers are not acquired by animals in paradigms involving a successor function (see Chapter 6) (Hauser, Chomsky, and Fitch 2002: 1577). The studies available overwhelmingly suggest that non-human animals do not have a natural ability to discriminate numerosity, attending to it only as a "last resort", if other bases for discrimination, such as shape, color, size, frequency, or duration of a stimulus, are eliminated (Brannon and Terrace 1998: 746). Rhesus monkeys (*Macaca mulatta*) can spontaneously represent the numbers of novel visual stimuli and they can extrapolate an ordinal rule to novel numerosities, but it seems that they do not do so using a counting algorithm (Brannon and Terrace 1998: 748). And while the parrot Alex showed remarkable skills in handling numbers, Pepperberg (1999b: 110) concludes that her data did not indicate if Alex could count; there is no evidence that he is capable of a counting process or of a recursive understanding of numbers.

3.2.4 *Functional items*

Unlike lexical categories, such as nouns and verbs, functional categories are closed-class items expressing grammatical functions relating to person, tense, number, case, etc.

Location Do animals understand spatial relations between objects? And do they have functional concepts that correspond in some way to locative markers such as adpositions in human languages? Premack (1976: 203) found that, after learning *red* and *dish*, chimpanzees understood the command 'Insert the apple (in) red dish', and the gorilla Koko is said to have acquired signs for the English prepositions *on*, *out*, and *up* (Patterson 1978a:79). Alex and several other grey parrots were found to "understand the *concepts* of *in* versus *on*: They understand that to obtain a desired item that is *in* another object, one type of physical manipulation must be used;

if the desired item is *on* another object, the birds' manipulation is very different" (Pepperberg 1999a: 15). And the bottle-nosed dolphin Phoenix was given the task of linking the action term for 'fetch' with a transport object, a destination object, and spatial terms, being asked to carry a frisbee through, over, or under a hoop, and he performed correctly on 53 percent of the trials for FRISBEE FETCH THROUGH/OVER/UNDER HOOP (Herman 1987, 1989). This might suggest that Phoenix was able to distinguish between different relational spatial functional concepts.[4]

Does that mean that these animals perceive a functional concept of space? The information that is available does not allow this question to be answered clearly in the affirmative. There is, however, a noteworthy observation made by Herman (1989: 24). In the tank of the dolphin Ake there were two fixtures: a window (WINDOW) to Ake's right and a gate (GATE) to her left. Without being taught so, the dolphin learned to use WINDOW and GATE as relational terms for RIGHT and LEFT.

Deixis While we have found no clear evidence that animals have form–meaning pairings corresponding, for example, to personal pronouns in human languages, there appears to be a concept of deixis, as has been shown especially in studies of chimpanzees who have been taught sign language: The animals studied by Gardner and Gardner (1969) display a distinction of personal deixis: The sign for 'me' was made by tapping one's own chest and the sign for 'you' by pointing away from the chest toward the addressee. Later on, analyzing the chimpanzees studied by the Gardners and Fouts, Rivas (2005) found that the animals had what he calls a "wild card sign", a frequently used pointing sign THAT/THERE/YOU having the appearance of a "polysemous" item in that it can be glossed as 'that' when pointing to an object, 'there' when pointing to a location, and 'you' when pointing to a person. But the gorilla Koko is claimed to have acquired signs for 'me' and 'you' (Patterson 1978a: 79), and the orangutan Chantek is said to have distinguished 'I' and 'you' but preferred

---

[4] Kako (1999: 10) proposes a different answer: "Does this mean that Phoenix has some knowledge of closed-class elements and their syntactic properties? I believe that the answer is no, as relational sentences with OVER/UNDER/THROUGH are really conjoined sentences in disguise. They were first taught as ACTION terms and are even glossed by Herman with the word GO preceding them (e.g. [go] UNDER). The upshot of this is that Phoenix appears to have interpreted FRISBEE FETCH UNDER HOOP as "Fetch the frisbee and go under the hoop." Her enactment indicates that these words continue to function like open-class words in human language."

to use a proper name rather than a pronoun when talking to a person (Miles 1990: 516).

An interesting case of spatial deixis is provided by the chimpanzee Sarah (Premack 1976: 320–1). One process concerns autonomous (e.g. nominal) participants (*an orange*) whose use is said to be extended to that of modifiers/attributes of other participants (*an orange dress*). Sarah appears to have extended the use of a pronominal demonstrative to attributive demonstrative without any training; her training was restricted to the pronominal distinction between 'this' and 'that'. She was asked to comprehend the difference between 'Sarah take this' and 'Sarah take that', or to produce 'Give Sarah this' vs. 'Give Sarah that'. Furthermore, she was asked to produce demonstratives not only pronominally but also attributively. When required to produce 'Give Sarah this cookie' vs. 'Give Sarah that cookie', she made only three errors in fifteen trials, with none on the first five trials.

Possible manifestations of a concept of deixis can also be seen in other traits of behavior, for example in the fact that in her signed utterances the female chimpanzee Washoe placed the addressee always before the addresser or speaker in her spontaneous constructions (McNeill 1974). Note further that the bonobo Kanzi was able to label differentially the contrast of deictic motion between 'come' and 'go' (Greenfield and Savage-Rumbaugh 1990; Kako 1999: 6); but there are no clues to suggest that he was aware of the deictic significance of these items.

Negation A question that has found some attention in animal studies concerns negation: Do animals have, or can they be trained to distinguish negative concepts such as (a) rejection, (b) non-existence, and (c) denial? Jackendoff (2002: 241) maintains that no animal call system includes a signal of generalized negation like no. On the other hand, the chimpanzee Washoe and the gorilla Koko are said to have acquired a sign for negation (Patterson 1978a: 79, 90), and Premack's chimpanzees have also been successfully taught to comprehend the concept of negation (Premack 1971, 1976: 156–60). However, among the five chimpanzees studied by Rivas (2005) on the basis of videotaped material, only one (Moja) had a sign NO, produced by shaking her head sideways, although it was not entirely clear whether this really was a negation marker.

That there is a concept of (a) seems to be well established: Most animal species that have been appropriately trained know how to handle the notion of rejection (Pepperberg 1999b: 80). And much the same applies to (b). Non-existence is a fairly advanced concept in cognitive and linguistic development: It presupposes knowledge about the expected presence of objects, events, and other phenomena, as well as the ability to recognize a discrepancy between the expected and the actual state of affairs. However, trained apes have apparently no problems understanding this concept. The chimpanzees of Gardner and Gardner (1978) used ASL to comment upon the absence of a familiar object at a customary location, and the chimpanzees studied by Savage-Rumbaugh (1984) responded correctly when asked questions about objects when the object of the question was not present for sensory reference. Similar observations have been made in some non-primate species. Trained marine mammals have in fact demonstrated some understanding of non-existence: The bottle-nosed dolphin (*Tursiops truncatus*) Ake was taught an interrogative form and to press a paddle to her right to indicate presence ('yes') of the named object or a paddle to her left to indicate absence ('no'), and she could handle the distinction productively (Herman and Forestell 1985), and the sea lion Rocky "balked" (did not respond) when asked to perform an action (e.g. a flipper touch) on an absent object.

The parrot Alex reacted to an object's absence by saying *nuh* to refuse an object offered in place of one he had requested (Pepperberg 1987c: 43–4, 1999b: 80, 83). After having been properly trained and tested, Alex showed at least limited use of the concept of non-occurrence or absence, and Pepperberg (1999b: 94) found that this use is directly comparable to that of other animals that have undergone similar training. Pepperberg warns, however, that such data provide little direct evidence for a general concept of non-existence, even if they indicate an awareness that something was not present.

We have found no conclusive evidence that animals comprehend (c). That non-human animals have difficulties understanding this concept might be suggested by the following example. When the chimpanzee Sarah was given a sentence such as "Red on green ?" (= 'Is red on green?'), referring to the relationship of two colored cards, on 30 percent of the occasions she altered the relationship to make it possible for her to answer 'Yes' rather than 'No' (Premack and Premack 1983: 116).

In concluding, mention should be made of a noteworthy finding made by Zuberbühler (2002) on wild Diana monkeys (*Cercopithecus Diana*) in the Ivory Coast. Male Campbell monkeys (*Cercopithecus campbelli*) have distinct alarm calls for leopards and for crowned-hawk eagles (A), and when hearing these calls, Diana monkeys respond with their own corresponding alarm calls. In addition, Campbell monkeys have a third call, consisting of a pair of low, resounding "boom" calls, B. This third call is used for disturbances that are not a direct threat, such as a falling tree, or a distant predator. Once Diana monkeys hear the sequence A-B of Campbell males, they ignore the alarm meaning B—that is, A-B does not trigger alarm calls in Diana monkeys. This situation might be taken as suggesting that a call A functions somehow like a negation marker for another call B; without further information, however, it is hard to tell how this behavior is to be understood.

In sum, trained animals are able to develop notions of rejection and refusal, and even of non-existence; but it seems that none of these non-human animals clearly has acquired a notion of denial, that is, the ability to deny the truth or falsity of a given assertion.

Questions What surfaces from the literature seems to indicate in fact that trained animals can acquire the ability to perceive questions. Premack's chimpanzees have been shown to handle the concept of a polar (yes–no) question in predications on same–different distinctions, as well as a WH-question when given two options and asked '*What* is A to A?' (Premack 1976: 147–55). Similarly, the chimpanzees Washoe and Sarah had acquired a well-established interrogative marker, hence Sarah had no problems understanding word questions like "Red ? apple" (What is the relation between red and apple?) (Premack 1971: 813). And the gorilla Koko appears to have created a sign for polar questions by using gestural intonation, retaining the hands in the sign position at the end of an utterance and seeking direct eye contact, although she did not acquire signs for word questions ('who?', 'what?', or 'where?'; Patterson 1978a: 78–80).

But an understanding of questions was not only found in primates. The grey parrot Alex did not only comprehend a vocal question and extract the relevant category from a word question but appears to have been able to produce questions: He used *what* to request information about objects in his environment, and since *what* never occurred in his responses to questions posed by his trainers (Pepperberg 1999b: 61; Kako 1999: 10), this might suggest that he had a concept of an interrogative marker. And much the same applies to the bottle-nosed dolphin (*Tursiops truncatus*) Ake, who was taught a polar interrogative form consisting of a sign for an object plus a question sign and responded productively to this form to signal presence vs. absence of an object (Herman 1989: 23).

3.2.5 Compositionality

Trained non-human primates learn to combine form–meaning pairings fairly early. The chimpanzee Washoe began combining signs after only 10 months of training, Moja at the age of six months, and the gorilla Koko is reported to have begun using sign combinations after four months of sign language training (Miles 1978: 105).

Can animals combine form–meaning pairings with each pairing retaining its meaning constant? Do they understand that utterances can be broken up into concepts and can be combined productively and in a principled way, and are they able to use at least two paradigms of linguistic forms productively in a sequence? With the term compositionality (or combinatorics) we refer to the fact that linguistic forms are, first, combinatorial, in that the meaning of forms remains stable across contexts: Meaningful units can be combined and still retain their semantic integrity—that is, combining is compositional, even if the meaning may be influenced by the context in which it occurs (for example, the form *big* as in *a big problem* does not have exactly the same meaning as in *a big car*). Second, we also use the term for the ability to productively combine novel forms and concepts.

There is fairly clear evidence that the apes that were studied as well as Alex, the grey parrot, and the two bottle-nosed dolphins, have the ability for discrete combinatorics. For Alex, for example, the expression *rose paper* refers to paper with the property of being colored rose, not to something intermediate between rose and paper; for Kanzi, the bonobo, the expression *Tickle ball*, meaning to tickle him by rubbing the ball all over his body, was perceived as consisting of two distinct units rather than of one unanalyzed entity, as is suggested by his ability to re-combine these units, and the bottle-nosed dolphins interpreted PIPE TAIL-TOUCH as a command to perform an action on an object, that is, as an expression consisting of two discrete entities (see Kako 1999: 6).

We noted above that Zuberbühler (2002) found that wild Diana monkeys (*Cercopithecus Diana*) in the Ivory Coast appear to have a combinatory rule that is not only suggestive of a concatenation of form–meaning pairings but also of a grammatical function. Male Campbell monkeys (*Cercopithecus campbelli*) have distinct alarm calls for leopards and for crowned-hawk eagles, and when hearing these calls, Diana monkeys respond with their own corresponding alarm calls. In addition, Campbell monkeys have a third call, consisting of a pair of low, resounding "boom" calls, which is used for disturbances that are not a direct threat, such as a falling tree, or a distant predator. Once Diana monkeys hear a sequence of the two calls of Campbell males, they ignore the alarm meaning. Note, however, that this sequence of two calls is not—and cannot be—productive since there are no other form–meaning pairings to which it could be applied.

A noteworthy ability is documented for the trained grey parrot Alex, who "recombines beginnings and ends of existent labels, rather than ends with ends or beginnings with beginnings. Thus there are utterances such as "banacker" (banana-cracker), but never "bancrack" or "ana-er". The meaning of the use of this behavior is unclear but suggests, as Pepperberg (1999a: 17) argues, some sensitivity to internal order and subunit structure that is not necessarily expected in non-humans.

3.2.6 Argument structure

Are animals able to form sentences that can be said to be homologs or analogs of what one finds in human language; in particular, do they have the ability to acquire an argument structure? We will say that an argument structure is built around a predication where there is a verb (or predicate) having at least one (nominal) argument.

Kanzi developed two kinds of productive combining: The first concerned a lexigram to specify an action (e.g. *tickle*), subsequently a pointing gesture was used to specify the agent. Savage-Rumbaugh and Lewin (1994: 161) found that Kanzi's "rule" of action first, using lexigrams, and agent second, using gestures, represented the opposite order of the spoken English that was used around him all the time, and they argue that this fact provides strong evidence for creative productivity in this animal. One might argue, however, that this deviance from the model that Kanzi was exposed to could be due to the different modalities that were recruited by the animal. Nevertheless, this "rule" appears to have been productive: Once Kanzi had learned the action–object "rule", he appears to have extended it to new situations. Such a situation obtained, for example, when Kanzi asked someone to play *tickle-ball*, meaning to tickle him by rubbing the ball all over his body. In this new combination, he followed exactly the ordering he had demonstrated earlier (Savage-Rumbaugh and Lewin 1994: 159).

A second kind of combining arrangement that Kanzi acquired was putting together two action units, such as *tickle-bite*, which he himself had created—caretakers almost never used such combinations (Savage-Rumbaugh and Lewin 1994: 161). Kanzi produced almost exclusively only two elements in a proposition; in the rendering of the authors concerned (Greenfield and Savage-Rumbaugh 1990; Kako 1999: 9), he comprehended "both the words *and* their relations to one another", as well as the influence of the action ("verb") meaning on the argument ("thematic role")—in other words, he appeared to have the ability to understand the nature of an elementary one-argument proposition.

An understanding of distinctions in participant roles has also been observed in other *Pan* species. The chimpanzee Dar was observed to vary the place of articulation of his signs to include agents, locations, and instruments (Rimpau, Gardner, and Gardner 1989; Casey and Kluender 1998). And the chimpanzee Sarah is said to have learned to distinguish between case functions ("subject" vs. "object"): Premack (1976: 322) says that Sarah comprehended 'name of' when it was used as part of an accusative phrase although, in all of her previous experience, it was confined to the nominative phrase. She was equally successful in transferring the quantifiers from the nominative phrase in which they were taught her to the accusative phrase in which they were later presented. For example, she had been trained to produce sentences of the form 'Some cracker is round,' 'All cracker is PL square,' etc., and later performed correctly when instructed 'Sarah take some cracker,' etc. Thus, although both 'name of' and the quantifiers were learned originally as parts of "nominative phrases", she comprehended them when she later experienced them as part of "accusative phrases". The problem with some animals is that the possible link between syntactic position and thematic role is not always clear: Are the animals able to predict some semantic properties of arguments from syntactic information?

Conversely, it has also been claimed that chimpanzees are not able to acquire argument structure. Rivas (2005) concludes that the five chimpanzees studied by him, who had been taught ASL, lacked a semantic or syntactic structure in combining signs.

Combining patterns suggestive of argument structure is not restricted to primate species. The grey parrot Alex learned to use the structures [*wanna go* + location unit] and [*want/wanna* + object unit] at least with a range of form–meaning categories (Pepperberg 1987c). While he distinguished volitional propositions (*want* X, *wanna go* X) and commands (e.g. *go* X, *you tickle me*), it is not entirely clear whether these come close to reflecting an argument structure involving stable arguments in combination with predicates.

Bottle-nosed dolphins have been taught to understand propositional structures whose constituents were object terms (O), nonrelational action terms (A), relational action terms (R), and the optional modifier (M) terms LEFT and RIGHT, and Herman (1989: 24, 36) proposes a set of five rules that the dolphin Ake learned, for example when being directed to take Object 2 (= O2) to Object 1 (= O1). Herman concludes that Ake had an understanding of argument structure because when, for example responding to LEFT OVER, with the object term O missing, she jumped over the ball to her left, and when given the sign FETCH alone with no object sign preceding it, she cleared the tank of objects, often bringing back two or three at once to her trainer.

3.2.7 *Linear arrangement*

Are animals capable of developing a way of consistently and productively ordering paradigms of form–meaning pairings—in other words, do they have something corresponding to word order in human language?

The data available suggest that the answer is essentially in the affirmative, even if consistent linear arrangement in animals does not seem possible without training. Thus, Premack (1976: 317) notes that chimpanzees can be taught word order, but only with explicit training, and they therefore differ from children, who acquire it on the basis of observational learning. Similar findings have been made on bonobos: The work of Savage-Rumbaugh and Lewin (1994: 160) suggests that acquiring linear arrangements is a matter of systematic training. The bonobo Kanzi "showed no particular ordering of such symbols during the first month; sometimes he put the action first, sometimes the object. *Hide peanut* occurred just as often as *peanut hide*, for instance. But thereafter he began to follow a rather strict order, that of putting the action first and the object second: *hide peanut, bite tomato,* and so on." The authors suspect that Kanzi learned this arrangement from his caretakers.

But there are also linear arrangements that this bonobo appears to have developed on his own: As we observed above, Savage-Rumbaugh and Lewin (1994: 161) found that Kanzi's arrangement [action (= lexigram)– agent (= gesture)] did not rely on any model that he may have been exposed to. Thus, Kanzi's arrangement has the appearance of a verb-initial syntax, in that ordering was determined by placement of an action lexigram before an argument, irrespective of the syntactic function of the argument (Greenfield and Savage-Rumbaugh 1990).

Kako (1999: 9) concludes that Kanzi observed a structure that comes close to a word order arrangement: Of the seven major relations involving different types of referents, four showed a statistically significant preference for the following orders: action–agent, entity–demonstrative, goal–action, and object–agent. Furthermore, Kanzi placed his smaller category of gestures (among them three action gestures) almost always after the larger category of lexigrams that he had learned to distinguish, irrespective of whether the combination was, for example, [action lexigram + demonstrative gesture] or a [goal lexigram + action gesture]; thus, his lexigram ordering superseded the ordering of argument functions (Greenfield and Savage-Rumbaugh 1990). While Kanzi thus showed an arrangement pattern that was not dependent on specific form–meaning pairings, it is characterized essentially only by one "rule" (Kako 1999: 8). Some patterns of linear order were also found in the gorilla Koko (Patterson 1978a: 91).

Regular patternings in the order of form–meaning pairings have been reported on a number of apes, and they tend to match word order in spoken English—the language used by the trainers. The chimpanzee Ally showed a preference for an order "demonstrative–noun" in 92 percent of his two-sign constructions and "subject–verb–object" order preference in 89 percent of his three-sign constructions (Miles 1978: 109–10), and in the two-sign utterances of the gorilla Koko, 75 percent of the 98 attributive phrases recorded had the order "adjective–noun" (Patterson 1978a: 91).

Sensitivity to linear arrangement was also found in the trained bottlenosed dolphins Phoenix and Ake, who could correctly enact relational sentences only by inferring the thematic roles of objects (transport vs. designation) from their syntactic position (Herman *et al.* 1984; Herman 1987). That they are sensitive to word order is suggested by the fact that they responded correctly above chance level to reversals of arguments when asked to distinguish between PIPE FETCH HOOP "Fetch the pipe to the hoop" and HOOP FETCH PIPE "Fetch the hoop to the pipe".

And there are also indications that the animal developed iconic ordering patterns. Savage-Rumbaugh and Lewin (1994: 161) observe that the action–action combinations that Kanzi developed were essentially iconic: The authors found that orders like *tickle–bite* and *chase–hide* were significantly more frequent than their inversions.

We mentioned earlier an interesting ordering distinction found in the orangutan Chantek: When the object referred to was present, then the form of an utterance would be object-GIVE, but when the object was not present, then the form was GIVE-object (Miles 1990: 519).

A number of the findings made in these studies suggest that linear arrangement is—at least to some extent—"lexically" rather than "rule" based. Terrace (1983: 23) and associates found that their chimpanzee Nim used the sign *more* in the first position in 85 percent of the two-sign utterances in which *more* appeared (e.g. *more banana, more tickle*), 78 percent had *give* in the first position, and 83 percent of the instances in which a transitive verb (e.g. *hug, tickle, give*) was combined with *me* and *Nim* had the transitive verb in the first position. Note that Nim's teachers had no reason to sign many of these combinations. That linear arrangement tends to be lexically based is an observation that has also been made of the language acquisition of children (e.g. Tomasello 2003b; Diessel 2005).

Analyzing the ninety-one sequences of signs that the chimpanzee Washoe, trained in ASL, was able to produce (Gardner and Gardner 1971), McNeill (1974) observes that this ape had nothing corresponding to syntactic patterns in human languages, but she displayed a constituent order where the addressee or party being addressed always preceded the addressor or speaker, and McNeill suggests that the animal showed a linguistic ability that contrasts with that of humans, namely one that emphasizes interpersonal or social interaction.

3.2.8 Coordination

While we have not found any kind of subordination in systems of animal communication, some patterns of coordinating concatenation have been reported (Fitch and Hauser 2004: 377). Premack's (1976: 324) findings suggest that the trained chimpanzee Sarah acquired the ability to conjoin nouns: Sarah could combine two and sometimes three nouns having the same syntactic function in the same sentence, all of them as objects of one and the same verb, for example, "Mary give Sarah apple banana orange." This finding may be taken as evidence that this chimpanzee acquired the ability of conjoining noun-like form–meaning pairings; whether it is suggestive of equi-verb deletion, as Premack suggests, is an issue that cannot be resolved without further information.

Another example is provided by bottle-nosed dolphins. The dolphin Phoenix was given the instruction to act on a sequence of two propositions each consisting of an object and an action. When told, for example, PIPE TAIL-TOUCH PIPE OVER she swam to the pipe, touched it with her tail flukes, and then jumped over it. Without any specific training or reinforcement history, Phoenix carried out instructions on such coordinated propositions successfully in eleven out of fifteen cases (Herman 1987: 24).

Serial verb constructions in chimpanzees? The chimpanzee Sarah was not only able to conjoin nouns but also verbal items. Early in training, Sarah was given only one action name at a time, one appropriate to the point in the sequence to which the action had progressed. Later, several verbs were made available to her at the same time, enabling her to write, "Wash apple, cut apple," or "Cut apple, give apple," etc. But Sarah did not write pairs of sentences of this kind; rather, she wrote instead "Cut give apple," and "Wash give apple," even if not always observing the correct order (Premack 1976: 244, 321). Such complex forms involving sequences of two verbs might be suggestive of verb serialization, which did not occur until after she had first produced equivalent outcomes by using multiple simpler sentences; for example, she did not write "Wash give apple" until she had written "Wash apple" and "Give apple" many times before (Premack 1976: 324). Much the same behavior was shown by the other two chimpanzees Peony and Elizabeth, who wrote, for example, "Elizabeth apple wash cut" in describing their own action or "Amy apple cut insert" in referring to the trainer Amy as an agent.

Once again, Premack (1976: 324) accounts for this behavior in terms of equi-noun deletion of (underlying) full sentences; but it is equally possible to maintain that these chimpanzees simply conjoined verbs in a fashion roughly comparable to serial verb constructions (Aikhenvald and Dixon 2006). However this may be, what is remarkable about this achievement is that it was created by these animals apparently without any model provided by their human trainers:

> Their use of conjunction was impressive, in part because of its several forms and the different contexts in which it occurred, but more important because it was invented by them. No aspect of conjunction was taught Sarah or the other subjects. The stringing together of object names in one case and action names in the other was their contribution. (Premack 1976: 321)

Causal relations? Can animals understand the nature of logical relations obtaining between different situations or propositional contents? Premack (1976: 336–7) argues that chimpanzees can make a causal analysis of their experiences. His subject was given an intact object, a blank space, and the same object in a changed or terminal state, along with various alternatives, and was encouraged to complete the sequence by placing one of the alternatives in the blank space. For example, the subject was given such items as an intact apple and a cut apple; a dry sponge and a wet sponge; a clear piece of paper and one with writing on it. In these cases, the three alternatives given the subject consisted of a knife, a bowl of water, and a writing instrument. Premack (1976: 337) found that three of the four chimpanzees tested in this way required no more than general adaptation to the test format before responding correctly. Their ability to place the knife between the intact and severed apple, the water between the dry and wet sponge, and the pencil between the unmarked and pencil-marked paper showed that they correctly identified the instrument or operator needed to change each object from its initial to its final state.

Can this be taken to substantiate Premack's claim that chimpanzees are capable of making causal analyses? It would seem that the answer is not clearly in the affirmative. The behavior described may simply suggest that these animals associate three different objects as belonging together on account of past experiences.

3.2.9 Taxonomic concepts

We have seen in the preceding section that non-human primates and a range of other animals exhibit a range of abilities in perceiving and producing structures that have at least analogs in human languages. In the present section we are concerned with linguistic features that have been claimed to distinguish human beings from their non-human cousins: Do animals have, or can they acquire relational categories, in the sense that they perceive and/ or describe one concept in terms of another concept? Tomasello and Call (1997: 134) maintain that at least in primates there is some fairly strong evidence that this is the case. As we saw above, a number of animals, including avian species, are able to both perceive and produce predications on the basis of the sameness and differences between objects.

Premack (1976: 354) suggests that his chimpanzees distinguish between what he calls first and second order relations: They are not only able to observe, for example, that the relation between 'red' and 'red' is same, as that between 'grey' and 'grey' is same, but also that the relation between 'red' and 'red' is the same as that between 'grey' and 'grey', the latter being a relation between relations, that is, a second order relation. But are animals also able to understand asymmetrical relations of dominance among concepts, where one concept presupposes, implies, or includes the other concept? More specifically, is there anything in animals that corresponds to hierarchical taxonomy or to head-dependent relations in human languages? The following discussion will be confined to relations as they manifest themselves between thing-like, referential concepts, that is, form–meaning pairings typically coded as nouns in human languages. To this end, we need to introduce a number of elementary taxonomic distinctions. We will use the term "taxonomy" in a general sense for conceptual structures describing relations among concepts in a principled way. Taxonomies may be "shallow", that is, not involving hyponymy, or hierarchical, that is based on some form of dominance or hyponymy, and our interest here is exclusively with the latter. For our purposes, the following kinds of taxonomic structures are immediately relevant:

(1) Hierarchical taxonomic relations
   a. Inclusion: A is a kind/type of B (e.g. *An apple tree is a kind/type of tree*).
   b. Property relationship: B has property A (*a red ball*).
   c. Partonomy (or meronymy): A is part of B (*A finger is part of a hand*).
   d. Social relationship: A is a relative of B (*Anne's father, husband*, etc.).
   e. Possession: A has B (*Anne's car, name*, etc.).
   f. Location: B is located at A (*the book on the table*).

These taxonomic relations are both alike and different from one another in a number of properties and, depending on which property one highlights, various alternative groupings are possible. And each of these relations subsumes a number of relations which we will not further differentiate (see Cruse 1986 for detailed treatment). For example, (1a) also includes simple hyponymy (*An A is a B*) and set membership (*Trees make a forest*), and (1d) can be catenary (i.e. it can form indefinitely long chains: A → B → C, e.g. *father of*) or non-catenary (A → B $\xrightarrow{*}$ C, e.g. *husband of*), or it may be transitive (A → B → C → A) or intransitive (A → B → C $\xrightarrow{*}$ A, e.g. *father of B is not father of A*). Not all of these relations, especially (1f), are commonly mentioned in treatments of taxonomy but are included here since they may have some bearing on the structure of head-dependent relationships.

We will now look into the question of whether, or to what extent, animals are able to perceive and/or produce taxonomic relations such as the ones listed in (1).

Tomasello and Call (1997: 133) note "that the natural categories of primates seem to be, like those of pigeons, on the concrete species level (e.g. crow) and the generic level (e.g. predator), not on the basic level (e.g. bird) which seems to be so important in human linguistically based cognitive processes." Does this mean that these animals have at least a modest capacity of comprehending inclusiveness on what Tomasello and Call call the generic level?

Inclusion A common domain where inclusion relations manifest themselves is that of modifying compounding, as in our example in (1a). We have not found any clear instances of it in non-human animals, nor of productive compounding in general. The combinations of form–meaning pairings that have been reported can be interpreted in different ways: They could stand for unitary, non-compositional meanings, or they could be combinations of free forms. The chimpanzee Washoe has been reported as describing a swan by signing WATER BIRD, but such combinations could as well be due to mechanisms other than compounding nor do they seem to be productive (Rivas 2005). To be sure, quite a number of animals have been shown to combine form–meaning pairings that can be interpreted as noun–noun compounds. The gorilla Koko labeled a stale sweet roll a COOKIE ROCK, a mask an EYE HAT, a ring a FINGER BRACELET, a zebra a WHITE TIGER, or a Pinocchio doll an ELEPHANT BABY (Patterson 1978a: 88, 1978b: 195); but there is nothing to establish that these are indeed modifying compounds.

But the study by Savage-Rumbaugh *et al.* (1980: 922–4) suggests that their chimpanzees may have some notion of inclusion (1a): These animals interpreted the symbols (lexigrams for specific foods or tools) in order to then label them with lexigrams for the hypernyms 'food' or 'tool'. The training of their chimpanzees Sherman and Austin led these animals to link the use of an object and the label of an object together. Sherman and Austin were then presented with fourteen food and fourteen tool items and they correctly categorized, respectively, twenty-four and twenty-five of the twenty-eight items, and the errors resulted with one exception from classifying tools used to prepare food (for example, knife and cutting board, which they often lick) as foods. Savage-Rumbaugh *et al.* (1980: 924) suggest that Sherman and Austin were able to treat 'food' and 'tool' as "representational labels, and to expand the use of these labels to novel exemplars because of training which encouraged the appearance of functional symbolic communication between chimpanzees."

Such observations are confirmed by research on other chimpanzees: Premack was able to establish that his trained chimpanzees Peony and Elizabeth could sort items on the basis of the labels plant vs. animal. The animals sorted leaves, stems, seeds, and flowers on the one hand, and fur, teeth, hair, and bones on the other into piles corresponding to the distinction plant vs. animal, respectively, at about the 78 and 80 percent level of accuracy. In a later step, erasers, pins, buttons, paper clips, and the like were added to the above items, and both subjects were led to re-sort to a new criterion. The previously separated plant and animal items were now combined and contrasted with hairpin, button, etc. (Premack 1976: 217–18).

Property relationship A paradigm instance of a property relationship would be one where there are nouns that take adjectives as modifiers. In fact, quite a number of animals have been taught what is described by their trainers as adjectives or modifiers; the gorilla Koko, for example, is said to have acquired distinct signs for sixteen modifiers, including 'big', 'clean', 'cold', 'good', etc. (Patterson 1978a: 79). But is this enough to develop a conceptual relationship between, for example, a noun and its modifier?

Trained African grey parrots (*Psittacus erithacus*) have been found to exhibit a remarkable ability to classify objects on the basis of physical properties, as is suggested by the following observations made by Pepperberg (1992): Using distinctions of color (blue, green, grey, orange, purple, rose, yellow), shape (2-corner, 3-corner, 4-corner, 5-corner, 6-corner), and material (chain, hide, key, paper, rock, wood, wool), her avian experimental subject Alex responded correctly in 63.3–83.3 percent of the thirty-four trials when asked for example: "What color is [item designated by shape and object label]?" This might suggest that Alex has some elementary concept of property relationship, in that he was able to assign different kinds of qualities to objects—that is, the expression *rose paper* referred to paper with the property of being colored rose. Furthermore, when presented for the first time with rose paper, Alex called it *rose hide*. After being allowed to chew on it, he called it *paper*. When prompted with "*What color?*" he replied "rose"; prompted further with *Rose what?* he replied "*rose... paper*" (with a brief pause between words) (Kako 1999: 7).

An equally impressive behavior exhibited by Alex is the following: He appears to have extended the word *rock* as the label to refer to the property 'hard'; he spontaneously produced the phrase *rock corn* and was given a kernel of dried corn. Subsequently, he continued to ask for dried corn with that phrase, and Pepperberg (1999a: 17) observes: "Alex did not learn *rock* as an adjective for hardness; rather he learned it as the label for a lavastone beak conditioner. Whether he abstracted the concept of hardness from the rock and intentionally applied it to the dried corn to form *rock corn* or simply hit upon the term in the course of his practice 'babbling' will never be known." However, *rock* did not acquire productivity, that is, it did not turn into a kind of modifier (Pepperberg 1987c: 46). That Alex might have acquired the taxonomic concept of inclusion is perhaps most clearly suggested by the following observations by Pepperberg:

> He understands categories in terms of hierarchical levels, so he knows that there's this weird (to him) sound called "color" and under that weird sound are grouped all these other sounds called "red," "blue," "green," "yellow," "orange," etc. that relate to a specific set of physical attributes of objects. Similarly, he understands there is another weird sound, "shape," and under that sound there are the other sound patterns "two-", "three-", "four-", "five-", and "six-corner" that relate to different physical attributes of the same objects. We can teach him new ways of categorizing items. If he's already learned to categorize items by color and shape, we can then ask him to categorize them by number. (Pepperberg 1999a)

The chimpanzee Sarah was given sentences of the target form "Sarah take red dish," and in taking the red dish, instead of either the green dish or the red/green pail, she is hypothesized to demonstrate comprehension of the attribute form. Premack (1976: 320) adds that Sarah's performance went beyond that of the child, for although she had been taught "Sarah take dish" and "Sarah take red", she had never been taught "red dish". Furthermore, having acquired a question marker, Sarah understood the questions "Red ? apple" (What is the relation between red and apple?) and "Yellow ? banana", where the only word available to her was "color of". She substituted this word for the negative marker, thereby forming the sentences, "Red color of apple" and "Yellow color of banana" (Premack 1971: 813). And when exposed for the first time to cantaloupe and strawberry, which were unnamed, referred to as "yellow fruit" and "red fruit," respectively, Sarah followed the instruction "Insert red fruit (in) dish" and correctly put the strawberry (not the cantaloupe) in the dish (Premack 1976: 323).

Another chimpanzee, Peony, was given sixteen blocks that differed in both color and form—red and yellow triangles and red and yellow squares, and she typically first sorted the blocks into two boxes on the basis of color. After the experimenter praised her, reassembled the blocks and gave them back, Peony shifted criteria, sorting them into the same boxes on the basis of shape, shifting criteria fourteen times in twenty trials (Premack 1976: 217).

Training modifier–object constituents also yielded some success with other animals. The orangutan Chantek began to use attributes, as in *red bird* and *white cheese food eat* (Miles 1990: 525) although there are no clues to the effect that he had a taxonomic understanding of the relationship. And it has also figured prominently in research projects on marine animals. The sea lions Rocky and Bucky studied by Schusterman and Krieger (1984) were taught modifiers for color (BLACK, WHITE, GREY), size (LARGE, SMALL), and locations (WATER, LAND), while the bottle-nosed dolphins Phoenix and Ake were given the locative modifiers LEFT, RIGHT, BOTTOM, and SURFACE (see below). The animals comprehended the meaning of these modifiers, even if there were combinations of modifiers; however, there is no indication that they understood these constituents in terms of a taxonomic relation object–modifier, let alone head–modifier.

To conclude, the abilities described above are remarkable but do not really establish that any of these animals clearly had the concept of a hierarchical relationship of the form [object [modifier]]. Still, there are a few indications that could be interpreted in favor of such a relationship. We noted above that the gorilla Koko acquired distinct signs for sixteen modifiers, and in her two-sign utterances recorded in 1974, 75 percent of the ninety-eight attributive phrases recorded had the "adjective" preceding the "noun" (Patterson 1978a: 91)—a fact that might be taken to suggest a modifier–head pattern.

Partonomy That apes are capable of understanding the nature of part–whole relationships has been claimed by a number of researchers, but the evidence adduced is, with few exceptions, not really convincing. Indications of a possible partonomy concept surfaced in the chimpanzee Sarah, trained by Premack:

At a later time, after a period in which fruit was routinely prepared in Sarah's view, the same tests were repeated, changed only by the addition of several new fruits—peach, grape, and fig—which in the meantime had been added to her vocabulary. On this second round of tests she was successful both in matching pieces of fruit to intact fruit and in matching names of fruit to intact fruit. There was a complete parallel, therefore, between the perceptual and the linguistic results. (Premack 1976: 138)

Sarah also comprehended the notion "name of" when it was used as part of a sentence object ("accusative") phrase although in all her previous experience it was confined to subject ("nominative") phrases. She had been trained to produce "X name of Y" statements (where X was the name of the object Y), yet subsequently she carried out instructions of the form "Sarah insert name-of-cracker in cup" versus "Sarah insert cracker in cup," where "name-of" occurred as part of the object phrase (Premack 1976: 321)

Furthermore, we observed above that the chimpanzees Peony and Elizabeth were able to sort plant parts (leaves, stems, seeds, and flowers) into a plant class and animal parts (fur, teeth, hair, and bones) into an animal class (Premack 1976: 217–18). Note also that spider monkeys, after having discovered a tree with fruit that had just become ripe, often proceeded immediately to other trees with the same type of fruit, "implying that the fruit trees in their environment were categorized on the basis of the particular type of fruit they bore" (Tomasello and Call 1997: 113).

Social relationship   Hauser, Chomsky, and Fitch (2002: 1575) observe that a wide variety of non-human primates have access to a rich knowledge of who is related to whom, as well as who is dominant and who is subordinate (see also Steklis 1988). In fact, primate understanding of relational categories is said to have evolved first in the social domain to comprehend third-party social relationships (Cheney and Seyfarth 1992; Tomasello 2000a: 173).

A number of non-primate species have been found to show understanding of at least some kind of kinship relationship; captive bat mothers (*Tadarida brasiliensis*), for example, recognize their pups, although pups do not recognize their mothers (Dechmann 2005: 487). Chimpanzees have a clear understanding of subordinate–dominant relations, and they can play different roles, acting as dominant in one situation but as subordinate in another situation—that is, they appear to understand a situation from a social-cognitive point of view (Hare *et al.* 2000: 783). Hare and his associates conclude that the fact that the same individuals adopted different strategies depending on the role they played in the experiment, subordinate or dominant, suggests that they were not following some blind behavioral contingencies or rules, but rather that they really did understand something of the situation from a social-cognitive point of view.

Tomasello and Call (1994; see also Tomasello 2000a: 166–7) found that there is strong evidence for an understanding of third-party social relationships in a number of different primate species, but this ability to understand and form categories of third-party social relationships appears to be confined to primates. But there are cases of redirected aggression, as when A1 (or A's kin) is attacked by B, A retaliates by attacking B's kin. Such cases, which are suggestive of some kind of analogical reasoning, are not restricted to apes; when a vervet monkey (*Cercopithecus aethiops*) A1 threatens an unrelated animal B1 following a fight between one of his relatives A2 and one of his opponent's relatives B2, A1 acts as if he recognizes that the relationship between B1 and B2 is similar to his own relationship with A2 (Cheney and Seyfarth 1992: 139).

Furthermore, while there is no clear evidence that non-apes have a concept of social relationship, there are some data on monkeys suggesting that there is at least a concept of a relationship between a mother and her offspring: Cheney and Seyfarth (1980) observe that adult female vervet monkeys, analyzed under natural conditions, can distinguish screams of their own 2-year-old juvenile offspring from the screams of other 2-year-olds in the group, and that the latter are able to associate specific screams with their mothers. Dasser (1988) trained two 5-year-old female Java monkeys (*Macaca fascicularis*), who were subsequently able to distinguish on color slides of members of their group mother–offspring relations from non-mother–offspring relations. Identification does not appear to have been achieved on the basis of perceptual features shared by the stimulus animals but rather on affiliation relations that the two subject animals had observed, enabling them to establish mother–child affiliation relations.

In sum, monkeys seem capable of classifying social relationships[5]—for example "mother–offspring bonds" or "bonds between the members of family X"—into types according to one or more abstract properties and to compare relationships on the basis of these representations (Cheney and

---

[5] Noteworthy observations on social relationship have also been made on some avian species. White-fronted bee-eaters (*Merops bullockoides*), for example, are communally breeding birds which seem to distinguish between kin and non-kin, close relatives and non-relatives, and between unrelated neighbors and non-neighbors (see e.g. Emlen and Wrege 1988).

Seyfarth 1992: 139, 142). Such findings, however, are not sufficient to establish that there is a taxonomic concept of social role relations such as 'mother-of', 'child-of', let alone 'father-of', in any of these animals.

Possession Fitch, Hauser, and Chomsky (in press: 9–10) note: "But there are many aspects of animal territorial behavior that are difficult to explain without some primitive notion of ownership, such as a 'home court advantage' effect that persists even when both contestants are equally familiar with the territory. Detailed experiments on animal 'ownership' show how it is influenced by dominance, priority of access, value of resource, and species-specific rules and exceptions.... Although one can always find differences, post hoc, between animal and human versions of behavior, these data suggest an 'ownership' concept in some animals with considerable overlap with our own."

In spite of these observations we have found no indication that there is a taxonomic relation possessor–possessee in any of the animals that have been studied. Such relations have been claimed; the gorilla Koko, for example, was able to produce combinations such as KOKO PURSE or HAT MINE (Patterson 1978a: 90), but there is nothing to show that they are suggestive of a possessive head–modifier construction.

Location Evidence for the presence of location relations is also poor. That the bottle-nosed dolphins Ake and Phoenix were able to understand a kind of locative modifier–object construction (Herman 1987, 1989) might be relevant but is certainly not sufficient evidence. The dolphins were confronted with a relational sentence with a modifier attached to either a transport word or a destination word, as in PERSON LEFT FRISBEE FETCH (for Ake) and FRISBEE FETCH BOTTOM HOOP (for Phoenix), and they were wholly correct on at least 63 percent of these tasks even though neither had been trained on this particular construction before. But perhaps more interesting is the following observation: Ake learned first arrangement patterns of the form [O1 + O2 + R], where O1 and O2 were object terms and R a relational action term. Later on she was taught to use LEFT and RIGHT as locative modifiers (M) before the object term, hence [M + O]. Without any training, she was exposed to the expanded structures [M + O1 + O2 + R] and [O1 + M + O2 + R], and she responded correctly to the first instance of a sequence of each type (Herman 1989: 24). This might be interpreted as implying that the dolphin was able to understand the nature of a hierarchical relation between [O] and [M + O], and possibly of a recursive structure; however, without any further evidence this remains no more than one possible interpretation.

Conclusion As we will see in Chapter 6, taxonomic relations such as the ones just discussed are a requirement for there to be noun modification—and also for noun phrase recursion.[6] The animals concerned are said to both comprehend and produce conglomerations of features and assign them to objects, such as 'red' and 'banana' in the case of the chimpanzee Sarah, or *rose* and *paper* in the case of the grey parrot Alex. And they appear to have abstract concepts such as 'red-ness' or 'banana-ness'. But are these cases really suggestive of taxonomic hierarchy?

Alex may well conceive the expression *rose paper* as consisting of a modifier and a head; still, it does not become entirely clear whether *rose paper* might not as easily stand for a rose thing having the quality paper, or for a combination of a rose thing and paper—in other words, the evidence provided is not sufficient to establish that there is some kind of inclusion, where the taxon *rose paper* is included in, or is an instance of the category *paper*. And there is also no compelling evidence to show, for example, that the chimpanzee Sarah's production "X name of Y" represents a taxonomic structure, where one is the hyponym and the other the hypernym. In other words, in most of the cases discussed, the evidence is not sufficient to establish a taxonomic relationship. Non-human primates appear to lack the ability to label relations between classes of objects (cf. Cheney and Seyfarth 1992: 143).

However, there are exceptions, that is, examples that seem to reflect taxonomic comprehension. Fairly clear cases were provided by our examples of inclusion relations and, slightly less convincing, of partonomy. The classification of food and tool by Sherman and Austin, and of plant vs. animal items, etc. by Peony and Elizabeth (see above) would seem to suggest that these chimpanzees can handle the taxonomic relation of inclusion and/or of part–whole relations.

3.3 Discussion

The data surveyed above raise a number of more general questions which we will now look into.

---

[6] Note that we are restricted in this book to linguistic manifestations of recursion.

3.3.1 Problems

In the preceding sections we ignored some specific problems associated with the significance of animal research. First, in our cross-species comparisons we ignored cladistic logic, that is, we did not take into account that there are relative degrees of genetic relationship, giving equal weight to comparisons between species of the same clade (or family), like *Pan* and *Homo*, as between species of different clades, such as *Pan* and *Zalophus* species, or *Zalophus* and *Psittacus* species. While we believe that such a procedure can be justified at the present stage of research (Hauser, Chomsky, and Fitch 2002), it goes without saying it has considerable drawbacks, as rightly pointed out by Jackendoff and Pinker (2005); for example, a property that is shared by humans with chimpanzees is possibly of a different nature and may have other implications than one shared with Californian sea lions or African grey parrots. Are the abilities exhibited by non-human animals homologs of corresponding human abilities, that is, are they due to common origin, hence genetically shared? Or are they analogs, having evolved independently? In the case of abilities to be found in apes one may be more readily prepared to attribute them to homology than to analogy. But, as we saw above, some of these abilities are found in much the same way in fairly distant relatives of the human species. Can form–meaning pairings such as the alarm calls of vervet or Diana monkeys, or the vocal accomplishments of parrots be related meaningfully to human words and phonology, respectively? We will not attempt to answer this question; but, considering all the evidence that is available, in particular the fact that these abilities are to be found in animals belonging to highly contrasting clades of animal evolution, a hypothesis in favor of homology does seem very plausible (Johansson 2001).

Another problem associated with some of the research reported above is that the animals analyzed were usually exposed to only a limited range of contexts and to tasks whose solution was predictable to some extent on the basis of the context and the training they had received. One may therefore wonder whether, or to what extent, the abilities claimed to exist in non-human animals may not be the product of a "Clever Hans phenomenon", that is, of the trainers' unintentional cues rather than of the tasks the animals were supposed to perform (de Luce and Wilder 1983: 5–8; Terrace 1983). Furthermore, is it really justified to compare knowledge arising as a result of explicit training with knowledge that is acquired via informal learning, as is characteristic of human children? And why do humans transmit their linguistic accomplishments to their offspring while animals do not really appear to have such an ability?

And there are further problems with the nature of animal cognition that have been discussed in biological, psychological, anthropological, and other works. Most of them were ignored here, such as the following: What is the nature of the short-term or "working memory" in animals? To what extent does animal behavior involve conceptual as opposed to specific learning, or abstract symbols at a representational level as opposed to contextually appropriate uses? And are categorization processes such as the ability to form hierarchies of inclusion relations entirely dependent on language, or could such skills have appeared prior to the emergence of language (Tomasello and Call 1997: 135)? Even languages that have been claimed to lack recursive phrase structures, like Pirahã of Amazonia (Everett 2005; Bower 2005), appear to have constructions that are suggestive at least of simple recursion (see Chapter 6), while no animal species has been found so far to produce such constructions.

A number of non-human animals, such as the grey parrot Alex (Pepperberg 1992), the chimpanzee Sarah (Premack 1976: 321), or the bottle-nosed dolphin Phoenix (Herman et al. 1984), have been claimed to have some understanding of recursion. With this term, the authors concerned do in fact refer to one manifestation of recursion, namely the ability to conjoin propositions or constituents (see Section 3.2.9 above); we propose to refer to this manifestation as iteration (see Chapter 6, (7)). But this is different from embedding recursion as it shows up, for example, in noun phrase modification and clause subordination: For this kind, which is widely held to constitute the paradigm manifestation of recursion, there does not appear to be any equivalent in animal communication (see also Hauser, Chomsky, and Fitch 2002); we will return to this issue in Section 7.2. On the basis of the evidence that is available, non-human animals lack the ability to form hierarchical phrase structure, to arrange propositional contents hierarchically, or to subordinate one clause in another clause, or to manipulate participants in discourse, for example by forming something corresponding to a topic-comment construction (Krifka 2005: 3).

Finally, why did non-human animals not develop language-related communication systems? In spite of all the abilities that have been found in animals, and in spite of all the "neural plasticity" that animals appear to dispose of in dealing with linguistic phenomena, Kako (1999: 12) rightly asks: What exactly are these animals doing in the wild with such plastic brains? While some cognitive abilities to develop a basic linguistic communication system appear to be in place in animals, no non-human animal species has developed anything that only remotely resembles systems of human linguistic communication. We have no conclusive answer to these questions other than observing that they need to be looked into in future research.

3.3.2 Language-like abilities in animals

In the introduction to this chapter we raised the question of whether non-human animal species show cognitive and communicative abilities that are a requirement for developing something corresponding in any significant way to human language abilities. In spite of all the problems and questions that we were concerned with in the preceding section, there is little doubt that animals exhibit a number of remarkable abilities. Note that our concern in this chapter was not with what animals cannot do, nor with how they acquired what they can do, but simply with what they can do. And according to some researchers, they can do a lot. Dogs have been found to be able to arrange elementary signs into 212 combinations of signs (Fleischer 1990), and Premack (1976: 331–2) claims that humans and chimpanzees share the following striking similarities: (a) Both species have "rich" conceptual structures, being able to distinguish between agent, object, action, etc.; (b) both species symbolize, that is, use one thing to represent another, and do so with regard to all possible kinds of things; and (c) both species are said to have syntax independent of semantics or general cognition, and their basic logical structure is the same, chimpanzee syntax differing only in detail.

While this depiction is not shared by most of the other researchers who have worked on animal behavior, we saw in fact that there is an impressive catalog of abilities to be found at least in some animals exposed to language training, especially the abilities listed in (2) below. Note that these abilities have not all been observed in one particular animal; rather, they represent the total of all abilities that we found, and they are confined essentially to animals that received some regimented teaching.

(2) Possible language-like abilities of some non-human animals
  a. to understand salient characteristics of concepts;
  b. to distinguish form–meaning pairings ("words");
  c. to acquire form–meaning pairings of more than 100 items, including items denoting objects, actions, and some numbers;
  d. to handle functional items for negation and interrogation;
  e. to have an elementary understanding of the notion of deixis;
  f. to use an elementary argument structure;
  g. to acquire some understanding of linear arrangement of form–meaning pairings;
  h. to conjoin propositions and/or form–meaning pairings;
  i. to acquire some basics of taxonomic hierarchy as it manifests itself in inclusion and part–whole relations.

On account of observations such as the ones made in this chapter, we concur with Fitch, Hauser, and Chomsky (in press: 13), who suggest that the safest assumption at present is that the mechanisms underlying human speech perception were largely in place before language evolved, and with Kako (1999: 12), who concludes that "several of the core properties of human syntax lie within the grasp of other animals."

3.3.3 Grammaticalization in animals?

In the preceding sections we did not have much to say about grammaticalization, and in fact there is not much to say. To be sure, we saw in Section 3.2.4 that some animals acquired form–meaning pairings that can be interpreted as equivalents of functional categories in human languages, such as markers for negation (non-existence), interrogation, spatial deixis ("demonstratives"), and personal deixis ("personal pronouns"). But the acquisition of such items was not based on parameters of grammaticalization as we defined them in Section 1.2.1; rather, these items appear to have been learned in the interaction between animal and human care-giver in much the same way as lexical form–meaning pairings.[7]

There are a few indications that animals are able to grammaticalize. For example, one salient cognitive strategy of grammaticalization consists in the transfer from concrete objects (e.g. body-part terms for 'back' or

---

[7] An anonymous referee of this work points out that this is predictably so since our focus is on the faculties of trained animals.

'head') to spatial relations (e.g. locative adpositions for 'behind', and 'on top of' or 'in front of', respectively) (see Section 2.2.1.1) on the basis of the extension parameter (Section 1.2.1.2). The bottle-nosed dolphin Ake might have used this strategy when extending the use of the signs for the object WINDOW (located to her right) and for GATE (located to her left), without being taught, to refer to the relational concepts 'right' and 'left', respectively (Herman 1989: 24). Such behavior could be suggestive of incipient grammaticalization, but more evidence is required to establish that it really is.

This raises the question of why grammaticalization is essentially absent in non-human animals. While it is not possible to propose a comprehensive answer, given the little information that is available on this issue, there is at least a partial answer: Grammaticalization requires a linguistic system that is used regularly and frequently within a community of speakers and is passed on from one group of speakers to another. Clearly, this does not apply to the language-related achievements of trained animals that we discussed in Section 3.2: The achievements they show in the course of their training are not transmitted to others. And since such achievements are not found in animals living in the wild, these animals are also barred from developing a grammaticalizing behavior. While there may be additional reasons for this inability, this reason in itself is sufficient to account for the lack of grammaticalization, which is a process usually extending over generations of speakers. As we will see later (Sections 6.2 and 7.2), this fact also has implications for other characteristics of animal behavior, such as the ability to express or conceptualize recursive structures.

3.4 Conclusion

The observations made above are largely in accordance with what has been argued for by a number of authors (e.g. Bickerton 1990: 23; Pinker and Bloom 1990: 726; Jackendoff 1990: 27; Newmeyer 1991: 10; Wilkins and Wakefield 1995: 179), namely that the structural features of human languages primarily concern apes' cognitive abilities rather than their communicative abilities (Fritz Newmeyer, p.c.). The main goal of the chapter was to determine how the scenario of grammaticalization proposed in Chapter 2 can be related to the behavior of non-human animals. On the basis of the animal abilities listed in (2), and in spite of the problems discussed in Section 3.3.1, it would seem that the two are not entirely mutually incompatible since there is clearly an overlap area: These abilities correspond at least to layer II of human language evolution as postulated in Figure 2.1 of Chapter 2, in some respects even to more advanced layers. We will return to this issue in Chapter 7 when we have looked at a number of other phenomena that have a bearing on this issue.

Animal behavior is one of the phenomena that have been taken in many works to provide a window on language genesis; in the next chapter we will deal with another possible window, namely pidgin languages, and we will relate animal communication to restricted linguistic systems created by humans.

4 On pidgins and other restricted linguistic systems

In discussions on language genesis and evolution, "degraded forms of language" are widely held to provide a primary source for proposing hypotheses, and among such systems, pidgin languages are taken to provide one of the main sources of evidence: They are believed to exhibit properties that shed light on earlier forms of human languages.[1] Accordingly, it has been claimed that human language began like an early stage pidgin without syntax (Calvin and Bickerton 2000: 137) or must have had characteristics of (unstable) pidgins (Hurford 2003: 53), and Bickerton (1990, 1995, 2005) and Givón (1995, 2002a, 2005) relate pidgins to what they call, respectively, "protolanguage" or "proto-grammar":

> Closer examination of pidgin communication reveals that it abides by several rather explicit syntactic rules that may be called **proto-grammar**. The common denominator of those rules is that they are extremely iconic—cognitively transparent, non-arbitrary—as compared with the considerably more arbitrary—symbolic—nature of grammatical morphology and syntactic constructions. (Givón 1995: 406)

The main goal of this chapter is to explore what the study of pidgins can contribute to reconstructing language evolution and to this end we will look in more detail at one particular pidgin (Sections 4.2, 4.3). But in addition, we will also be concerned with the question of how pidgins relate to other forms of restricted linguistic systems (Sections 4.7, 4.8). Unlike most of the other authors who have explored the potential of pidgins for reconstructing language evolution, we will not be restricted to pidgins in their "jargon" stage but will also deal with pidgins that have achieved a more advanced stage of development.

---

[1] We are grateful to Philip Baker and Peter Bakker for critical comments on this chapter.

4.1 Introduction

Are degraded forms of language (Jackendoff 1999) or restricted linguistic systems (Botha 2003b, 2005, 2005/6) really windows that may shed light on presumed earlier stages of language evolution, as has widely been assumed or implied (e.g. Sankoff 1979; Bickerton 1990, 1995; Romaine 1992b: 234; Aitchison 1996; Jackendoff 1999: 275; Comrie 2000: 1000; Calvin and Bickerton 2000: 137; Givón 2002a, 2002b: 35, 2005; Hurford 2003: 53; Bakker 2003; Botha 2005/6)? We will be dealing with this question by looking in more detail at one kind of such languages or systems. Pidgin languages (in short: pidgins) are paradigm instances of restricted linguistic systems, but they are also a heterogeneous type of language (Bakker 1995: 39). The social and linguistic space covered by the term "pidgin" is commonly classified into jargons ("unstable" or "rudimentary pidgins"), stable pidgins, and extended or expanded pidgins (or pidgin/creoles)—a classification that has never been properly defined (Philip Baker, p.c.). Extended pidgins are either first languages for some of their speakers or the main language of a community of speakers; all other categories are spoken essentially only as second languages. In discussions on language evolution, pidgins tend to be described as representing fairly "structureless" means of communication that are formed ad hoc and do not exhibit any marked degree of consistency—in other words, as something commonly associated with "jargons". As we hope to demonstrate in this chapter, such a description is hardly appropriate when one looks at a wider range of pidgins, especially pidgins that developed in Africa (Heine 1973).

In a number of relevant studies, pidgins are treated in tandem with creoles. In fact, it is frequently not easy to trace a clear boundary between the two, and Mufwene (1996) uses the term "creole" in a wider sense to include pidgins that serve as vernaculars or primary means of communication for at least a portion of their speakers. For the purposes of the present discussion we will keep the two apart and ignore creoles and the issues concerning the relationship between the two, including all the controversies surrounding creole studies (see, for example, Mufwene 1996; McWhorter 1998; DeGraff 2000). As Bakker (1995) has shown, there are a number of structural properties that justify keeping pidgins and creoles apart.

On the basis of a survey of pidgins in different parts of the world (Heine 1979; Romaine 1992a; Boretzky 1983; Tosco and Owens 1993; Holm 2000;

Bakker 1995, 2003), we use the term "pidgin" as a relative notion, meaning that pidgins are defined with reference to the languages from which they are historically derived, that is, their respective source (or "lexifier") languages: They are reduced languages that result from extended contact between groups of people with no language in common (Holm 2000: 5). Thus, English is the source language of Nigerian Pidgin, Ghanaian Pidgin, Solomons Pijin, Tok Pisin, Bislama, etc., while Zulu (or Nguni) is the source language of Fanagalo, and Kikongo that of Kituba. Except for extended pidgins, pidgins have essentially no native speakers, and in the development from their source languages many of them have undergone what, following Romaine (1992b: 232), we will call a "stripping" process that involves most or all of the changes listed in (1). Note that this process tends to be confined to the early stages of pidginization, in their more advanced stages, pidgins develop new grammatical structures; we will turn to this issue in Section 4.3.

(1) The "stripping" process of pidginization
   a. In phonology, the number of phonemic distinctions is reduced, distinctions in vowel length and tone tend to be given up, and consonant clusters tend to be simplified.
   b. Inflectional and derivational morphology tends to be drastically reduced or lost.
   c. Grammatical distinctions of gender, number, valency, and of tense, aspect, and modality tend to be lost, typically in that order (see Bakker 2003).
   d. Mechanisms of clause subordination are also likely to disappear.
   e. Words tend to become multifunctional.
   f. The lexicon shrinks to a fraction of the size it has in the source language.

The result of pidginization is a form of language having little affixal morphology, hence it has hardly any distinction between morphemes and words, which means that words tend to be unanalyzable entities. Grammatical functions are likely to be expressed either by lexical material or by word order, or else are not formally expressed; but word order is flexible (Bakker 1995: 31). There are only a limited number of lexical items, and context plays a central role in utterance construction and interpretation. With these generalizations we are ignoring the many particular properties exhibited by individual instances of pidgins (see below).

Pidgins may show some unexpected properties; for example, a number of pidgins have been found to have affixes separating nouns from verbs (Bakker 1995: 32); the Arabic-based pidgin Turku of north-central Africa distinguishes morphologically between verbs (-*u*) and nouns, whereas the varieties of Arabic that provided the source of Turku lack such a morphological distinction (Tosco and Owens 1993).

In spite of such observations, we do not propose a synchronic-typological definition of pidgins, for the following reason: Most of the structural properties exhibited by pidgins can also be found in some form or other in languages that clearly are not pidgins. For example, pidgins are commonly considered to have only a minimum of inflection and derivation, but this also applies to many languages of West Africa or Southeast Asia, including Chinese; note further that all pidgins known so far have at least some derivational morphology (Bakker 1995: 32). Much the same applies to the restricted phonological inventories that tend to characterize pidgins: There are also "fully-fledged" languages that have severely restricted phoneme inventories, and Bakker (2003) observes that those of pidgins are not necessarily smaller than the smallest inventories of the languages spoken as first languages by pidgin speakers. To be sure, the size of vocabulary characterizing pidgins amounts only to a fraction of what can be found in "fully-fledged" languages; however, there are also other linguistic communication systems that clearly are not pidgins but have a restricted lexicon. One salient sociolinguistic characteristic of pidgins can be seen in the fact that their use tends to be restricted to relatively few domains of social interaction, such as trade or labor; but again, this also applies to other linguistic systems, such as lingua francas that are not pidgins.

4.2 Kenya Pidgin Swahili (KPS)

In order to illustrate the nature of pidgins we will now look in more detail at one particular pidgin, namely Kenya Pidgin Swahili (henceforth: KPS), a variety of the East African language Swahili that was spoken in up-country Kenya in the 1960s and 1970s, when the first-named author had a chance to work on it (Heine 1973, 1991). Its most representative speakers were people having no or hardly any formal education, living on what was then known as the "White Highlands" of up-country Kenya, in and around urban centers such as Nairobi, Thika, Nakuru, Eldoret, Kitale, and Nanyuki. The first languages of the pidgin were Bantu languages such as Kikuyu, Kamba, Luyia, and Gusii, or Nilotic languages such as Luo, Kalenjin, Maa (Maasai), Turkana, as well as Cushitic languages such as Somali and Oromo, thus representing three of the four African language phyla (Niger-Congo, Nilo-Saharan, and Afroasiatic, respectively). As far as we know there were never any speakers using KPS as their L1 (first language), and its speakers tended to refer to it as "nobody's language".

KPS is elusive to the traditional taxonomy of pidgins that we introduced in Section 4.1. It has properties of a jargon for many speakers of northern Kenya, approaches what is commonly assumed to be a stable pidgin, but— as we will see in Section 4.3—also shows properties of extended pidgins. It may best be called a post-pidgin continuum in that it contains a large range of varieties extending along a scale from full pidgin to varieties approaching Coastal Swahili (CS),[2] that is, the Swahili dialects spoken natively along the East African coastal belt (Duran 1979). Its speakers do not normally consider it to be a distinct linguistic form; rather, they maintain that they simply speak "Swahili". In its genesis, KPS was to some extent a "workforce pidgin", being the medium of communication on the European-owned plantations in the "White Highlands" of Kenya, but it was also used in other multi-lingual settings, especially in the urban centres.

The pidgin variety discussed here is nearly, though not entirely, identical with what Muthiani (1974) describes as "Kisetla". The name *Kisetla* (*ki-* is a prefix of noun class 7, denoting 'language', and *setla* 'European settler') is due to the use of Swahili as a lingua franca by British settlers in colonial Kenya; hence, its original meaning was "the Swahili variety spoken by British settlers in up-country Kenya". However, the way the term is used by Muthiani (1974: 32) refers primarily to the variety spoken as a pidgin spoken "by many Africans". Neither of the terms "Kisetla" nor "Kenya Pidgin Swahili" was accepted by the speakers of this variety when we worked on it in the late 1960s and the 1970s; rather, the existence of KPS tended to be denied both by its speakers and by speakers of Standard Swahili. KPS differs from *Kihindi* ('Indian language'), the Swahili pidgin variety spoken by a large segment of the Indian population in up-country Kenya, especially shop-keepers, mechanics, etc.; to our knowledge, no description of this variety exists (Heine 1973: 59–69).

[2] These varieties include but are not restricted to Kimvita, the Swahili dialect spoken in and around Mombasa.

There is no clear information on whether, or to what extent, KPS still exists the way it was recorded more than thirty years ago. With the introduction of Standard Swahili[3] as a compulsory medium of education since the 1980s and as the medium of radio and other mass media there has been a development leading to the decline of the pidgin in favor of the standard variety.

The history of this pidgin began with the arrival of British and later on of Indian immigrants since the end of the nineteenth century and the use of varieties of Coastal Swahili as a lingua franca in up-country Kenya among the European and Indian immigrants and Africans. Being virtually the only linguistic medium for communication among these communities in the emerging administrative and commercial centers and plantation areas of up-country Kenya, the lingua franca underwent massive restructuring. On the basis of our knowledge of the "native" L1 varieties of Swahili (henceforth: CS, i.e. Coastal Swahili varieties which form the basis of Standard Swahili) that served as a source for KPS and of written documents, it is reasonable to assume that at the latest by the 1930s, a form of KPS had emerged that was the result in particular of the following processes (see Heine 1983 for details):

(2) The "stripping" process in KPS
    a. In phonology, the number of phonemic distinctions is reduced. For example, the dental fricatives *th* and *dh*, and the velar fricative *gh* (all three being restricted to Arabic loanwords) were entirely lost, the voiced fricatives *z* and *v* tended to be replaced by the corresponding voiceless fricatives *s* and *f*, respectively, and the distinction in vowel length was also lost.
    b. Inflectional and derivational morphology was largely lost: The system of verbal derivational extensions and nominal inflections was drastically reduced.
    c. Grammatical distinctions of gender, number, valency, and of tense, aspect, and modality largely disappeared: Both the noun class and the tense–aspect systems collapsed, with only few relics left.
    d. All inflectional mechanisms of clause subordination disappeared.

---

[3] Standard Swahili is based on the Kiunguja dialect of Zanzibar rather than on coastal varieties of Kenya.

e. Words tend to become multifunctional. For example, the noun *mguu* 'foot, leg' was extended to 'car wheel' or to any item supporting another item.
f. The inventory of lexical items shrank drastically, to 500 to 1000 with some speakers and to a few hundred with others.

In more general terms, the rise of KPS has been characterized in the following way:

One major result of this process is that almost the entire inflectional and derivational morphology characteristic of Bantu languages was given up, the agglutinating structure of Swahili was replaced by an analytic-isolating type of structure, and that hypotaxis gave way to parataxis as a means of structuring texts. With the breakdown of the noun class system and gender–number agreement, the language lost its major means of marking syntactic relations.... Thus, KPS in its earlier forms can be reconstructed as a "jargon" consisting of several hundreds of lexical items, a handful of affixes, and a few word order rules. (Heine 1991: 32)

An example illustrating the result of this process is provided by Muthiani, where (3a) illustrates the source structure and (3b) the pidginized end-product:

(3) Swahili
   a. Coastal Swahili (CS)
      Ni-   ta-   ku-   piga.
      1.SG- FUT- 2.SG.O - beat
      'I'll beat you.'
   b. "Kisetla" (Muthiani 1974: 37)
      mimi piga wewe   kesho.[4]
      I   beat you   tomorrow
      'I'll beat you tomorrow.'

The structure of KPS, as we found it to be in the late 1960s, is that of an isolating language where grammatical functions were either unmarked or expressed by lexical means. Exceptions included most of all the following:

(a) Properties of the noun class system were not lost entirely but survived in some way or other among all speakers of KPS, even if

---

[4] This example is suggestive of Indian KPS speakers, who generally tend to omit tense–aspect markers. According to our informants of KPS, the verbal future prefix *ta-* would be used in this example.

the only semantic distinction characterizing this system is for the most part that between human and non-human participants.
(b) Some of the verbal derivational extensions survive as lexicalized relics. For example, the CS verb -*nyonga* 'strangle' survives in KPS in two extended forms: *jinyonga* (with the reflexive prefix *ji-*) 'hang oneself, commit suicide' and *nyongwa* (with the passive suffix -*w-*) 'be hanged'.
(c) Tense–aspect markers are mostly absent, but the future marker *ta-* and the present/progressive prefix *na-* tend to be retained, even if the function of the latter has been extended to function as a general marker for non-future situations.
(d) There are essentially no personal inflectional prefixes, even though human subject referents are occasionally cross-referenced on the verb and on demonstratives.
(e) There is no inflectional negation; rather, the invariable particle *hapana* 'no, there is not' is used (see below). Still, inflection survives in some lexicalized forms; thus, instead of the CS form (4a), there is the KPS phrase (4b), but more commonly the CS form is used as an invariable expression.

(4) Swahili
   a. CS
      si-   ju-   i.
      NEG.1.SG-   know-   NEG
      'I don't know.'
   b. KPS
      Mimi   hapana   jua.   or   Sijui.
      I   NEG   know
      'I don't know.'

(f) Clause coordination by means of the conjunctions *au* 'or' and *na* 'and' is retained.
(g) The conjunctions *kama* 'if', *lakini* 'but', and *(kwa) sababu* 'because' are the only productive means of clause subordination.

Like other pidgins, meaning relations in KPS are to a large extent determined by context pragmatics, which means that semantic or syntactic functions tend to be unmarked (see Heine 1991: 118 for an example illustrating this structure). However, it has retained the essential word order arrangements of CS: It places modifiers after the head, and the basic word order is verb-medial (SVO). As long as there are sufficient contextual clues, however, there is some range in word order variation. For example, without any further contextual information, sentence (5) has the meaning (5a), where *watu hii* 'these people' is the clausal subject. But in a situation where the relevant people were presented as topical in the preceding context it can as well mean (5b), where *watu hii* is a topicalized object.

(5) KPS
Watu   hii   bado   ona.
people   this   not.yet   see
a. 'These people have not yet seen (it).'
b. 'These people, (I) have not yet seen (them).'

The extent to which meaning is context-dependent can be illustrated with the following example from an informant whose first language was Maasai: The hearer can only establish on the basis of the pragmatics of the speech situation that, for example, the subject referent of the verb *-wacha* 'leave' is *hawa* 'they' whereas in the case of the verb *-kufa* it is *mtu* 'person, man'. As long as the hearer can be expected to be able to reconstruct the referential relations among participants, there are few constraints on the building of discourses. Accordingly, in (6b) the basic SVO order is ignored in favor of an OVS order since the hearer can be expected to infer from the meaning of the items employed who the agent and the patient are—that is, that the object *redio* 'radio' is topicalized and the subject *mimi* backgrounded in the sentence-final position.

(6) KPS
  a. Hawa   hapana   jali   na-   toa   mtu   huku
     they   NEG   mind   NF-   remove   person   there
     na-   peleka   huku   na-   kufa   halafu   na-
     NF-   take.to   there   NF-   die   then   NF-
     wacha   huku [...].
     leave   there
     'They didn't care, they pulled the man out, took him there until he died, then they left him there [...].'
  b. Redio   hii   kwisha   ona   mimi.
     radio   this   PFV   see   I
     'This radio, I have (already) seen it.'

To summarize, with few exceptions, the main morphosyntactic characteristics of CS have disappeared in KPS, which can be defined as a restricted system, largely devoid of inflectional and derivational morphology. The few relics that have survived concern salient conceptual distinctions such as marking future vs. non-future, or human vs. non-human referents, and the skeleton of linear arrangement of the source language CS.

There is, however, one important observation to be made on the syntactic structure of KPS in its early stages. While essentially all morphology for clause subordination was eliminated in the "stripping" process, there nevertheless was syntactic recursion (see Chapter 6): Data from older speakers who had preserved the older structure of KPS to some extent suggest that there was productive noun modification, both adjectival and possessive modification, for example *kitu mbaya* (thing bad) 'a bad thing', *watu ya taun* (people of town) 'urban dwellers'. Accordingly, there is little doubt that at least noun phrase recursion was present at all stages in the development of this pidgin.

4.3 The rise of new functional categories

But there is also a second phase in the history of this pidgin. After the "stripping" process in the early stages of their development, pidgins will, under appropriate conditions, develop new grammatical structures, as we observed in Section 4.1, and this is what happened in KPS.

Being impoverished vis-à-vis the coastal varieties of Swahili (CS), from which it is historically derived, KPS experienced a number of developments within its short lifespan of roughly forty to fifty years. These developments almost invariably involved linguistic material that was available in the language but that was put to new uses of two kinds: Either it concerned free forms, that is, independent words that assumed functional roles corresponding to some of the lost inflections. For example, the person inflections *ni-* (1.SG), *u-* (2.SG), etc. were lost (except for lexicalized fossils), and their function was taken over by the self-standing pronouns *mimi* 'I', *wewe* 'you (SG)', etc. But more generally, pidgin speakers searched for ways of expressing functional concepts by drawing on universal strategies of grammaticalization, using existing lexical structures in contexts that invited novel interpretations of these structures in terms of functional concepts. We will now deal with these structures in turn.

Copular structures Neither CS nor KPS require a copula in equational propositions; still, CS has a copula *ni* (or *ndiyo*) used for both classificational (*John is a teacher*) and identificational (*The teacher is John*) predications. This copula has largely disappeared in KPS, which developed a new copular use pattern by grammaticalizing the locative copula *iko*[5] 'be somewhere, exist' to a general copula: Not only has *iko* retained the locative and existential functions of CS; as a result of its desemanticization (or generalization) it has become an all-purpose copula. The following examples illustrate its uses.

(7) KPS
   a. Watu   ya   Uganda   iko   watu   pole   sana [...].
      people of   Uganda   COP people kind very
      'The people of Uganda are very kind people [...].'
   b. Ugonjwa   hii   iko   mbaya.
      disease   this   COP   bad
      'This disease is dangerous.'

This copula has also given rise to a marker of predicative possession ('to have'), where it frequently takes the comitative preposition *na* 'with'. Since the use of *na* is optional, the equational and the possessive use patterns are frequently ambiguous, as in the following example:

(8) KPS
    Mtu   hii   hapana   iko   baba.
    person this   NEG   COP   father
   a. 'This person is not a father.'
   b. 'This person has no father.'

To conclude, via grammaticalization of the locative-existential copula *-ko* of CS, KPS has developed new copular and possessive constructions that have no homolog in the source language.

Tense–aspect Among the universal strategies used to create new categories for immediate perfects or past tenses there is one whereby verbs meaning 'to come from' are grammaticalized to tense–aspect markers; cf. French *Il vient d'aller à Paris* (he comes from to.go to Paris) 'He has just

---

[5] *iko* is an invariable marker that is derived from the CS locative copula *-ko* taking noun class agreement prefixes. The frozen form *iko* contains the prefix *i-* of noun class 9.

gone to Paris.' This strategy has also been employed by KPS speakers by developing their verb *toka* 'come from' into an immediate-past auxiliary (Heine 1991: 42): (9a) illustrates the lexical and (9b) the functional use of *toka*.

(9) KPS (Heine 1983: 92)

a. Mimi   na-   toka   taun.
   I   NF-   come.from   town
   'I am coming from town.'

b. Mimi   na-   toka   andika   barua.
   I   NF-   come.from   write   letter
   'I just wrote a letter.'

Another tense–aspect use pattern characterizing KPS concerns the grammaticalization of the verb *kwisha* 'be finished' to a perfect auxiliary. This process occurred in CS, where it has given rise to an *already*-perfect (denoting an event that occurred earlier than expected) in combination with the perfect prefix -*me*-, for example:

(10) CS

A-   me-   kwisha   ku-   ja.
3.SG-   PERF-   *kwisha* INF-   come.
'He has come already.'

What is new in KPS is that, first, *kwisha* is not supported by the perfect prefix, which has disappeared in the pidgin, and second, that the use of *kwisha* has been extended to serve as a general perfect marker. Thus, in examples such as the following, the meaning 'already' no longer makes sense.

(11) KPS

Mimi   na-   jaza   petroli   huku   Nairobi   na   mimi
I   NF-   fill   petrol   there   Nairobi   and   I

kwisha   kwenda   maili   mia   moja   tu!
PERF   go   mile   100   one   only

'I filled the tank in Nairobi but drove only one hundred miles (still, the tank is already empty)!'

The future prefix -*ta*- belongs to the few inflections that have survived the transition from coastal L1-Swahili (CS) to pidgin. However, -*ta*- is restricted to the positive future, while the negative future is a new creation within the pidgin. An evolution from markers of ability ('be able, can') to future tenses is crosslinguistically not very common but is attested.[6] KPS speakers appear to have used this strategy to form a negative future, using the phrase *hapana weza* (NEG can) or *hawezi*, derived from CS *ha-wez-i* (NEG.3.SG-can-NEG), for this purpose. Thus, the following sentence may have its literal meaning 'The children cannot work', but more likely is understood to express negative future meaning.

(12) KPS
   Watoto   hapana   weza   fanya   kazi.
   children NEG   can   make   work
   'The children will not work.'

For further examples of new tense–aspect categories in KPS, see below.

Modality   One way of creating new expressions for deontic modality is to draw on a universal strategy whereby predicates of the kind 'It is enough/fitting/suitable/good (that)' are grammaticalized to markers for necessity or obligation (Heine and Kuteva 2002a). KPS speakers have done so by developing the expression *mzuri* '(it is) good (that)' into a marker of deontic modality, cf. (13).

(13) KPS
   a. Chakula   hii   iko   mzuri.
      food   this   be   good
      'This food is good/tasty.'

   b. Mzuri   nyinyi   na-   ona   daktari.
      good   you.PL   NF-   see   doctor
      'You (PL) should see a doctor!'

Another universal conceptual process concerns verbs meaning 'leave', 'abandon', or 'let' that are grammaticalized to hortative markers or related concepts of deontic modality (Heine and Kuteva 2002a). This process has given rise in KPS to a hortative use pattern: Example (14a) illustrates the lexical use of *wacha* 'leave' (< CS *-(w)acha* 'leave, abandon'), that we had already in (6), and (14b) that of a hortative marker:

(14) Kenya Pidgin Swahili
   a. Yeye   kwisha   wacha   kazi.
      3:SG   PERF   leave   work
      'He has left work.'

---

[6] See Bybee, Perkins, and Pagliuca (1994: 266), who hypothesize that this evolution involves the following stages: ABILITY > ROOT POSSIBILITY > INTENTION > FUTURE.

  b. Wacha yeye na- letia sisi biya.
    HORT s/he NF- bring us/we beer
    'Let him bring us beer!'

Negation One crosslinguistically common pathway leading to the rise of verbal negation markers has been described by Croft (1991); we gave a sketchy rendering of it in Section 2.2.1.2 (Verb > negation). In accordance with this pathway, negative existential verbs ('not to exist') develop into general markers of verbal negation. In CS, verbal negation is expressed by a discontinuous structure consisting of the inflectional prefix *h*V- (*si-* in the first-person singular) and the negative suffix *-i*, cf. (15a). This negation structure has been eliminated in KPS except for a few, mostly lexicalized relics. Instead, early KPS speakers have drawn on the existential pathway by grammaticalizing the existential construction of CS, consisting of a negation prefix *h*V-, one of the three locative noun class markers *pa-* (C16), *ku-* (C17), or *mu-* (C18), plus the existential copula *-na* 'exist, be with', illustrated in (15b).

(15) CS
  a. H- a- sem- i kitu.
    NEG- 3.SG- say- NEG thing
    'He doesn't say anything.'
  b. Ha- pa- na chakula.
    NEG- C16 exist food
    'There is not food around.'

In accordance with this pathway, the CS negative existential phrase *ha-pa-na* illustrated in (15b), using the noun class marker of the locative class 16, has been grammaticalized to an invariable verbal negation marker *hapana* in KPS, occurring immediately before the verb. Accordingly, the meaning of the CS sentence in (15a) would be expressed in KPS typically as in (16).

(16) KPS
  Yeye hapana sema kitu.
  s/he NEG say thing

The results of grammaticalization in this process were the following: Via extension, the negative existential verb acquired a new context, namely serving as a pre-verbal particle. Desemanticization had the effect that the existential verb lost its existential and its predicate functions, thereby being reduced to the function of negation, and decategorization meant that the erstwhile verb lost the ability to be inflected and became restricted to the pre-verbal position.

Relative clauses With the collapse and disappearance of the CS relative clause construction, based on the relativizer -*o*-, KPS speakers were left with no relative clause construction; what in other languages would be expressed by relative clauses was encoded in the emerging pidgin simply by juxtaposing clauses, as in (17a).[7] But these speakers gradually created a new structure, drawing on universal principles of grammaticalization. Crosslinguistically the most common way in which relative clause markers arise is via the grammaticalization of demonstrative pronouns, English *that* being a case in point (see Section 2.2.3). Like earlier speakers of English, KPS speakers drew on their distal demonstrative *ile* 'that' to mark both restrictive and non-restrictive relative clauses (see Heine 1991: 47–51 for more details). At the initial stage, *ile* appears to have been used only in contexts where it could also be interpreted as a demonstrative, that is, where it was ambiguous, as in (17b). Eventually, the use of *ile* was extended to contexts where it functions unambiguously as a relative clause marker, as in (17c). As (17d) shows, the relative marker is no longer restricted to definite referents. That the grammaticalization process from demonstrative to relative clause marker has been completed is suggested by examples such as (17e) where the relative pronoun can co-occur with the demonstrative. Note that (17e) is an instance of a center-embedded construction, in that the relative clause is placed *within* the main clause; we will return to this example in Section 6.6.

(17) KPS

    a. Iko   chupa   ine   tu   na-   bakia.
       be   bottle   four   only   NF-   remain
       'There are only four bottles (of beer) [which are] left.'

    b. Wewe   na-   weza   ona   Fort Jesus   **ile**   na-   jeng-   wa
       you   NF-   can   see   Fort Jesus   DEM/REL   NF-   build-   PASS
       na   watu   ya   Portugal.
       by   people   of   Portugal
       i 'You can see Fort Jesus; that one has been built by the Portuguese.'

---

[7] One might argue that *na-bakia* in (17a) is "underlyingly" a relative clause but one that is not formally marked. We did not find compelling evidence in support of such an analysis.

ii 'You can see Fort Jesus, which has been built by the Portuguese.'

c. [...] lazima wewe na- leta barua kutoka kwa
　　　　 must　 you　 NF- bring letter from　 at
watu **ile**　 wewe na- fanya kazi.
people REL　 you　 NF- make work

'[...] you must bring a letter from your employer.'

d. Na　 yeye na-　 lal-　 isha　 yeye ndani ya sanduku
　 and　 s/he NF-　 lie-　 CAUS s/he inside of box
ile　wanyama likuwa　　　 na- kula chakula yao [...].
REL　 animals　 PAST.PROG NF- eat　 food　 their

'And she placed him in a crib [...].' (Lit.: 'And she layed him in a box that animals eat their food from [...].')

e. **Ile**　 mtu　 [**ile** iko　 watoto tatu] kwisha fung- wa.
　 DEM person REL have children three PERF　 shut- PASS

'That man, who has three children, has been jailed.'

All these stages co-occur in KPS (at least when we studied it in the late 1960s). As Table 4.1 shows (see also Heine 1991: 50), the largest number of uses concern situations where *ile* is in some way or other ambiguous between a relative and a demonstrative reading. The use of the relative marker is in most contexts optional; however, there is a strong tendency to omit it with indefinite head nouns but use it with definite head nouns.

The following example shows that the relativizer *ile* has been grammaticalized to the extent that its use is fully recursive:

TABLE 4.1. *Text analysis of relative clause structures in KPS*

Stage	Type of relative structure	Number of occurrences	Percentage (%)
0	No relative marker	13	23.2
1	*ile* functions simultaneously as a demonstrative and a relative marker	34	60.7
2	*ile* is exclusively a relative marker	9	16.1
Total		56	100.0

Note: Sample size: 4,000 words.

(18) KPS

Lete	biya	ile	mimi	na-	nunua	kwa	duka	ile
bring	beer	REL	I	NF-	buy	at	shop	DEM
ya	Muhindi	ile	mama	yake	na-	uza-uza		
of	Indian	REL	mother	his	NF-	sell-sell		
kila	kitu	ile	wewe	na-	penda.			
each	thing	REL	you	NF-	like			

'Get (me) the beer that I bought yesterday in that shop of the Indian whose mother sells everything you like.'

But *ile* has not only been grammaticalized to a relative clause marker; in combinations with specific nouns it also turned into a more general clause subordinator, thereby taking over many of the functions characterizing the inflectional prefix -*o*- of CS (or of Standard Swahili).

Adverbial clauses Virtually all of the developments characterizing KPS can appropriately be portrayed as being drastic reductions vis-à-vis CS. But, as the examples presented above suggest, the pidgin has somehow acquired a new typological profile that is characterized by new functional distinctions for which there is no exact equivalent in CS. In some cases this means that KPS developed grammatical distinctions that are absent in CS. For example, CS has an inflectional verbal prefix -*po*- introducing temporal adverbial clauses (built on the relativizer -*o*-), as in (19). KPS has lost this inflection and has created a new use pattern based on a lexical structure which is *siku ile* ('day which') when referring to events more distant in time, cf. (20a), and *saa ile* ('hour which') for events that happened recently, cf. (20b). Thus, KPS speakers dispose of a more differentiated grammatical distinction than CS speakers, in that for the latter there is no grammaticalized way of distinguishing between (20a) and (20b).

(19) CS

Ni-	ambi-	e	(wakati)	u-	taka-	po-	fika.
1.SG-	tell-	SUBJ	(time)	2.SG-	FUT.REL-	TEMP-	arrive

'Tell me when you'll arrive.'

(20) KPS

a.
Mambia	mimi	siku	ile	wewe	ta-	fika.
tell	I	day	REL	you	FUT-	arrive

'Tell me when (i.e. which day, month, or year) you'll arrive.'

b. Mambia mimi saa   ile wewe ta-   fika.
   tell   I   hour REL you   FUT-   arrive
   'Tell me when (i.e. which time of the day) you'll arrive.'

Question words In many cases the new grammatical structures that evolved in the pidgin were already present as minor use patterns for specific purposes in CS. What grammaticalization achieved is that these patterns were activated by the pidgin speakers, used more frequently, extended to novel contexts, and developed into new functional categories. The rise of new question words illustrates this process (see Heine 1991: 38–9): CS has a paradigm of unanalyzable disyllabic interrogative pronouns, which are listed in Table 4.2. With the exception of *wapi* 'where?', these pronouns are rarely used in the pidgin, where the analytic phrases listed in Table 4.2 tend to be used instead, consisting of a noun plus the attributive *gani* 'which?', see also (21c). This strategy, recruiting generic nouns that refer to some specific ontological domain plus an attributive interrogative to create new question words, is crosslinguistically widespread, and it is also typical of creoles (Peter Bakker, p.c.); as Baker (2006) demonstrates, interrogatives formed in this way became productive in the history of many, if not all creoles; with their semantic transparency and vocabulary-building capacity such interrogatives provided a valuable attribute in the evolution of creoles. At the same time, analytic (bimorphemic) interrogatives were already available as a use pattern in CS; thus, (21a) and (21b) are equivalent questions in many contexts. What happened in KPS is that these use patterns were generalized, being used more frequently in a wider range of contexts.

(21) CS and KPS
    a. u-   na-   taka   nini?   (CS)
       2.SG- PRES-   want   what
    b. u-   na-   taka   kitu   gani?   (CS)
       2.SG- PRES-   want   thing which
       'What do you want?'
    c. wewe   na-   taka   kitu gani?   (KPS)
       2.SG   NF-   want   what
       'What do you want?'

As we observed in our discussion of adverbial clause marking, in the present case as well the pidgin speakers dispose of a grammatical distinction that is absent in CS (even if it can be expressed lexically): Where

TABLE 4.2. *Interrogative pronouns in Coastal Swahili and Kenya Pidgin Swahili*

Meaning	CS form	Preferred KPS form	Literal meaning of KPS form
'who?'	nani	mtu gani	'person which?'
'what?'	nini	kitu gani	'thing which?'
'where?'	wapi	mahali gani	'place which?'
'when?'	lini	siku gani	'day which?'
		saa gani	'hour which?'
'how?'	vipi	namna gani	'manner which?'

CS has only one temporal interrogative (*lini* 'when?'), the pidgin distinguishes between specific (*saa gani*) and more general time periods (*siku gani*).

4.4 Discussion

As we pointed out above, pidgins tend to be portrayed as fairly "structureless" forms of communication that are formed ad hoc and do not exhibit any marked degree of consistency. While this may be true for "jargons", that is, pidgins in their earliest stage of development, it does not apply to any of the pidgins that we are familiar with, including KPS. The KPS structures that we presented in Section 4.3 are new grammatical use patterns in this pidgin that must have evolved within roughly half a century; there are essentially no equivalents in CS. Their development is fully in accordance with parameters of grammaticalization, as they can also be observed in other languages.[8]

In many of their uses these patterns are ambiguous, as they can still be interpreted with reference to their lexical meaning. For example, the immediate-past marker *toka* can still be understood in its lexical sense. Some of these use patterns, however, have acquired new contexts where the lexical meaning no longer makes sense—hence, where they can be

---

[8] In our entire corpus of grammatical developments we came across only one example of a possible "degrammaticalization" (see Section 1.2.2): The CS adverb *katikati* 'between' is occasionally used by KPS speakers as a noun meaning 'center, inside'.

interpreted meaningfully only with reference to the new grammatical function. Thus, the erstwhile verb *toka* has also been extended to contexts where its lexical meaning 'come from', cf. (22a), is ruled out, that is, where its immediate-past tense function provides the only reasonable interpretation, as in (22b), where the meaning of the verb *ingia* 'enter' is incompatible with that of *toka* (which also means 'come out'):

(22) KPS

   a. Yeye na- toka nyumbani.
      s/he NF- come.from house[9]
      'He is coming from the house.'

   b. Yeye na- toka ingia nyumbani.
      s/he NF- come.from enter house
      'He just entered the house.'

Furthermore, the process leading to the rise of these structures was not necessarily one where speakers intended to introduce a new functional category; rather, speakers used existing constructions in new contexts for specific communicative purposes that made sense in these contexts, and in some of these contexts, a new grammatical function emerged. In many cases, this context extension did not affect the structure of grammatical categorization but rather remained a contextually restricted option. In some cases, however, it had the effect that new grammatical use patterns emerged serving specific functional purposes and acquiring a new status as distinct functional categories. The result is that KPS acquired a profile that contrasts with that of CS on the one hand, and with that of a maximally restricted linguistic system on the other.

That it is the manipulation of existing forms in new contexts that determines the emergence of new use patterns and functional categories is suggested by the following example. The CS adverb *bado* has two different, though related, meanings, which are 'still' in affirmative and 'not yet' in negative constructions, and both meanings are retained in KPS. But in KPS this adverb has served as a basis for new contrasting functional use patterns: When preceded by the non-future (NF) marker *na-* it denotes an event that continues longer than expected (= 'still'), cf. (23a). When used without the NF marker it denotes an event that was supposed to have started but has not started as yet (= 'not yet'). However,

---

[9] The CS noun *nyumba* 'house' is frequently used in its locative form *nyumbani* 'in the house' in KPS to denote both 'house' and 'in the house'.

in the latter use pattern, *bado* has undergone further desemanticization, in that it tends to be generalized as a negative perfect marker, as in (23b), where the meaning 'not yet' does not appear to make much sense.

(23) KPS
    a. Ndito   yangu   bado   na-   kwenda   skuli.
       girl   my   *bado*   NF-   go   school
       'My daughter is still attending school.'
    b. Hii   mzee   bado   kwenda   skuli.
       this   old.person   *bado*   go   school
       'This old man has not attended school (= is illiterate).'

There are also other conclusions to be drawn from the findings presented in this section. The first concerns the question of how much time is minimally required for new functional categories to evolve or, more generally, for a language to acquire a new typological profile. As we noted above, KPS is a product of British colonialization in Kenya, its genesis can be dated back to around the beginning of the twentieth century; we are not aware of any pre-colonial pidgin varieties in up-country Kenya. Since the mid 1930s there are records to the effect, first, that what was once a "natural" language (CS), had turned into an unquestionable pidgin (KPS).[10] That the pidgin acquired a set of new grammatical use patterns within less than five decades suggests that grammaticalization proceeded much faster in this pidgin than in what we know from "natural" languages.

Another issue concerns the agents of language change. That the creators of KPS were L1 speakers of the languages spoken in up-country Kenya is fairly uncontroversial, such languages being Kikuyu, Kamba, Luo, Luhya, Maasai, and Kalenjin, as well as roughly a dozen other languages. The contribution of native speakers of CS was at all times modest: There never was any substantial community of native Swahili speakers in the rural areas of up-country Kenya. The contribution made by the colonizers in this process, that is, British farmers, government officials, teachers, etc., is unknown. The agents must have been primarily male adults working on the plantations of European settlers and traders in commercial centers such as Nairobi, Thika, Nakuru, Eldoret, Kitale, Nanyuki, etc., and women working as household servants may also have

[10] We have no information on any possible "jargon" stage that may have pre-dated the pidgin variety described here.

played some role. There is no indication that children played a role in this process (see Section 7.1.6).

4.5 Grammaticalization in other pidgins

That a pidgin, after having undergone a "stripping" process, may acquire new grammatical structures is well documented, especially for extended pidgins or creoles such as Tok Pisin and other English-based pidgins in Oceania (e.g. Sankoff and Laberge 1973; Sankoff and Brown 1976; Sankoff 1979; Keesing 1991; Romaine 1992a, 1992b, 1995, 1999; Aitchison 1996; Meyerhoff 2001); in fact, some of the best described cases of grammaticalization in the literature on this field are devoted to Tok Pisin (Romaine 1995, 1999). Sankoff (1979) shows how in Tok Pisin new grammatical categories evolved, such as markers for number, tense, and causativity, or complementizers and relative clause markers, and how meaningful morphemes, such as personal pronouns, may develop into largely obligatory and redundant elements of clause structure. Keesing (1991) describes how English-based extended pidgins such as Solomons Pijin or Bislama of Vanuatu developed new grammatical profiles as a result of the influence from Austronesian languages,[11] and similar processes have been reported for Ghanaian Pidgin English (Huber 1996, 1999). What these processes have in common is that they are essentially always based on grammaticalization. In the remainder of this section we provide a few examples to illustrate these processes; the reader is referred to the works listed above for a wealth of additional examples.

In Chapter 2 we discussed a process leading from adverbs to demonstratives and another process from demonstratives to relative clause markers, in accordance with the universal pathway in (24).

(24) Locative adverb > demonstrative > relative marker

An example of this pathway is provided by the English-based Tok Pisin and other varieties of Melanesian PE, where the English adverb *here* was

---

[11] Tok Pisin, Solomons Pijin, and Bislama are all to a very large extent continuations and further developments of East Australian Aboriginal Pidgin English. This is due to the "labour trade" whereby men from these islands were "recruited," often against their will, to work in Queensland (from 1863) for three years. A great many features of these pidgins, some of them to be discussed below, are already attested East Australian Aboriginal Pidgin English (Philip Baker, p.c.).

grammaticalized to a demonstrative and a relative clause marker *ia* (or *ya*). Since its origins around 1880, strategies of relativization have been gradually arising in Tok Pisin. According to Romaine (1992a: 146, 170), zero-marked relative clauses may have been the original strategy in Tok Pisin, and relative clauses are comparatively recent developments in the history of the language, and this applies especially to "bracketing relative clauses" (Romaine 1992a: 166–7). Sankoff and Brown (1976) discuss how the new relativizer *ia* developed from a demonstrative or generalized deictic particle in discourse. The following examples illustrate the various stages in this development, where (25a) shows the use as a place adverbial, (25b) as a postposed deictic or demonstrative, and (25c) as a double-bracketed relativizer.

(25) Tok Pisin (Sankoff 1979: 32; Sankoff and Brown 1976; Romaine 1992a: 162, 166)

a. yu   stap   hia.
   (you   stay   here)
   'You stay here.'

b. Ee!   Man   ia   toktok   wantaim   husat?
   hey   man   ia   talk   to   whom
   'Hey! Who's this guy talking to?'

c. Disla   liklik   anis ia   [em i   bin   dens wantaim em
   (this   little   ant   ia   he   TAM   dance   he
   festaim ia]   em   go   nau.
   first   REL   he   go   now)
   'This little ant that he danced with the first time left.'

Sankoff describes this process in the following way:

We argued that relativization grew out of the need in discourse for "bracketing" devices for use in the organization of information, and that as the use of Tok Pisin expanded, the slot occurring after *ia* (itself always postposed to the noun it qualifies) was a strategic location for the insertion of further qualifying information. As this information came more and more to take the shape of a full sentence, *ia* became available for reanalysis as a relativizing particle. (Sankoff 1979: 32)

And Tok Pisin provides further information on the rise of clause subordination. One concerns the grammaticalization from question word to clause subordinator—a process that is particularly common in European languages (see Section 5.3.1.2). There appear to have been two different ways of using the interrogative-to-relativizer grammaticalization in Tok Pisin. One concerns the marker *husat*, presumably derived from a propositional meaning *who is that?*. Romaine (1992a: 159) says that the emergence of the relativizer *husat* is more characteristic of written Tok Pisin and does not normally occur in the spoken language:

(26) Tok Pisin (Romaine 1992a: 159)
Em man [**husat** i drawim] em i go lapun tru na em i dai pinis.
'The man [who drew (it)] got very old and died.'

The second is the one we discussed above, namely *we*, historically derived from the English place interrogative adverb *where*. In accordance with its conceptual source, it is used mainly for marking locative relatives, although its use has been extended marginally to subject and object relatives (Romaine 1992a: 159).

On the basis of the data available it is possible to reconstruct the development of the relativizer *we* (see Romaine 1992a: 160–1): At stage 0, it was used exclusively as a question word, as it is up to today; cf. (27a). At stage 1, its use was extended to locative relatives, as in (27b). Stage 3 marks the transition from locative to object relative, where the head noun *wok* 'work' is repeated, cf. (27c). At the final stage 4, *we* no longer allows a locative interpretation; it is now a general relativizer, as in (27d).

(27) Tok Pisin (Woolford 1979: 121; Romaine 1992a: 160–1)
    a. Dispela tupela man i stap **we**?
       this   two   man be located where
       'Where are those two men?'

    b. Em la go antap long hul [**we** ol pasim long tupela ain].
       'He wanted to go on top into the hole [where they shut (him) behind two iron bars].'

    c. Ol mas go na painim gutpla, gutpla wok [**we** displa wok i ken i kambek na helpim ol pipol].
       'They have to go and look for very good work, where this work can come back and help the people.'

    d. Mipela ol **we** i save kaikai saksak em i putim long mipela tasol.
       we (pl.) REL know eat sago he put to us only
       'We *who* are used to eating sago, he gave it to us only.'

The development of new modes of clause subordination is virtually predictable in the later development of pidgins. The following is an example from Ghanaian Pidgin English of West Africa, illustrating the process from a speech act verb 'say' to a complementizer. The following sentence can be interpreted in the same way as a direct-speech utterance, where the erstwhile verb *se* 'say' functions as a quotative marker (28a), or as a [main clause–complement clause] structure, where *se* has the function of a complementizer[12] after the perception verb *si* 'see' (28b).[13]

(28) Ghanaian Pidgin English (Huber 1999: 189)

a si   se   (")yɛa, dɛ   tin   we   mì. (")
(I see CPL yeah   DEF thing weigh me)

a. 'I saw, "Yeah, the thing weighs me down".'
b. 'I saw that the thing weighed me down.'

Ghanaian Pidgin English illustrates another component of the development from 'say' to clause subordinator. Complementizers may grammaticalize further into purpose clause markers (see Section 5.3.4), and *se* also serves as a purpose clause marker (29a), but this pidgin has gone one step further, in grammaticalizing the 'say'-verb to a marker of cause clauses, cf. (29b).

(29) Ghanaian Pidgin English (Huber 1999: 190)

a. Dos   trɔks   we   dè   briŋ   fɔ   dʒɛmɛni
(DEM   trucks CPL they bring   for   Germany

se   mek   dè   kam   kari   dɛ   bɔla [...]
se   CAU   they come   carry   DEF   refuse)

'Those lorries that they brought from Germany in order that they remove the refuse [...]'.

b. sɔmbɔdi   dè   pe   hjudʒ   sam   ɔf   mɔni
(somebody   PROG   pay   huge   sum   of   money

---

[12] Note, however, that there is a caveat with the complementizer *se* of Ghanaian Pidgin English: "[...] *se*, one of whose main functions is to introduce subordinate clauses after verbs of cognition, perception, or saying. This particle appears to be cognate with English *say* but its position in WAPE [West African Pidgin English; a.n.] has probably been consolidated by the near-homophonous Akan form *sɛ* 'that, whether, if' which in turn may go back to *se* 'say' [...]" (Huber 1999:188).

[13] Huber comments on this example thus: "The ambiguity of these sentences attests to the fact that the complementizer results from a grammaticalization of the main verb *se* 'say', and shows that intermediate stages of this process are still observable in GhaPE [Ghanaian Pidgin English; a.n.]." (Huber 1999: 189).

      se   ì   wã   kã   ste   nima.
      se   3.SG INT come stay Nima)
      'People are paying huge sums of money because they want to stay in Nima.'

More examples of grammaticalization can be found in other English-based extended pidgins. In Bislama, a pidgin of Vanuatu, speakers used an expression commonly recruited crosslinguistically to develop progressive and durative aspect markers (Heine and Kuteva 2002a: 127, 198), namely the verb *stap* 'stay, be present, exist', historically derived from English *stop*, to develop a durative aspect marker:[14]

  (30) Bislama (English-based pidgin; Keesing 1991: 328)
       em   i   stap   pik- im   yam.
       he   he-   DUR   dig- TRS yam
       'He's in the process of digging yams.'

The following example from Solomons Pijin illustrates the magnitude of grammaticalization that can be observed in the English-based pidgins of Melanesia and Oceania: The future (FUT) marker *bae* is a grammaticalized form of the English adverb *by-and-by*, the particle *das* (or *des, tes*), which is presumably a grammaticalized form of English *just*, expresses an immediate future tense when combined with the future marker *bae*, the transitivizing suffix *-im* (TRS) arose via the grammaticalization of the English object pronoun *him*, and the inclusive personal pronoun *iumi* 'we (including hearer)' is diachronically a combination of English *you* and *me*, being one of a number of examples where the pidgin has grammaticalized new categories of personal deixis for which there are no equivalents in the English source language. These four categories are all creations that arose in the process of pidgin development.[15]

  (31) Solomons Pijin (English-based pidgin; Keesing 1988: 215)
       bae   iumi   das   luk-   im.
       FUT   we. INCL just   see-   TRS
       'We'll see it in a while.'

---

[14] Our interpretation of this case rests on Keesing (1988). The grammaticalization of verbs meaning 'stay, be present, exist' as durative or progressive markers is crosslinguistically widespread (see Heine and Kuteva 2002a). Note further that this grammaticalization is not confined to Bislama, it can also be observed in Tok Pisin and Solomons Pijin, and it remains unclear whether the process looked at here took place independently of these other cases.

[15] That in all of them language contact played some role (Keesing 1991; Heine and Kuteva 2005) is irrelevant for the present purposes.

In Nigerian Pidgin English, a number of verbs were grammaticalized to prepositions and markers of argument structure. Accordingly, the verb *tek* 'take' (< English *take*) is used for instrumental (32a) and a number of other participants, such as time in (32b), *giv* 'give' (< *give*) for benefactive participants (32c), and the motion verbs *go* 'go' (< *go*) and *kom* 'come' (< *come*) for directed motion, cf. (32d).

(32) Nigerian Pidgin (Faraclas 1996: 76–9, 171)
    a. A tek nayf kọt dì nyam.
       (I take.FACT knife cut the yam)
       'I cut the yam with a knife.'
    b. A tek nayt kọt dì nyam.
       (I take.FACT night cut the yam)
       'I cut the yam at night.' (Lit. 'I took the night cut the yam.')
    c. A bay nyam giv yù.
       (I buy.FACT yam give you)
       'I bought you the yam.'
    d. A gò tek dì chudren go makẹt.
       (I FUT take the children go market)
       'I will take the children to the market.'

These are but a few examples showing that the situation that we found in KPS is by no means unusual: Once pidgins have passed the "stripping" phase, they will—under appropriate conditions—undergo the same kind of grammaticalization processes as any other languages. Indeed, these cases are frequently young or incipient grammaticalizations, but this would be predicted by grammaticalization theory on account of the young age of the linguistic systems concerned. A wealth of additional cases have been documented for Tok Pisin and related pidgins; suffice it to mention the rise of a proximative aspect category (Romaine 1999), or of pronominal dual and trial markers out of the numerals *tu* 'two' and *tri* 'three' (see Heine and Kuteva 2005). The examples provided suggest that grammaticalization is not restricted to extended pidgins: At no stage in its short history would KPS qualify as an extended pidgin; all the processes that we described in Section 4.3 took place within an extremely short timespan; we will return to this fact in the next section.

Even in pidgins that have been claimed to "have no structure" (Bickerton 1990: 120–2) there is clause subordination: Both pidgin varieties discussed by Bickerton, one spoken by immigrants to Hawaii and the other (Russo-norsk) used in trading contacts between Russian and Scandinavian sailors, have conditional subordinate clauses. And there are also indications that grammaticalization is at work, as the following example suggests:

(33) Hawaiian pidgin (Bickerton 1990: 120)
Ifu   laik   meiki, mo   beta   *make* time, mani   no
if   like   make   more   better die   time   money no
kaen   *hapai.*
can   carry
'If you want to build (a temple), you should do it just before you die—you can't take it with you!"

Bickerton observes that *mo beta* (more better) roughly corresponds to the English auxiliary *should*. There is a crosslinguistically attested grammaticalization process whereby expressions like "It is enough/fitting/suitable/good (that)" are grammaticalized to deontic modals of necessity or obligation, meaning 'must' or 'should' (Heine and Kuteva 2002a). We saw one instance of this process in Section 4.3, and it would seem plausible that *mo beta* of Hawaiian pidgin is another instance, whereby something like '(it is) better (to) do' is reinterpreted as '(you) should do' in an incipient process of grammaticalization.[16]

4.6 A pidgin window on early language?

We observed in the introduction to this chapter that pidgins have been used by a number of authors to draw inferences on early human language. The question now is whether, or to what extent, such a procedure is justified. On the basis of the observations made in this chapter, one may hesitate to answer this question in the affirmative. Ignoring the fact that the evolution of early language may have involved a cognitive and socio-cultural ecology that contrasts with that in which KPS and other pidgins arose, there are also a number of other factors that make it hard to establish a reasonable analog between the two kinds of phenomena, such as the following:

---

[16] The corresponding English construction, as in *You'd better go now* appears to be another instance of this process, which may have served as a model to these speakers of Hawaiian pidgin English.

(a) Pidgins, at least in their "stripping" phase, have been described as being the result of a process from grammatically complex to less complex forms of language. According to our findings on grammaticalization that we presented in Chapter 2, which are in accordance with what many other authors have argued for, the evolution of early language must have proceeded in the opposite direction from less complex to more complex grammar.

(b) While pidgins are morphosyntactically impoverished vis-à-vis the languages from which they are derived, there are usually quite a number of "non-pidgin" features that survive the pidginization process: Pidgins are—at least to some extent—derived from other languages and inherit a number of properties of these languages. For example, Fanagalo, Kituba, and KPS are pidgins that grew out of Bantu languages spoken natively, that is, of Zulu and other varieties of Nguni languages, Kikongo, and Coastal Swahili, respectively. This is reflected, for example, in the fact that all three exhibit structural properties bearing witness to their origin as Bantu languages: They have retained some relics of Bantu noun class systems, of Bantu verbal derivational extensions, and of Bantu word order arrangements. There is no conceivable analog of such a situation in early language.

(c) In their genesis, pidgin speakers had at least some linguistic material to build on: They had at all times what Bakker (2003) calls a "functionally adequate language", namely their first language, which in some way also provided a model for developing the pidgin.

(d) The development of pidgins has also been shaped in some way or other by language contact and bilingualism. For example, speakers of KPS had, in addition to their respective mother tongues, other languages to draw on; many of these speakers knew several other Kenyan vernacular languages. Furthermore, they had Coastal Swahili in the form of its standardized variety constantly accessible to them via speakers of Standard Swahili, as a medium of education, broadcasting, in speeches held in more formal situations such as public gatherings, election campaigns, etc. Accordingly, pidgin speakers were exposed to multiple models provided by other languages or language varieties, and it would be surprising if the structure of the pidgins would not have been influenced by these models.

(e) Irrespective of the fact that early-language speakers may not have been able to dispose of the cognitive endowment that modern-language speakers do, pidgin speakers were able to draw on cognitive skills that were available to them on the basis of the communicative networks they were exposed to—in particular skills that would enable them to establish and express relations among different concepts.

(f) A paramount sociolinguistic characteristic of pidgins can be seen in the fact that their use tends to be restricted to very few domains of social interaction, such as those of trade or labor (e.g. Bakker 1995; Holm 2000). There is no reason to assume that early language was similarly restricted in its use.

The situation of early language is therefore likely to have contrasted sharply with that of pidgins: There was no language on which early language was built, there were no other languages around to serve as possible models and, accordingly, there were no models that could have served as cognitive templates to shape early language, etc. This raises the question of what justification there may be to draw on pidgins as possible analogs of early language. The only possible justification that we are aware of is that some pidgin properties, such as the ones listed in (1), may resemble properties that are hypothesized to have characterized early language, as has been argued by a number of authors (especially Sankoff 1979; Bickerton 1990; Romaine 1992b: 234; Aitchison 1996; Jackendoff 1999; Calvin and Bickerton 2000: 137; Givón 2002a, 2005). For example, with reference to the scenario of grammaticalization that we proposed in Chapter 2 (Figure 2.1), the structure of KPS and a number of pidgins can be located largely within layers IV and V, that is, it tends to lack the properties of the final layer VI.

Any hypothesis to the effect that pidgins provide an analog of early language needs to account for the differences that we listed above, by showing that these differences are either irrelevant or epi-phenomenal. So far, such an account does not exist.

Nevertheless, there are also reasons why a pidgin like KPS is relevant for a reconstruction of early language. These reasons all relate to the second component of pidgin development that we dealt with in Section 4.3, namely to the way in which a language having little in terms of grammatical complexity gains in complexity.

The first reason for arguing in favor of a systematic link between pidgins and early language concerns the rise of functional categories. From all we know, there was a stage in the history of KPS and other pidgins where structures such as the ones discussed in Sections 4.3 and 4.4 were absent, that is, where these languages were morphosyntactically poorer than they are now. Disposing of hardly any distinct grammatical morphology, of no formal apparatus to signal syntactic relations, nor of any productive means of clause subordination, the speakers of these languages created a range of novel functional categories by combining existing material in new ways. On the basis of our bridge hypothesis (see Section 1.1.3) we argue that the strategy employed by KPS speakers was the same as the one that characterized the behavior of early-language speakers. This strategy can be characterized in the following way:

(34) A salient strategy used to develop new functional categories
   a. Use existing linguistic forms in new contexts to suggest novel meanings.
   b. Use "concrete" (physically definable, easily accessible, clearly-delineated, etc.) concepts to describe more "abstract" concepts.
   c. Describe functional distinctions concerning semantic and syntactic relations by means of lexical concepts and constructions.[17]

In accordance with this strategy, we observed that speakers of Kenya Pidgin Swahili extended, for example, the use of the verb *toka* 'come from' from its canonical position before nouns to occur before verbs, whereby *toka* lost its "concrete" meaning of physical motion [To come from a place] and assumed a more "abstract" function [To come from an action], namely that of denoting immediate-past tense (see example (9)). In this way, lexical concepts were pressed into service to express specific functional concepts.

The second reason concerns the speed required for the rise of new functional categories. Under "normal" circumstances, this may involve one or more centuries to take its course—even if there is a great range of variation from one case to another. The pidgin evidence that we surveyed

---

[17] We are concerned here with pidgins; otherwise, grammaticalization is by no means confined to lexical concepts, as we observed above.

suggests that essentially the same process can take place within a few generations. The case of KPS is particularly noteworthy since it shows that under appropriate conditions a language can acquire a whole range of new grammatical use patterns and categories within half a century. This suggests that the speed of grammaticalization is contingent upon the linguistic and sociolinguistic environment in which it takes place; but more research is required on this issue.[18]

The third reason is perhaps the most relevant one with reference to any hypothesis on language genesis. Mühlhäusler (1986: 135) describes "structural expansion" in pidgins as being roughly synonymous with the processes of grammaticalization and addition of lexical resources. Ignoring lexical expansion, which is an issue beyond the scope of the present book, our observations on KPS suggest that structural expansion is in no way restricted to extended pidgins. Once there is a sufficient stock of lexical items used regularly by a group of speakers and this is passed on to other speakers, the ground is cleared for functional use patterns and categories to evolve: In regularly used collocations of two lexical items there is some probability that one of them will assume an "auxiliary" function, and that some of these collocations will turn into grammatical use patterns. Thus, pidgins can show us how a communication system relying primarily on lexical items, disposing only of a minimal apparatus of derivational and inflectional trappings and of marking clause subordination, gains in structural complexity, developing towards "fully-fledged" languages.

Fourth, while the newly evolving patterns and categories may be influenced by models provided by the source language or other languages that are available to the pidgin speakers, this is not necessarily so. We saw in Section 4.3 that the pidgin speakers created some novel functional categories which do not appear to have analogs in the source languages concerned, such as an immediate-past tense form by means of the verb *toka* 'come' or a category of deontic modality built on the adjective *mzuri* 'good'. Since these are grammaticalizations that can also be found in non-pidgins across the world (Heine and Kuteva 2002a),

---

[18] Philip Baker rightly reminds us that with regard to the speed required for the rise of new functional categories one needs to take account of the possibility that a particular pidgin derives in part at least from an earlier pidgin, as is certainly the case in the southwest Pacific area.

it is reasonable to assume that they are based on universal strategies of grammaticalization.

And finally, the study of pidgins offers a convenient laboratory for studying language evolution: Most modern pidgins arose within the last two centuries; in many cases we know roughly the time when and under which sociolinguistic circumstances they arose, when specific grammatical structures emerged, and how they evolved. Accordingly, pidgins provide a useful tool for reconstructing the emergence and development of grammar.

To conclude, there appears to be justification to link the ontological domain of pidgin speakers to that of early-language speakers—in spite of all the caveats that we pointed out above. But, as we hope to have shown in this chapter, little would be gained by simply equating the two domains—that is, by arguing for example that the former can be used as a readily available template to understand the latter.

4.7 Other restricted systems

The hypothesis that grammaticalization, and hence the development of a more elaborate grammar, requires, first, regular and frequent communication and, second, a system that is passed on from one group of speakers to another can be tested by looking at other kinds of restricted linguistic systems. In the present section we provide a summarizing discussion of a number of systems that lack these properties.

Homesigns One testing ground is provided by homesign systems (in short, homesigns).[19] Homesigns, which have been documented in at least thirteen different countries, are developed by congenitally deaf children born to hearing parents who did not expose these children to a conventional sign language. The children spontaneously use and conventionalize gestures (= homesigns) in order to communicate even if they are not exposed to a conventional sign language model, and thus lack a usable signed or spoken input (e.g. Goldin-Meadow 1982, 2002; Goldin-Meadow

---

[19] Our interest here is exclusively with systems developed by children; the behavior of adult homesigners would seem to require a separate treatment. The term "homesign" is not entirely satisfactory since the linguistic units making up such systems are to be described more appropriately as gestures rather than as signs (Bencie Woll, p.c.).

and Mylander 1990; Goldin-Meadow, Mylander, and Butcher 1995; Morford and Goldin-Meadow 1997; Morford 2002; van den Bogaerde 2004).

Homesigning children were found to have stable gesture vocabularies distinguishing nouns and verbs, they could combine gestures productively into sentences, and they could form sequences of sentences by combining gestures into at least two propositions, mostly but not exclusively involving a temporal sequence of events, and they could refer to objects, actions, attributes, and locations that were not in the here and now (Goldin-Meadow and Mylander 1990; Morford and Goldin-Meadow 1997). But we have found no clear evidence that there is explicit clause subordination, nor do there appear to be grammaticalized distinctions relating, for example, to tense, aspect, number, adpositions, or to nominal and verbal modification, even if there may be some means for signaling aspectual notions by means of movements (Beppie van den Bogaerde, p.c.); nor did we find any clues that grammaticalization parameters such as desemanticization and decategorialization had been drawn on by the homesigners.

There are occasionally gestures that can be interpretated as being suggestive of functional categories. For example, Morford and Goldin-Meadow (1997: 424–5) found the homesigner David to use a gesture for the question word 'where' (a shrug with upturned hands), and another gesture for the habitual location of a non-present object (patting the top of his head), and Goldin-Meadow (2003: 144) describes what she calls a narrative gesture 'away' (a palm or point hand extended or arced away from the body) marking a piece of gestural discourse as a narrative. And homesigners also appear to use strategies, or create structures, that are a prerequisite for grammaticalization, involving the extension mechanism. This concerns the ability to productively combine word units; for example, in the homesign variety of Nicaragua, homesigns for 'fruits' are a sequence where preparation of the fruit for eating is followed by the sign EAT, for example, PEEL EAT 'banana', RUB-ON-SHIRT EAT 'apple', SLICE-OFF-TOP-WITH-MACHETE EAT 'pineapple' (Morford 2002: 333). All this, however, is not sufficient to argue that homesigners do grammaticalize.

Homesign systems are not normally transmitted from one generation to the next. This suggests that homesigners do not dispose of an appropriate sociolinguistic environment that appears to be a requirement for grammaticalization to take place; hence, they do not grammaticalize.

Goldin-Meadow (1982) argues that homesigners use recursion in structuring their discourses. However, in the data available we found no clear evidence that there is embedding recursion.[20] For example, Goldin-Meadow interprets the utterance (35a) as being suggestive of subordinate linkage analogous to relativized sentences in English. Example (35a) was produced by the homesigner David to request Heidi to give him a toy grape. While we do not wish to exclude the possibility that this example reflects recursion, the evidence available is not sufficient to establish that it does. In a similar fashion, example (35b) might be suggestive of recursive noun modification [[adjective]–noun], but once again, more data would be required to show that such a structure really exists.

(35) The homesigner David (Goldin-Meadow 1982: 58, 59)[21]
    a. MOVE-palm-EAT. ('You/Heidi *move* to my *palm* grape which one *eats*.')
    b. ROUND-penny-me. ('You/Heidi give *me* the *penny* which is *round*.')

The situation is different when a homesign is transmitted to another group of people, as has happened in Nicaragua in the transition from a homesign of the first cohort to Nicaraguan Sign Language of the second cohort around 1985 (see Section 7.2), and presumably also in some other community homesigns (Beppie van den Bogaerde, p.c.). Such situations are likely to lead to some form of grammaticalization (see Sections 7.1.6; 7.2). Morford describes the transition from homesign to Nicaraguan Sign Language thus:

Homesigners started with inconsistent gesture as input, and innovated structure. Nicaraguan signers started with structured input (i.e., homesign) and grammaticized elements of the input. (Morford 2002: 333)

Twins' languages[22] These are forms of communication that are created and spoken by two (sometimes even more) hearing children who acquire language simultaneously and, while having access to normal spoken

---

[20] Note that Goldin-Meadow (1982: 54) uses the term recursion in a wider sense—one that includes iteration (see Section 6.1.4).

[21] In these examples, the following conventions are used: lowercase = referents of deictic signs; capitals = glosses for the referents of characterizing signs (Goldin-Meadow 1982: 7).

[22] The following account is based entirely on Bakker (1987a, 1987b, 1990, 2006). We are grateful to Peter Bakker for having made his materials available to us. In earlier publications, Bakker preferred the term "autonomous language" to twins' language.

language in their critical stage of language acquisition, do not find much parental attention, receiving more language input from one another than from their parents, and using each other as models of speech formation (Bakker 1987a, 1987b, 1990, 2006). The children differ neither mentally nor physically from other children; it would seem that it is on the one hand a strong mutual psychological bond and on the other hand relative social isolation from other children and family members that are the main contributing factors for the emergence of twins' languages. Note that the children are not necessarily twins, they may be close siblings or close friends; but the probability that a twins' language will arise is highest with monozygotic and with same-sex twins, and even higher with triplets. Being spoken mostly between the ages of 4 and 6, the languages tend to disappear gradually thereafter, in that more and more words and grammatical rules of the matrix language (i.e. their parents' language) are adopted.

Twins' languages are neither "invented" nor are they intended to be secret languages since their speakers are as a rule monolingual; they usually arise as first languages, believed by some to be fossilized forms of child language. The children understand their matrix language, which in the thirteen cases documented was Danish, English, Estonian, German, Swiss German, Icelandic, or Russian. It has been claimed that twins' languages, also known as secret languages or autonomous languages, or as manifestations of idioglossia or cryptophasia, can be found in some way in 40 percent of twins in early childhood, but documented cases are few.

The cases that have been documented are spoken systems with little gesturing or signing. They show drastic phonological, morphological, and syntactic simplification vis-à-vis their respective matrix languages. There is a stock of mostly lexical items which can be combined and remain stable in combinations. These items are overwhelmingly noun- or verb-like, but there are also a few other kinds of items, such as terms for colors or numbers. While onomatopoetic forms are not uncommon (e.g. *tutu* 'train' in an English, *hihi* 'horse' in an Icelandic, or *düdüdüt* 'post coach' in a Swiss German twins' language), the lexicon is derived mainly (at least 90 percent) from lexical items of the matrix language, being basically arbitrary and multi-functional, not uncommonly belonging simultaneously to several grammatical categories. The form *hapn* of a German twins' language, for instance, means both 'food' and 'to eat'.

Utterances consist mostly of nouns and verbs strung together, they are mono-propositional, although occasionally sequences of propositions do occur, as in (36). Word order is fairly free, but old information, or information that is uppermost in the speaker's mind, tends to be presented first, while new information is placed later. A common feature across twins' languages is that utterances are frequently introduced by vocative nouns, cf. (37). Compounds are occasionally encountered, for example *koko-dach* 'store' (< German *Kauf* + *Dach* 'buy + roof') in a German twins' language, but are not necessarily suggestive of a head–modifier construction. No form of grammaticalization has been found in any of the twins' languages.

There is no inflectional morphology; grammatical forms such as copulas, plural markers, case markers, adpositions, tense–aspect markers, or conjunctions are absent, but reduplication is not uncommon. The few functional elements that do exist, in particular distinct words for negation and yes–no questions, appear to be derived in some way or other from the matrix language that the children are exposed to rather than having been created via grammaticalization, cf. the negation marker in (37) and (38) or the question marker in (39). There is no clause subordination of any kind, nor is there any kind of recursive structure. But a possibly noteworthy characteristic is that some of these systems show complex predicates consisting of combinations of two verbs, as in example (37).

(36) English twins' language (Bakker 2006)
 Gaän odo migno-migno, feu odo.
 God take.away rain heat send
 'God, take away the rain and send the sun.'

(37) Russian twins' language (Bakker 1990: 88)
 Liulia, ne nata potet.
 Yura, NEG want look
 'Yura, I don't want that you look.'

(38) Danish twins' language (Bakker 1990: 85)
 hun mis bie enaj.
 dog cat bite no
 'Dogs may not bite the cat(s).'

(39) Swiss German twins' language (Bakker 2006)
 Bobby hemma há.
 Bobby home Q
 'Is Bobby home?'

Exactly what the status of twins' languages is vis-à-vis other forms of first language acquisition remains a question that cannot be answered conclusively on the basis of the limited data available. We follow Bakker (1987a, 1987b, 1990, 2006) in assuming that they represent a specific structural type: First, they share a number of sociolinguistic and linguistic properties, such as the ones mentioned above, second, they are fairly efficient media of communication for their speakers, spoken fast and fluently, usually being unintelligible to non-speakers, third, they are apparently not transmitted to other groups of speakers, fourth, it seems that they cannot be reduced to being impoverished variants of the respective matrix languages or representing a frozen stage of first language acquisition. Even if the lexicon is overwhelmingly derived from the respective matrix language, the grammar is not, showing principles of discourse organization that are not the same as those underlying the latter. And finally, twins' languages do not exhibit any of the prerequisites for grammaticalization that we mentioned in the introduction to this section, hence there is no grammaticalization—in particular no inflection, derivation, clause subordination, or recursive structure.

Isolated children But what about feral human beings who in their childhood have been raised intentionally in social and linguistic isolation—in short, "isolated children"? There are three cases that provide a reasonable basis for answering this question, the persons concerned being Kaspar Hauser, Genie, and Chelsea (Curtiss 1977, 1994: 225–31; MacSwan and Rolstad 2005; for more information on feral children, see Aitchison 1989, and Candland 1993). The following sketchy remarks are restricted to language-like behavior that allow for comparisons across the different kinds of human and animal behavior.

All the available data suggest that Kaspar Hauser was kept in total isolation in a small room from the age of 3 or 4 until about 15 or 16 years of age, being released and discovered in Nuremberg in 1828, but he was assassinated in 1833 at age 21. Within this short period of five years he learned to read and write, and he mastered semantic aspects of the German language.

A few months after his discovery, Kaspar acquired a sizable vocabulary and began to combine words into "sentences." While being able to actively participate in intellectual discussions, there was little in terms of morphology or syntax. Kaspar's verbs were restricted mostly to infinitive forms, and there were almost no conjunctions, participles, or adverbs; there is no evidence of grammaticalization of any kind. And he rarely used the first person pronoun *ich*, referring to himself in the third person (*Kaspar*). The linguistic categories he created were not necessarily those that he learned from the people who looked after him; for example, he developed a distinction between human and non-human concepts, the former expressed by *bua* (from Bavarian *bua* 'boy', High German *Bub*) and the latter by *ross* (from German *Ross* 'horse'). The semantic over-generalizations that we observed in animals were also characteristic of Kaspar's speech. His ability to create novel expressions by re-combining existing ones is illustrated by these examples: He referred to a fat man as 'the man with the mountain' and to a woman whose shawl dragged on the ground as 'the lady with the tail.'

Genie was isolated from age 1;8 to 13;7—that is, for a period of twelve years she was imprisoned by her father in a small bedroom in the back of the family home in California. She had little linguistic input; her brother and father were her primary caretakers but did not speak to her. Her blind mother managed to escape with her from the home when Genie was 13, and when Genie was discovered in 1970 she could barely walk, chew or bite, and she did not understand nor speak language.

Genie began to produce words within a few months, and three to four months later she had acquired an expressive vocabulary of 100 to 200 words, including words for colors, numbers, and lexical taxonomic hierarchy (superordinate, basic, subordinate), and she could combine two words. She produced increasingly longer strings of words, but her utterances remained what Curtiss (1994: 228) describes as "agrammatic and hierarchically flat." As the following utterance illustrates, propositions are simply juxtaposed, and there are essentially no personal pronouns, with distinctions of personal deixis being expressed by means of proper names.

(40) Genie (MacSwan and Rolstad 2005: 232)
   Father hit Genie cry long time ago.
   'When my father hit me, I cried, a long time ago.'

She was able to talk about persons and objects in their absence, and about events to come. There remained a discrepancy between her lexical and semantic achievements on the one hand and her grammatical ones on the other. After eight years, her speech was still essentially without closed-class morphology, there were no categories such as prepositions or pronouns, and hardly any syntactic devices; accordingly, there was no grammaticalizing behavior, and no recursion.[23] Nevertheless, Genie's competence for effective communication was well in place but her grasp of the linguistic skills that are important for social participation in conversations (e.g. use of social rituals) remained generally poor. And she relied heavily on simple propositions or on repetition for expressing pragmatic functions.

The case of the hearing-impaired Chelsea is of a different nature; she was an adult who attempted first language acquisition only in her thirties. Unlike Genie, she acquired a reasonable command of social rituals and conversational operators. But as in the cases of Kaspar and Genie, there is a marked disparity between lexical achievements and the ability to combine vocabulary into appropriate and grammatical utterances: While her ability to acquire a lexicon progressed steadily, her multi-word utterances were for the most part grammatically ill formed and remained frequently propositionally unclear. But conceivably she acquired the use of noun modification, as examples such as (41) might suggest. Whether such examples reflect the ability to understand the concept of taxonomic inclusion (see Section 3.2.9), however, remains unclear.

(41) Chelsea (MacSwan and Rolstad 2005: 233)
Missy girl same both girl.
[Comparing the gender of two animals]

Combining semantically relevant nouns was all Chelsea could achieve in terms of expressive language. There are no indications that Chelsea used grammaticalization in her utterance production. Curtiss (1994: 230) notes that while lexical and propositional semantics developed with apparent ease in the cases of both Kaspar and Genie, Chelsea experienced problems with propositional form.

4.8 An elementary linguistic system?

From all we saw in the preceding section there are remarkable resemblances among the structures dealt with in Section 4.7: When comparing the

---

[23] Here we differ from Goldin-Meadow (1982: 53), who argues that Genie learned recursion. That Genie was in fact able to use adjective–noun combinations productively is suggested by utterances such as the following: *Ask teacher yellow material. Blue paint. Yellow green paint. Genie have blue material* (Bickerton 1990: 117). But whether this should be interpreted as being suggestive of noun phrase recursion is a question that is hard to answer on the basis of the evidence available.

behavior of homesigners, twins' language speakers, and isolated children with that of trained animals as described in Section 3.2, there is a catalogue of properties they share, namely most or all of the ones listed in (42).[24]

(42) An elementary linguistic system
- a. There is a well-marked ability to create a limited stock of form–meaning units[25] ("lexical items").
- b. The units are, at least to some extent, arbitrary form–meaning pairings.
- c. These units can be combined to produce new meanings.
- d. The units are compositional, that is, their meaning remains stable in combinations.
- e. The units include items for objects ("nouns") and actions ("verbs"), but also some more abstract items, such as items for colors and numbers.
- f. The units can be arranged into propositional structures ("predicate frames"), even if this ability appears to be fairly elementary.
- g. There is also an ability to form sequences of propositions; but there is little to suggest that these sequences are structured in terms of grammatical rules other than simple juxtaposition.
- h. There are linguistic means to express questions and negation.
- i. There appears to be an elementary ability to comprehend conceptual taxonomic hierarchy
(superordinate vs. subordinate category).
- j. There are hardly any non-lexical units; communication is achieved overwhelmingly or entirely without functional categories.
- k. The system allows for successful linguistic communication among its speakers.
- l. There are no clear indications of grammaticalization.
- m. If there are any functional categories, they are not created by using parameters of grammaticalization.
- n. There is essentially no form of clause subordination.
- o. There are no phrase structures that are clearly suggestive of recursion.

[24] We concur with Fritz Newmeyer (p.c.) that the database for all these restricted systems is fairly small and that any generalizations should be taken with care.
[25] The form these units may take differs from one system to another; they may be spoken words, gestures, lexigrams, or plastic labels.

p. The system is not normally transmitted from one group of speakers to another (or from one generation to the next).

Taken together, the properties listed in (42) are suggestive of some distinct kind of elementary linguistic system. Note that Casey and Kluender (1998: 74–6) also argue that both deaf children with impoverished language input and some trained non-human primates may represent an intermediate form of communication in the evolution of language, and comparing twins' languages with homesigns, Peter Bakker characterizes the latter thus:

These languages appear to be different from adult sign languages, but very similar to the AL's [twins' languages; a.n.]. They have relatively free order of subject, verb [and] object, they have many vocatives and sentence final and sentence initial negations and they lack morphology (contrary to adult sign languages). This may point to the fact that we have to do with a 'natural' type of language, which emerges in certain situations of deprived language input. (Bakker 1990: 92)

Thus, it appears that the linguistic abilities exhibited by the humans discussed in Section 4.7 do not seem to be dramatically different from what we found in trained animals in Chapter 3, both primates and some non-primates. They all have a capacity to develop a lexically based communication system that allows them to interact fairly successfully with others. And second, these observations also suggest that neither the humans nor the animals show any clear traces of grammaticalization, subordination, and recursion. There are a few cues suggesting that the ability to understand the concept of hierarchical taxonomy is in place, not only in the humans but also in some trained non-human primates (see Section 3.2.10), but there is no clear evidence that there is recursive syntax.

Kaspar and Genie use third-person reference, that is, personal names, for all distinctions of personal deixis (cf. example (40)). This might be seen as an indication that they had a concept both of personal deixis and of incipient grammaticalization; nominal or pronominal expressions having third-person reference do not uncommonly provide a conceptual template for creating categories for second- or first-person reference— the French impersonal pronoun *on*, which has acquired the significance of a first-person plural pronoun in modern colloquial French being a case in point (see Section 2.2.1.1). But very likely, this behavior is due to factors other than grammaticalization.

But whether the properties listed in (42) do in fact reflect a specific type of elementary linguistic system is a question that we leave open to future research. A clear difference between animals on the one hand and homesigners and isolated children on the other consists in their respective non-linguistic abilities, and this difference is suggestive of a dissociation of linguistic and non-linguistic cognitive abilities, as argued for by MacSwan and Rolstad (2005). Kaspar is reported to have become competent in mathematics and to participate actively in philosophical discussions, Genie showed fully developed abilities in the domain of visual-spatial organization, and Chelsea could perform mathematical operations, and she could even keep a balanced checkbook. No corresponding abilities have been reported for trained animals. Since our concern is exclusively with manifestations of language-like behavior, we are not able to offer any interpretation for this discrepancy (but see Curtiss 1994; MacSwan and Rolstad 2005).

4.9 Conclusion

A number of different forms of linguistic communication have been grouped together as "degraded forms of language" (Jackendoff 1999) or restricted linguistic systems (Botha 2003b, 2005, 2005/6), most of all child language, late untutored second language aquisition, pidgins and related contact varieties, language use of agrammatic aphasics, homesigns, twins' languages, historically early non-grammaticalized language, and language use by isolated children like Kaspar Hauser, Genie, or Chelsea. With reference to language evolution, it would seem that these systems can be divided into two groups. The first comprises non-grammaticalized systems, such as those of homesigners, twins' language speakers, and isolated children, but also those of some trained animals (see Section 3.3); in the second group are the weakly grammaticalized systems such as (extended) pidgins, and presumably also the Nicaraguan Sign Language. Systems of the first group are characterized by the presence of many or all of the properties listed in (42): They lack salient properties of human languages, in particular grammaticalization, subordination, and recursive syntactic structures, while the latter have such properties, at least in some basic form.

The reasons for this typological distinction are probably complex, but there is one reason that is probably crucial—one that is sociolinguistic in nature: As we observed in Section 3.3, grammaticalization requires a linguistic system that is used regularly and frequently within a community of speakers and is passed on from one group of speakers to another. This condition is somehow met in the case of the second group, but not in the case of the first group: Homesigners, isolated children, twins' language speakers, and trained animals do not dispose of an appropriate sociolinguistic environment that would have made grammaticalization possible. This leaves us with the question of why systems of the first group also lack subordination and recursive syntactic structures. The answer has to be deferred to Chapters 5 and 6, where we will argue that grammaticalization is causally responsible for these two properties to arise.

5 Clause subordination

If there is one question that has excited students of language genesis and language evolution perhaps more than any others then it is the one that we listed in Chapter 1 as (1n), namely: How did the properties believed to be restricted to modern human languages arise, in particular syntax and the recursive use of language structures? We will deal with this question in more detail, devoting the present chapter to syntax (more precisely, morphosyntax) and the next chapter to recursion. In a recent publication, Bickerton observes:

Moreover, the question—as one has to keep reminding non-linguists—is not just how did syntax come to be, but why it took the form that it did, rather than some other form.... clearly we have neither understood syntax nor produced an adequate theory of language evolution until this last question has been answered (Bickerton 2005: 9)

The present chapter is concerned with this last question, focusing on a core area of syntax, namely complex sentences—more narrowly on how clause subordination came to be. As elsewhere in this book we will use grammaticalization theory to reconstruct some salient patterns of evolution; and as elsewhere we argue that synchronic language structure can be viewed as the frozen product of cognitive and communicative processes that happened in the past. Accordingly, we argue that underlying the evolution of clause combining there is a strategy whereby existing grammatical means are recruited for novel discourse functions, and that there are two distinct mechanisms whereby the forms used for these discourse functions are transformed into new syntactic structures. The subject matter of this chapter is far from new; after Givón's (1979c) pioneering work it has received some attention in recent years (especially Harris and Campbell 1995; Bybee and Noonan 2001; Hopper and Traugott 2003: 175–211; Givón 2006; see also Diessel 2005), and our discussion will build on these foundations.

Clause subordination is a complex subject matter and the typological diversity of structures it exhibits in the languages of the world is enormous. It therefore goes without saying that we will not be able to do justice to this vast field; rather, we will be restricted to a few salient manifestations of it that appear to be crosslinguistically common, ignoring the many other manifestations that have come to be known.

5.1 Introduction

Selection of the data to be discussed below was guided by the question that is central to this book: What can insights on documented language change reveal about early language? We will once more adopt a reductionist perspective, leaving aside all the fascinating conceptual and pragmatic processes that have jointly contributed over centuries to the crystallization of new grammatical structures; in accordance with this reductionist perspective, we will not deal in any detail with syntax per se but will focus on its formal exponents, more precisely on the markers of clause subordination; for a comprehensive syntactic analysis, see Givón (2006). Harris and Campbell (1995: 284) maintain that the historical sources of such markers do not necessarily imply the sources of the structures of which these markers are a part; however, it would seem that, as a rule, they do; we will return to this issue in Section 5.4.

Following other treatments of this subject we will distinguish between coordination, involving two (or more) relatively independent clauses, and subordination, that is a combination of two (or more) clauses where one can be defined as being the main (or matrix, or nucleus) clause and the other as being subordinate to (or dependent on) the main clause on the basis of morphosyntactic properties. As has been pointed out by many authors, and as will also become obvious in subsequent sections, the boundary between the two kinds of clauses is not really neat. In our treatment we will ignore cases where two clauses are reduced to one. This process can be the result of at least two different kinds of processes. One concerns the loss of a main clause whereby the subordinate clause is reinterpreted as the new main clause. Paradigm cases are provided by cleft constructions [It is X that S], where the copular main clause [It is X] is grammaticalized to a focus (and sometimes further to a subject) noun phrase and the subordinate clause [that S] turns into the new main clause, as has happened in a number of languages (see e.g. Heine and Reh 1984: 147–82; Harris and Campbell 1995: 151–68). Another kind concerns subordinate clauses that are grammaticalized (or lexicalized) to adverbials, conjunctions, or parenthetical markers within the main clause. For example, in the Central Khoisan language Khwe (Kxoe) of Namibia, the temporal adverbial clause in (1), meaning roughly 'when it is like that', has developed into the temporal conjunction *taátenu* 'then', which has become one of the most frequently used discourse markers of the language. Other cases concern clausal structures that assume functions as modal or evidential markers (see Hopper and Traugott 2003: 207–9; Thompson and Mulac 1991). Such cases concern situations where there already exists an established form of clause subordination, while our interest in this chapter is with situations where new forms of subordination arise.

(1) Khwe (Kxoe, Central Khoisan, Khoisan)
ta-   á-   te-   nu   xavána ǁé   kúùn-
be.thus- JUNC- PRES- when again 1.M.PL go-
à-   tè [...].
JUNC-   PRES
'Then we went again [...].'

On methodology We argue in this chapter that clause subordination is the product of grammaticalization of non-subordinate sentence structures. The crucial question that this hypothesis raises is the following: What is the evidence in support of this hypothesis?

As may have become clear in the preceding chapters, there are two kinds of evidence on which our reconstruction work is based. The first is historical in nature: Since grammaticalization theory rests on generalizations on historical processes, its hypotheses must be verifiable, or falsifiable, by means of historical data. For example, if we hypothesize that the English complementizer *that* or its German equivalent *dass* are grammaticalized products of the demonstrative *that* or *das*, respectively, then we need to demonstrate—as has in fact been done (see e.g. Hopper and Traugott 1993: 185–9)—that at some stage in the history of English there was a demonstrative *that* but no complementizer *that*, and that the latter is historically derived from the former. Consequently, we need to demonstrate that there was a situation in earlier forms of English where there were no complement clauses introduced by *that* (or *dass* in German). We will refer to this component of our methodology as the **diachronic criterion**.

But our methodology has a second component. Since grammaticalization is a regular process, instances of it can also be identified by means of structural properties—that is, properties that bear witness to the process concerned. We will refer to this as the **structural criterion**: Grammaticalized expressions differ from their non-grammaticalized counterparts with reference to the four parameters distinguished in Section 1.2.1: These expressions are the result of extension, desemanticization, decategorialization, and erosion—hence, they can be identified in terms of these parameters.[1] Accordingly, we also need to demonstrate that, for example, the complementizer *that* differs from the demonstrative *that* with reference to these parameters.

With regard to the present example, the structural criterion yields the following outcome: Extension means that the use of *that* was extended to a new context where it came to be used preceding complement clauses—with the result that a new function emerged, namely that of expressing a syntactic relation. Desemanticization had the effect that the item lost a central semantic property that the demonstrative had, namely expressing spatial deixis. Effects of decategorialization can be seen, for example, in the fact that whereas the demonstrative *that* is sensitive to number (taking the plural form *those*) and spatial deixis (distal *that* vs. proximal *this*), and can be used either as a nominal determiner (*that story*) or a pronoun (*I know that*), the complementizer lacks such morphosyntactic properties, and it also lacks the ability to have participant status that the pronoun *that* has, as in *I know that*. Additional effects characteristic of decategorialization can be seen in the German equivalent of English *that*: As a demonstrative (*das*), it is sensitive to gender, number, and case, while as a complementizer (*dass*) it is an invariable particle. Finally, erosion means that *that* as a complementizer lacks a phonetic property that the demonstrative has: It has lost the ability to receive stress.

To conclude, both the diachronic and the structural criteria lead to the same conclusion, namely that *that* as a complementizer is the product of grammaticalization of the demonstrative. Ideally, both criteria should apply in order to prove cases of grammaticalization. But, as we will see below, in many cases there is no appropriate historical data to draw on the diachronic criterion, and in such cases we will be confined to the structural criterion. It

---

[1] As we observed there, not all of these parameters are necessarily relevant in a given case.

goes without saying that a reconstruction based on both criteria is empirically stronger than one based on only one criterion.

Is this process necessarily unidirectional? The answer is essentially yes: There are quite a number of languages where a change from demonstrative to complementizer has been documented, while we are not aware of any language where a complementizer changed into a demonstrative. Nevertheless, a few cases have been adduced where subordinate structures gave rise to non-subordinate ones (see Hopper and Traugott 1993: 184–5; Harris and Campbell 1995); however, such cases are not only uncommon but also idiosyncratic, that is, they do not seem to correspond to any crosslinguistically regular pattern of grammatical change. Accordingly, provided that the structural criterion applies, we will henceforth assume that the unidirectionality principle is observed.

Two channels As we argue in this chapter, there are crosslinguistically two main ways in which clause subordination arises: Either via expansion, that is, the reinterpretation of a nominal as a clausal (propositional) participant, or via the integration of two independent sentences within one sentence; the two ways are sketched in (2).[2] This argument entails a strong claim, namely that clause subordination is historically derived from non-subordinate sentences. The same claim has been made independently by Givón (2006): Analyzing a wide range of languages of worldwide distribution, he concludes that there are two main diachronic sources or channels leading to complex sentences (or clause union), namely via embedded verb phrase complements (type A) and clause chaining (type B). His type A relates to expansion, while type B corresponds to integration.[3]

(2) The two main channels in the rise of clause subordination
    a. S [NP] > S_1 [S_2]   Expansion
    b. $S_1 + S_2$ > S_1 [S_2]   Integration

The formulas in (2) show that once either of these two processes has taken place there is recursion, in that one clause (S_2) is embedded in another

---

[2] Note that (2a) is not really a satisfactory way of rendering the actual process. First, "[NP]" need not be a noun phrase but may as well be an adverbial phrase, and, second, the formula applies if "[NP]" is an adjunct, that is, a peripheral participant of S, but not if it is an argument of S.

[3] The reader is referred to this study by Givón (2006), which discusses a much wider range of processes than we are able to cover here and provides a coherent syntactic account of these processes.

clause (S_1); we will return to this issue in Chapter 6. Hopper and Traugott (2003: 176) propose a cline of clause combining leading from parataxis to subordination. This cline concerns only integration (2b); as we will see below, this is not the only way in which clause subordination arises.

Section 5.2 is devoted to expansion, while the subsequent sections deal with integration in its major manifestations: Section 5.3.1 is concerned with certain aspects of relative clauses, that is, with clauses that function as attributive modifiers of nouns (or noun phrases).[4] Section 5.3.2 deals with complement clauses, that is, with subordinate clauses that are complements (or arguments) of another clause (i.e. the main or matrix clause),[5] and Section 5.3.3 with adverbial clauses, that is, with clauses that have adverbial functions, being modifiers or adjuncts of matrix clauses or predicates. Problems associated with our analysis, as well as some implications that this analysis has, are discussed in Section 5.4. Historical information on grammatical change in the languages of the world is scanty, and many of the reconstructions proposed are based on applying the methodology of grammaticalization theory to linguistic data that are only accessible via our structural criterion; however, a number of the reconstructions are also supported by attested historical evidence (see below).

One major theme of this chapter is to demonstrate that devices that first served to structure independent sentences come to assume functions of subordination, and to show how this process happens. This, however, does not necessarily mean that there was no previous form of subordination. Since the methodology of historical linguistics covers only a fairly narrow time depth, we can never be sure whether there did not exist such a form in the earlier history of that language; as Harris and Campbell (1995: 282 ff.) rightly emphasize, the rise of a new form of subordination may simply mean that an existing form was either modified or replaced. In

---

[4] This description is far from satisfactory and takes care essentially only of restrictive relative clauses (see de Vries 2002: 14–15 for discussion).

[5] This usage differs slightly from that of Thompson and Longacre (1985: 172; see also Hopper and Traugott 2003: 177) who restrict such clauses to "those which function as noun phrases (called complements)." Even if the majority of complement clauses function as noun phrases, not all really do; for example, in our understanding the English phrase *that he's a liar* is a complement clause of the sentence *It seems that he's a liar* although it does not appear to be able to function as a noun phrase. Our definition is similar to that of Noonan (1985: 42) according to which a complement clause is a notional sentence or predication that is a subject or object of a predicate.

some language families, no subordinate structures can be reconstructed; for example, no specific relative clause marking can be reconstructed for the Germanic languages. But this does not mean that previously in the relevant families there were no corresponding subordination structures.

5.2 Expansion

While complement and adverbial subordinate clauses differ in their structures, we will treat them together in this section since the mechanism of expansion operates in much the same way in both. In accordance with (2a), expansion is based on a conceptual strategy whereby clausal (propositional) participants are treated like nominal participants, and the strategy has the effect that—over time—nominal structures acquire the properties of subordinate clauses.[6] There do tend to be some nominal properties that survive this process, however, such as the following (but see also below):

(3) Structural properties commonly found on subordinate clauses arising via expansion
   a. The marker of subordination resembles a grammatical form associated with noun phrase structure, such as a marker of case, gender, definiteness, or an adposition.
   b. The verb of the subordinate clause is frequently non-finite, encoded like an infinitival, gerundival, participial, or a nominalized constituent and takes the case marking of a corresponding nominal participant.
   c. The arguments of the subordinate clause are encoded in a form that tends to differ from that of the main clause.
   d. The agent or notional subject takes a genitive/possessive or other case form, typically having the appearance of a genitival modifier of the subordinate verb.
   e. The patient or notional object may also take a genitive/possessive or other case form.

---

[6] Grammaticalization frequently involves a reduction of structure (Fritz Newmeyer, p.c.), as is suggested by the parameters proposed in Section 1.2, which, except for one, involve loss of properties. In cases such as the present, where there is expansion from nominal to clausal structures, the formal exponents include the loss of nominal meaning (desemanticization) and cardinal nominal morphosyntactic properties (decategorialization); we will return to this issue below.

f. There are severe restrictions on distinctions such as tense, aspect, modality, negation, etc. that can be expressed—in fact, such distinctions may be absent altogether.

Note that these are not definitional but rather diagnostic properties meant to identify instances of expansion, and not all of the properties are necessarily present in a given case. To be sure, nominal encodings such as the ones listed in (3) are in no way restricted to specific languages; rather, they are found in some way in quite a number of languages. For example, English *He witnessed the enemy's destruction of the city* largely corresponds to (3), being a nominal version of the largely equivalent sentence *He witnessed that the enemy destroyed the city*.

With reference to the four parameters of grammaticalization, expansion tends to have the following effects in particular: Extension means that an existing morphological device is extended from nominal to clausal structures, with the effect that a new function, that of presenting subordinate clauses, emerges. This has the effect that the nominal function associated with this device is lost in the relevant contexts (desemanticization), and also that the ability associated with nominal structures to take determiners and modifiers is lost (decategorialization). Erosion, which may but need not be involved, means that the marker of subordination loses in phonetic substance, becoming shorter or phonetically simplified vis-à-vis the corresponding nominal marker (see Section 1.2.1.5).

Expansion from noun to complement clause   Givón (2006: 12) describes this channel as one where "the complement-clause event is treated analogically as a **nominal object** of the main clause." Complement clauses arising via expansion are not uncommonly restricted to a limited spectrum of main clause (matrix) verbs, most of all speech-act, cognition, and/or verbs of volition. These verbs typically take both clausal and propositional complements, such verbs being for example 'see', 'hear', 'feel', 'want', 'finish', 'start', 'know', 'tell', 'remember', 'say', etc. Furthermore, this structure may be confined to complement clauses where the subject of the complement clause is coreferential with that of the main clause subject. We will now look at a few examples illustrating the effects of expansion (for further examples, see Givón 2006).

In the following example from the Tungusic language Evenki, the complementizer is an accusative case marker (ACC), that is, the complement

clause is introduced by a case suffix, in accordance with (3a), the verb əmə-
'come' of the complement clause is presented in the resultative participle
(PART), cf. (3b), and the agent of the complement clause appears as a
possessor suffix (-s 'your') on the participle form of the verb (cf. (3d)).

(4) Evenki (Tungusic; Comrie 1981: 83)
    ənii-   m   əə-   ćəə-   n   saa-   rə si   tənəwə
    mother- my NEG- PAST- 3.SG know- ?[7] you yesterday
    əmə-   nəə-   wəə-   s.
    come- PART- ACC- 2.SG
    'My mother doesn't know that you arrived yesterday.'

A similar structure is exhibited by the Estonian example in (5), which illustrates one of the two ways in which complement clauses having speech-act or mental-state verbs as main verbs are expressed in this language: The verb is non-finite, constructed in the present tense of the active participle (3b), and the subject/agent appears in the genitive case (GEN) (3d):

(5) Estonian (Finno-Ugric; Harris and Campbell 1995: 99)
    sai   kuul-   da   seal   ühe   mehe   ela-
    got   hear-   INF   there one.GEN man.GEN live-
    vat.
    PRES.ACTIVE.PTCPL
    'S/he came to hear that a man lives there.'

Similarly, in the Ik language of northeastern Uganda, the complement verb 'eat' in (6) is non-finite and appears in the accusative case (ACC) (3b) and the patient 'food' of the complement clause in the genitive case (GEN) (3e).

(6) Ik (Kuliak, Nilo-Saharan)
    bɛɗ-   a   atsꞌ- és-   íka   ŋkáká-   é.
    want- 3.SG eat- INF- ACC food/meat- GEN
    'He wants to eat meat.'

With regard to (3f), expansion may entail that the complement clause may not be able to express grammatical distinctions commonly found in other clauses. For example, in the Ik complement clause illustrated in (6) no tense–aspect distinction can be expressed, and in the Chibchan language Jacaltec, complement clauses "of the gerund type" cannot be negated while finite complement clauses can (Craig 1977: 242).

[7] No gloss is provided by the author.

Expansion from noun to finite clause But there is a second variety of the expansion strategy whereby the subordinate clause is also treated like a nominal participant but instead of a non-finite structure it takes the form of a fully-fledged finite clause. Accordingly, instances of this variety conform to property (3a) but not to the other properties—that is, the structure of the subordinate clause does not differ dramatically from that of the main clause—other than taking a marker of subordination that is homophonous with, or similar to, a nominal case or other marker. The following examples illustrate this variety.

In the Nigerian language Kanuri, the dative case suffix (or enclitic) *-rò* (DAT) is attached to the finite clause in (7a) to form complement clauses (7b), and in the Andean language Quechua (8) and the Khwe language of Namibia (9) it is the accusative case marker (ACC), cf. (9a), which is attached to the finite complement clause (9b).

(7) Kanuri (Saharan, Nilo-Saharan; Noonan 1985: 47)
    a. Sávā-   nyī   íshìn.
       friend- my   come(3.SG)
       'My friend is coming.'
    b. Sávā-   nyī   íshìn-   **rò**   təmăŋə́nà.
       friend- my   come(3.SG)- DAT   thought.1.SG.PERF
       'I thought my friend would come.'

(8) Imbabura Quechua (Cole 1982: 43)
    Pedro ya-   n   [ñuka Agatu- pi kawsa- ni] **-ta**.
    Pedro think- 3   I   Agato- in live-   1-   **-ACC**
    'Pedro thinks that I live in Agato.'

(9) Khwe (Central Khoisan, Khoisan)
    a. doá-   m̀   ꞌà |xꞌún- á-   hān.
       kudu- M.SG ACC kill- JUNC- PERF
       '(They) killed a kudu antelope.'
    b. xàcí tcà-   á-   tè   ꞌà   tí |xꞌân
       she   be.sick- JUNC- PRES ACC I   very
       qāámà- à-   tè.
       regret- JUNC- PRES
       'I am a lot sorry that she is sick.'

The expansion strategy, whereby a clausal proposition is treated like a noun, is found in headless relative clauses in some languages, where the clausal proposition receives a marker otherwise reserved for nouns; almost invariably, this marker is a definite article. In the Khwe language just mentioned this is a person-gender-number marker, which is used primarily to signal that a noun or noun phrase is referential or definite (10a), but whose use has also been extended to present relative clauses (10b).

(10) Khwe (Central Khoisan, Khoisan)
    a. |óán   mà
       child M.SG
       'the (male) child'

    b. tcà-   á-   tè   **mà**   ún-   á-   ǁòè   béé.
       be.sick- JUNC- PRES M.SG hunt- JUNC- FUT NEG
       'He who is sick will not go hunting.'

In languages lacking formal means of marking nouns, expansion can simply mean that a complement noun phrase is replaced by a clause, and in such cases it remains unclear whether there is in fact clause expansion, or else integration (5.3). For example, in the Khoisan language !Xun, one way of forming complement clauses after cognition verbs is by simply using a clause instead of a complement noun phrase. Thus, in example (11a), *mí dàbà ǁ'àn* 'my child is sick' has the form of a main clause. That it is a subordinate clause can be seen in the fact that it lacks the topic marker *má*, which is an obligatory part of the main clause, as in (11b).

(11) !Xun (North Khoisan, Khoisan)
    a. mí   m-   é   bhȁlì   mí   dàbà   ǁ'àn.
       1.SG TOP- PAST dream 1.SG child be.sick
       'I was dreaming that my child is sick.'

    b. mí   dàbà   má   ǁ'àn.
       1.SG child TOP be.sick
       'My child is sick.'

Adverbial clauses Expansion from noun to finite clause is particularly common in the rise of adverbial clauses, to which we now turn (see also Section 2.2.6). An incipient stage in the development where an adposition or case marker is used via the extension mechanism to also mark adverbial clauses is provided by the following English example volunteered by Hopper and Traugott (2003: 184–5). They observe that, fairly recently, the prepositional phrases *on the basis (of)* and *in terms of* have come to be used to link clauses together, for example, *They're a general nuisance **in terms of** they harass people trying to enjoy the park*. A fully grammaticalized English example—which we briefly touched upon earlier—can be seen in the development of *for*, which was a preposition of location and purpose in Old English, cf. (12a), but came to be used as a purpose clause subordinator (and a complementizer) by early Middle English, cf. (12b) (van Gelderen 2004: 30).

(12) English (van Gelderen 2004: 30)
   a. Old English (*Beowulf*, 358)
      þæt   he   **for**   eaxlum   gestod
      that he   before   shoulders   stepped
      'that he stood in front of [...].'
   b. Early Middle English (Layamon, Otho, 1113–14)
      Locrin   &   Camber   to   þon   scipen   comen.
      Locrin   and   Camber   to   the   ships   came
      for to   habben   al   þa   æhte.
      for   to   have   all the   goods
      'Locrin and Camber came to the ships to take all the goods.'

Presumably this is an example of a development leading straight from adposition (or case marker) to clause subordination. But perhaps more common is a development as sketched in (13), where there is an intermediate stage where the complement is neither a noun phrase nor a clause but rather a verb having some nominalizing morphology on it, such as an infinitive, participle, or nominalization marker.

(13) Three stages in the development from adposition to clause subordinator
   a. noun phrase   — adposition
   b. nominalized verb   — adposition/subordinator
   c. clause   — subordinator

There is a wealth of data dealing with the process leading from prepositions or postpositions, being heads of noun phrases, to adverbial clause markers, that is, heads of adverbial clauses. The mechanism involved is described by Genetti for Bodic languages of Tibeto-Burman in the following way:

In Tibeto-Burman languages it is common for case postpositions, markers of concrete spatial or social relationship between entities, to develop into adverbial subordinators, markers of temporal and logical relationships between propositions. The syntactic mechanism by which such a development occurs is nominalization, which allows for verbs to be inflected with nominal morphology. (Genetti 1991: 227)

Example (14) illustrates the extension from locative postposition (LOC), cf. (14a), to conditional subordinator (14b) in Classical Tibetan, and Table 5.1 lists the main channels of grammaticalization.

(14) Classical Tibetan (Tibeto-Burman; Genetti 1991: 229)
    a. gyas   **na**   bsgyur.
       right   LOC   turn
       'Turn to the right.'
    b. me   yod   **na**   du-ba   ʼbyung.
       fire   be   if   smoke   become
       'If there is fire, there is smoke.'

TABLE 5.1. *Grammaticalization patterns from postposition to clause subordinator in Bodic languages*

Function as postposition	Function as clause subordinator
Locative	'if/although', 'when/while/after'
Ablative	'when/while/after', 'because', non-final
Allative	purpose
Dative	purpose
Ergative/instrumental	'because', 'when', 'while'

Source: Genetti (1991: 229)

In the Kathmandu dialect of the Tibeto-Burman language Newari, the development of postpositions into subordinators occurred repeatedly over the last several centuries and "the morphosyntactic mechanism by which the development occurred was nominalization, followed by a reanalysis of originally nominal morphology as verbal morphology, via the reanalysis of unmarked deverbal nominals as erstwhile finite verbs" (Genetti 1991: 228).

As we observed above, in the transition from adposition to clause subordinator there tends to be an intermediate stage where the relevant form can be interpreted with reference to both. The following example from Classical Newari exhibits this intermediate stage: (15a) illustrates the postposition stage with a clearly nominal morphology suffixed to a noun; (15b) shows the intermediate stage, where the instrumental morpheme (INSTR) is ambiguous, in that the clause can be analyzed either as a nominal clause marked with a case suffix, or as a finite verb followed by a causal subordinator.

(15) Classical Newari (Tibeto-Burman; Genetti 1991: 246)
  a. thva   deśa-   s   utpatā-   **naM**   khvayāva conaM.
     this   country- LOC   disaster- INSTR   cry   stay
     '(Somebody) was weeping on account of a public disaster.'
  b. āva   chan   daya-   **n**   jin   rājy   li
     now   you   have-   INSTR   I   kingdom   back
     kāya   dhuna.
     take   finish
     'Because you are here I have now won back my kingdom.'

For a similar example from the Omotic language Maale, see Section 2.2.6. Further examples from Laz, a South Caucasian language of the Kartvelian family, are provided by Harris and Campbell (1995: 291–3): the genitive case marker -*ši* and several adpositions are also used to present subordinate clauses. That this situation is the result of a unidirectional process is suggested by the diachronic criterion (5.1): In their nominal uses the genitive marker and the adpositions can be reconstructed back to earlier stages of Kartvelian while this is not possible in the case of their subordinator functions.

While the development from adposition or case marker to subordinator has given rise in many languages to finite subordinate clauses, it was restricted to non-finite subordinate clauses in other languages. In the Jaminjung language of Northern Australia, the TIME case marker -*mindij* may be added to finite (inflecting) verbs to form temporal subordinate clauses, but otherwise the use of the case suffixes (or clitics) of this language, including -*mindij*, is extended only to non-finite clauses. Thus, in (16a), -*mindij* heads a noun whereas in (16b) it heads a non-finite subordinate clause.

(16) Jaminjung (Jaminjungan; Schultze-Berndt 2000: 60, 115)
  a. jalang=guji   na-   ruma-   ny,   buru
     now=FIRST   2.SG-   COME-   PST   return
     na-   jga-   ny   gugu-   **mindij**=na.
     2.SG-   GO-   PST   water-   TIME- NOW
     'You just came now (i.e. recently), you had gone back in the wet season.'
  b. gulban   burrb-   **mindij**   nga-   w-
     ground   finish-   TIME   1.SG-   FUT-
     ijga   buru   Kununurra.
     GO   return   Kununurra
     'After the funeral I'm going back to Kununurra.'

To conclude, there is appropriate evidence showing that adpositions or case markers may give rise to adverbial clause subordinators while it is hard to find examples of an opposite directionality, that is, where a clause subordinator developed into a preposition or postposition. Since adpositions are frequently derived from nouns (or noun phrases), various developments discussed in this section can eventually be traced back to nouns (see Section 5.3.3.2). A paradigm example is provided by the Old English noun phrase [*in stede* + genitive] 'in the place of', which developed into a preposition and later on into the clause connective *instead of* (Traugott 2003).

5.3 Integration

Expansion is a ubiquitous mechanism, but it appears to be crosslinguistically less widespread than integration, whereby two independent sentences, or events, are conflated into one sentence and one of the two turns into a subordinate clause of the other—in accordance with (2b), reprinted here as (17a). We follow Givón (2006) in assuming that the integration of two events tends to entail in particular the features listed in (17b).

(17) Integration
  a. $S_1 + S_2 > S_1[S_2]$
  b. Features commonly found in event integration
    (i) Referential integration: The sharing of referents between the two events.
    (ii) Temporal integration: Simultaneity or direct temporal adjacency of the two events.
    (iii) Spatial integration: The sharing of the same location between the two events.
    (iv) Phonological integration: There tends to be only one intonation contour for the integrated sentence.

Unlike the detailed account of Givón, the present section is restricted to integration as it manifests itself in relative clauses (5.3.1), complement clauses (5.3.2), and adverbial clauses (5.3.3).

5.3.1 *Relative clauses*

In the introduction to this chapter we defined relative clauses as clauses that function as attributive modifiers of nouns (or noun phrases). In the more familiar languages, relative clauses are commonly introduced by relative clause markers or, in short, relative markers or relativizers, for example English *who*, *which*, *that*, etc. But they may as well involve no formal marking, for example *The book I read yesterday* (for a survey of the ways the relative clause construction is marked in the languages of the world, see Comrie and Kuteva 2005). Since our interest is primarily, though not exclusively, with the morphosyntax of clause subordination, we are concerned most of all with relative clauses marked by some kind of morphological expression; accordingly, we will not be able to do justice to other kinds of relative clauses. We will deal primarily with restrictive relative clauses, even if much of what we have to say also applies to non-restrictive relative clauses. Furthermore, we are confined to finite relative clauses. While most languages have them, some languages have only non-finite ones; most modern Lezgian languages, in the same way as Proto-Lezgian, for example, lack finite relative clauses (Harris and Campbell 1995: 311).

5.3.1.1 The demonstrative channel
Presumably the most frequent source of markers introducing (restrictive) relative clauses is provided by demonstrative pronouns.[8] The history of English *that* (Old English *þæt*, *þe*, later replaced by *that*), illustrates one major way in which this process may proceed (see O'Neil 1977; Hopper and Traugott 1993: 190 ff.; van Gelderen 2004: 81 ff. for detailed description). Note that both proximal and distal demonstratives may be used in this process. The process is in accordance with the main mechanisms of grammaticalization: desemanticization leads to a loss of the spatial deixis of the demonstrative, and decategorialization means that the demonstrative pronoun loses its inflectional variability and its freedom to occur on its own as an argument of the matrix clause and is restricted to introducing subordinate clauses, usually in a fixed position vis-à-vis the noun phrase to which it refers.

In Old Swedish, the transition from paratactic to hypotactic forms of clause combining is well documented. It was the masculine form *sā* of the demonstrative pronoun of Old Norse that was reinterpreted as a relative

[8] Frajzyngier (1996: 336–42) claims a development from relative clause marker to demonstrative in some Chadic languages. The evidence presented, however, is such that we do not consider this to be an exception to the generalizations proposed here but rather as one that is in need of a more appropriate analysis.

pronoun (Zeevaert 2006: 20–1). In the following example, there is still ambiguity between the (paratactic) demonstrative and the (hypotactic) use of *sā* as a non-restrictive relativizer:

(18) Old Norse (ca. 800 AD; Zeevaert 2006: 21)
stikuR karþi kubl þau aft auint sunu sin **sa fial austr.**
(i) 'Stig made these monuments after his son Eyvind. **He died in the east'.** *or*
(ii) 'Stig made these monuments after his son Eyvind, **who died in the east.**'

The rise of (restrictive) relative clauses based on this channel typically takes a form as paraphrased in (19), where two juxtaposed sentences, S_1 and S_2, are combined within one sentence and S_2 is reinterpreted as a relative clause,[9] roughly as illustrated in (19). The likely effects of this process are listed in (20). Note that the transition from demonstrative pronoun to relative pronoun is gradual, involving an intermediate stage where the relevant item is ambiguous in that it can be interpreted both with reference to a demonstrative and a clause subordinator; we saw this previously in our example in Section 4.3, another example will be provided in Section 6.4.1.1.

(19) From $[S_1 + S_2]$ juxtaposition to S_1 $[S_2]$ relativization
 a. *There is the car;* **that** *(one) I like.*
 b. *There is the car [that I like].*

(20) Reinterpretation processes in demonstrative-derived relative clauses
 a. The demonstrative pronoun (DP) of S_2 refers anaphorically (or cataphorically) to some participant of S_1.
 b. The DP is grammaticalized to a relative clause marker.
 c. S_2 is grammaticalized to a relative clause.
 d. The grammaticalization of the DP entails desemanticization (e.g. loss of spatial deixis) and decategorialization (e.g. loss of morphosyntactic properties, or of the ability to be used as a demonstrative attribute).
 e. The DP tends to undergo erosion (e.g. loss of ability to receive stress).
 f. The two clauses tend to be united under one intonation contour.

---

[9] Fritz Newmeyer (p.c.) argues that *that* in (19b) should be interpreted as a complementizer rather than a relativizer. For an account in the Minimalist framework, whereby the demonstrative starts out to occupy the specifier of a complement phrase and later, as a relativizer, becomes a head, see van Gelderen (2004: 81 ff.).

With the grammaticalization from demonstrative pronoun to relative clause marker in Old English, the latter could retain the case marking of its head noun, as in the following example, where the relative pronoun *þone* is accusative even though it functions as the subject of the relative clause:

(21) Old English (*Matthew*, 8.5; quoted from van Gelderen 2004: 82)
Ic   wat   wytodlice þæt   ge   secaþ **Þone**   haeland
I   know   truly   that you   seek   that.ACC   savior
Þone   þe   on   rode   ahangen   waes.
that.ACC   that   on   cross   hung   was
'I know that you seek the savior who was crucified.'

Hopper and Traugott (2003: 197–9) discuss a number of additional properties characterizing the English *that*-construction, but these properties are crosslinguistically not generally relevant. For example, the pronominal element in the relative clause standing for the noun phrase that is coreferential with the matrix noun phrase may disappear in the course of grammaticalization, as it did in English, but there are languages where this did not happen even in tightly subordinated relative clauses. Furthermore, there are languages where the newly created relative clause marker has to be placed next to the noun phrase to which it refers, but there are other languages where this is not a requirement. This means that the relative clause marker can occur at quite some distance from its head noun phrase.

The Ik language of northeastern Uganda provides a typical example of this channel: The proximal demonstrative *na*, plural *ni* 'this' has developed into a relative clause marker; example (22) illustrates the former and (23) the latter function. As a demonstrative, it immediately follows the noun it determines, while as a relative clause marker it is placed at the end of the noun phrase, that is, after other modifiers. Thus, in (23c), it follows the numeral 'two'.

Unlike its English counterpart *that*, the Ik item is less grammaticalized in that it has retained its number distinction, while the English item is decategorialized to the extent that it does not show number agreement with its head noun, hence (24a) is possible but not (24b).

(22) Ik (Kuliak, Nilo-Saharan; König 2002)
a. cek-   a   **ná**
   woman-   NOM   DEM.PROX
   'this woman'

    b. kʊrʊuɓáa   ní
       things.NOM   DEM.PROX.PL
       'these things'

(23) Ik (Kuliak, Nilo-Saharan; König 2002)
    a. cek-   a   **na**   wicé-   á   bɛɖᵃ   bíraa
       woman-   NOM   REL.SG children- ACC   like   be.not
       nɛɛ́   nᵃ.
       here.DAT   DEM
       'The woman whom the children like is not here.'
    b. ena   **na**   kʊrʊɓaɖi- a   **ní**   ɪlíɓ-   atᵃ.
       see   ENC   things-   ACC   REL.PL be.green- 3.PL
       'He sees things which are green.'
    c. cɛm-   í-   a   bɛɖ-   ɛ̀s-   o   de-
       fight-   1.SG-   *a*   want-   INF- ABL   foot-
       ɪk-   ɛ   léɓetse   **ní**   ze-   ikᵃ.
       PL-   GEN   two.OBL   REL.PL big-   PL
       'I am looking for two big feet.' (Lit.: 'I am wanting two big feet.')

(24) English
    a. The bottle **that** you bought yesterday is empty.
    b. *The bottles **those** you bought yesterday are empty.

Another example is provided by Classical Chinese. The item *zhi* was used exclusively as a demonstrative pronoun in the inscriptions of tortoise shells of the Shang Dynasty (c. sixteenth–eleventh century BC). Later on it was grammaticalized to a marker of relative clauses and associate and genitive phrases. The non-grammaticalized and the grammaticalized uses co-existed for several centuries (Shi and Li 2002: 6). But *zhi* is not the only Chinese demonstrative to have undergone a grammaticalization from demonstrative to relative clause marker.[10] Another example is provided by *di*. Originally a demonstrative pronoun, it was first grammaticalized to mark relative clauses in the ninth century AD, undergoing erosion by losing its tone and changing to *de*. Shi and Li (2002: 13) describe the process as one of reanalysis, whereby the structure (25a) was reinterpreted as (25b), with *di/de* changing from forming a constituent with the following noun to forming a new constituent with the preceding relative clause.

---

[10] We are ignoring here that, in addition to relative clause marking, the demonstratives also assumed other functions such as marking genitive phrase, associate phrases, etc.

(25) The grammaticalization of *di/de* in Chinese (Shi and Li 2002: 13)
 a. [relative clause + [$di_{\text{demonstrative}}$ + noun]]
 b. [[relative clause-*de*]   + noun]

To conclude, wherever there is historical evidence it shows that there is a unidirectional development from demonstrative to relative marker; conversely, we are not aware of any language where a relative clause marker has given rise to a demonstrative. And the process has the salient characteristics of grammaticalization, such as the bleaching out of demonstrative semantics (desemanticization) and loss of categorial properties characteristic of demonstratives.

5.3.1.2 The interrogative channel

The development from question word to relative marker (e.g. English *who, which*) is most centrally found in European languages in particular and Indo-European languages in general (Heine and Kuteva 2005). In addition, it is also found in a number of languages having a history of contact with Indo-European languages; see Heine and Kuteva (2006) for details. However, it is not restricted to these languages, as the following example may show (see also Harris and Campbell 1995: 285 ff.).

In the Chadic language Gidar, one of the two relative markers, *án* (26b), is identical with the interrogative pronoun for 'what?' (26a), and we hypothesize that this is an instance of the interrogative-to-relative channel.

(26) Gidar (Chadic, Afroasiatic; Frajzyngier 1996: 450)
 a. tìzí   à   ttókə́- k   án   nə̀   lbá-hə́.
    Tizi   3.M   ask-   PERF   what   1.SG buy
    'Tizi asked what I bought' (i.e. he asked the question).
 b. ɗə́f   án   ná   gə̀mə́- t
    man   REL   FUT   take-   3.F
    'man who will choose her'

As we will see below, question words do not only give rise to relative markers, they may as well develop into complementizers (see Harris and Campbell 1995: 298).

5.3.2 *Complement clauses*

We have defined complement clauses in Section 5.1 as subordinate clauses that are complements (or arguments) of another clause (the main or matrix clause). Complement clauses are typically though not necessarily introduced by a morphological expression called the complementizer. This term has been used in a number of works in an extended sense, referring to various types of clause subordinators; in the present work, we restrict its use to complement clauses. There is a large variety in the range of different complement clause structures in the languages of the world (see, e.g., Noonan 1985) and we will not be able to do justice to this variety in this section. What we want to achieve is to provide an outline sketch of the evolution of some common types of clause complementation arising via integration.

5.3.2.1 Introduction
As we saw in Section 5.2, expansion constitutes an important mechanism for developing complement clauses; although, integration of two independent clauses appears to be more common.

5.3.2.2 The noun channel
One major source for complementizers consists of some generic noun (N) having argument status in the main clause and at the same time forming the head noun of a relative clause. The grammaticalization process involved changes structure (27a) into one like (27b), whereby the noun (N) plus the relative clause marker (REL) jointly give rise to a complementizer and the relative clause turns into a complement, or free relative[11] clause (the English examples are meant to illustrate the nature of the process concerned).

(27) Generic Noun—REL developing into headless relative/complement clause
  a. *I don't know the thing [which he wants]*
  b. *I don't know [**what** he wants]*

Nouns serving as complements normally have a generic meaning, standing for ontological categories such as PERSON, THING, PLACE, TIME, or MANNER; Table 5.2 lists the kinds of nouns most commonly used cross-linguistically. There is some variation in the way this process affects the structure of the emerging complementizer. The following are the most common developments:

---

[11] We follow Fritz Newmeyer (p.c.) in assuming that (27b) is an instance of a free relative, rather than a headless relative clause.

TABLE 5.2. *Nouns commonly used as complements in the rise of complement clauses*

Meaning of noun	Meaning of the resulting complement clause
'person'	Subject, object clause
'thing', 'matter'	Subject, object clause
'place'	Locative clause
'time'	Temporal clause
'manner', 'kind', 'way'	Manner clause

(28) The structure of noun-REL derived complementizers
   a. The noun and the relative clause marker merge into a complementizer (merger)
   b. The relative clause marker is lost (REL-loss)
   c. The noun stem is lost (N-loss)

This process differs from the other processes discussed in this chapter in one fundamental way: It implies that there is already clause subordination, and that one type of subordination (relativization involving generic nouns as heads) is "exapted" for another type (complementation).

The Ewe language illustrates the use of merger (29): *amé-si* (person-REL) 'who(m)', *nú-si* (thing-REL) 'what', *afi-si* (place-REL) 'where', *ɣe-si-ɣi* (time-REL) 'when', for example:

(29) Ewe (Kwa, Niger-Congo; Westermann 1930: 148)
   nye-   mé-   nyá   amé-si   tsɔ́-   e   o.
   I-   NEG-   know   who   take-   it   NEG
   'I don't know who has taken it.'

The rise of new complementizers can be demonstrated with the following examples from Ik, a language of northeastern Uganda having verb-initial (VSO) basic word order and an elaborate case system distinguishing seven case categories. What makes Ik an interesting case for our purposes is that it exhibits a typologically unusual property: It has an extremely conservative case system which is retained even if a noun is grammaticalized into some other word category. The result is that in this language there are case-marked adverbs, adpositions, adverbial conjunctions, and complementizers (König 2002). This fact allows us to reconstruct a situation where all the items that are inflected for case are historically derived from nouns. For example, *mɛna* is a plural noun meaning 'matter', taking plural agreement. Sentence (30a) illustrates its nominal and (30b) its complementizer use, where *mɛna* is followed by a complement clause. Note that the object noun is inflected in the nominative case (NOM): There is a rule in Ik according to which objects are coded in the nominative when the subject has first- or second-person reference. Consequently, the object appears in the accusative in (30c) since the subject has third-person reference. One might argue that *mɛna* is not a complementizer but a noun in all these sentences. But this analysis would raise a problem: When a noun is followed by a clause, that clause must be introduced by the relative clause marker *na*, plural *ni*. This is in fact the case in (30b) and (30c); however, the relative marker can be omitted (optional REL-loss) whenever *mɛna* functions as a complementizer, as in (30d), but never when it is used as a noun. In other words, *mɛna* has the function of a complementizer but, with one exception, the morphosyntax of a noun.

(30) Ik (Kuliak, Nilo-Saharan; König 2002)
 a. ńtá ye- í- í mɛn- a.
    NEG know- 1.SG- NEG matter- NOM
    'I don't know the matter.'
 b. ńtá ye- í- í mɛn- a ni
    NEG know- 1.SG- NEG matter- NOM REL.PL
    tód- at[a].
    say- 3.PL
    'I don't know what they say.'
 c. ńtá ye- í mɛn- á ni tód- at[a].
    NEG know- NEG matter- ACC REL.PL say- 3.PL
    'He doesn't know what they say.'
 d. ńtá ye- í mɛn- á tód- at[a].
    NEG know- NEG matter- ACC say- 3.PL
    'He doesn't know what they say.'

The items *kɔrɔɓáa* 'thing' and *na* 'place' behave exactly like *mɛna*: They are nouns but at the same time also complementizers, as can be seen in (31a). But there is one difference: Unlike *mɛna*, they can occur sentence-initially when functioning as complementizers, and in this position they are not inflected for case, cf. (31b).

(31) Ik (Kuliak, Nilo-Saharan; König 2002)
 a. ńtá ye- at- í kɔrɔɓádí- a itiy- at[a].
    NEG know- 7.PL- NEG thing- ACC do- 3.PL
    'They don't know what they do.'

b. kɔrɔɓáa   ítíya-   ídᵃ   ńtá   ye-   í-   í.
   what   do-   2.SG   NEG   know-   1.SG-   NEG
   'What you are doing I don't know.'

These properties are shared by the remaining items *tʊmɛda na* and *tóimɛn*, with one exception: The latter have no more nominal meaning, that is, they are exclusively complementizers. But they are still inflected for case. Thus, the verb *en-* 'see' in (32a) requires a complement in the accusative (ACC), whereas the verb *Itɛt-* 'reach, notice' in (32b) must have a complement in the dative case (DAT).

(32) Ik (Kuliak, Nilo-Saharan; König 2002)
   a. na   kɔnto.óɗowi   én-   ío   oɲor-   a
      when   one.day   see-   NAR   elephant-   NOM
      toimɛní-   a [...].
      that-   ACC
      'When one day the elephant saw that [...].'
   b. ɪtɛ́t-   ía   ná   **tóimɛní-**   kᵉ   ńɟa
      notice-   1.SG   ENC.SG   that-   DAT   eat
      nyɛ́ɟ-   a   bi-   kᵃ
      hunger-   NOM   you-   ACC.
      'I noticed that you felt hungry.' (Lit: 'hunger eats you')

We observed above that the relative clause marker *na* (in the case of *mɛna* it is the plural marker *ni*) is used optionally with *mɛna*, *kɔrɔɓáa*, and *na*, that is, it can be used or omitted with no difference in meaning. In this respect, *tʊ́mɛda na* and *tóimɛn* show a different behavior: With the former, *na* is obligatory, it has become a frozen part of the complementizer (merger), while *tóimɛn* cannot take *na*, that is, the relativizer has been lost (REL-loss).

Table 5.3 provides a list of the five complementizers and some of their properties. On the one hand, the situation of these complementizers is typologically unusual; it is probably hard to find other languages where functional items such as clause subordinators consistently take case inflections. On the other hand, it illustrates in a nutshell the evolution of one salient type of complementizers, namely that derived from nouns, which allows for the following generalizations:

(a) There are nouns (or noun phrases) pressed into service as functional categories to code complement clauses (or headless relative clauses).

(b) In this process the nouns are desemanticized, losing their lexical meaning, and decategorialized, losing most of their nominal properties, such as the ability to take plural markers or modifiers.

(c) However, they tend to retain some properties bearing witness to their lexical origin. In the present case, such properties include the ability to be case-inflected, and the optional use (in the case of *tűmɛda na* the obligatory use) of the relative clause marker.

(d) As a result of the process from noun to complementizer, nominal properties are extended to also become properties of subordinate clauses. For example, case inflections are crosslinguistically a characteristic of nouns (in some languages also of noun phrases), but in Ik they have become part of clause subordination as well (see König 2002 for details).

(e) Clause subordinators in a given language are frequently not a homogeneous class, some are more strongly grammaticalized, others less so. Thus, Table 5.3 suggests that the Ik complementizers can be arranged along a cline of grammaticalization: At one end there is *mɛna*, which largely still behaves like a noun, and many of its uses can be interpreted with reference to its nominal meaning. At the other end there is *tóimɛn*, which has lost most of its associations with nouns, having no function other than introducing complement clauses.

As a result of decategorialization and erosion, complementizers tend to become increasingly dissimilar from their lexical sources, losing both

TABLE 5.3. *Complementizers in Ik*

Complementizer	Functional domain	Gloss	Inflected for case	Nominal meaning	Relative clause marker required	Cannot occur sentence-initially
mɛna (ni)	Object	'what'	+	'matter'	+/−	+
kɔrɔɓáa (na)	Object	'what'	+	'thing'	+/−	−
na (na)	Space	'where'	+	'place'	+/−	−
tűmɛda na	Space	'where'	+	−	+	−
tóimɛn	Object	'that'	+	−	−	−

Source: Based on König (2002).

nominal properties and phonetic substance. The following example (33) from the !Xun language of southwestern Africa illustrates the grammaticalized output of the process: The complementizer *tcé* is the compressed version of a noun phrase, consisting diachronically of the noun *tcí* 'thing' and the relative clause marker *-è*, hence *tcí-è > tcé*. Thus, (33) means historically 'I don't know the thing that he wants'.

(33) !Xun, W2 dialect (North Khoisan, Khoisan)
mí   má |ōā   ǃǃèhī   **tcé**   hà   kàlè.
1.SG TOP   NEG   know   what   N1   want
'I don't know what he wants.'

Conceivably, some languages have experienced a process straight from relative marker to complementizer (Lehmann 1995: 1213–14; Heine and Kuteva 2002a). For example, the relative marker *xa=* of the Oto-Manguean language Chalcatongo Mixtec is said to have given rise to a complementizer (Macaulay 1996: 153, 160), and the same is claimed for the Thai relative marker *thîi* (Bisang 1998: 780) and the relative pronoun *she/asher* of Early Biblical Hebrew (Cristofaro 1998: 64–5). Note, however, that—as pointed out already in Heine and Kuteva (2002a: 254)—more research is needed on the structure, and the genetic and areal distribution of this development. In other languages again there is no evidence of relativization; the following examples suggest that the complement clause is added to the complement noun without any formal marking. The result is that the complement noun is grammaticalized to a complementizing pronoun; thus, instead of (27) we have (34).[12] The following examples, both involving verb-final languages, illustrate this channel; note that in these languages the modifier precedes its head. In the Namibian Khoisan language Nama, the noun *!xái-s* (*!xái-sà* oblique case) 'matter, story' has given rise to the object clause complementizing pronoun *!xái-'è, !xái-sà* 'that', 'whether', and, as such, is still inflected like a noun, cf. (35).

(34) From complement noun to complementizer
 a. [S₁ N] [S₂]   e.g. *[I don't know the thing] [he wants (it)]*
 b. S₁ [CPL + S₂]   e.g. *I don't know [**what** he wants]*

(35) Nama (Central Khoisan; Krönlein 1889: 206; Hagman 1977: 138)
tííta   ke   kè |'úú   'íí   !úū-   ts
1.SG   TOP   PAST not.know   PAST   go-   2.SG.M

---

[12] Fritz Newmeyer (p.c.) maintains that *what* in (34b) is a pronoun rather than a complementizer, being the thematic object of *want*.

	ta	!xái-	sà.
	IMPFV	COMP-	3.SG.M

'I didn't know that you were going.'

In a similar fashion, the Japanese nominalizer or complementizer *koto* has the etymological meaning 'thing' (Lehmann 1982: 65); the following example illustrates its use as a complementizer. Like Nama, Japanese is an SOV language, hence the complementizer follows its complement clause.

(36) Japanese (Kuno 1973; quoted from Lehmann 1982: 65)

Ano	hito	ga/no	hon	o	kai-	ta	**koto**
that	person	NOM/GEN	book	ACC	write-	PART	NOMIN
ga	yoku	sirarete	iru.				
NOM	well	known	is				

'That that person has written a book is well known.'

5.3.2.3 The verb channel

There are two main kinds of verbs that crosslinguistically provide conceptual sources for complementizers. The most common kind is provided by speech act verbs meaning 'say', the second one being similative verbs meaning 'be like', 'be equal', or 'resemble'. We will look at both sources in turn (for further examples, see Heine and Kuteva 2002a: 257–8, 261–5, 273–5).

On the basis of crosslinguistic evidence, a grammaticalization channel typically involving the following main stages has been proposed (see e.g. Lord 1976; Heine, Claudi, and Hünnemeyer 1991: 158; Ebert 1991: 87; Frajzyngier 1996; Klamer 2000; Crass 2002; see also Section 6.4):

(37) Main stages in the evolution from verb for 'say' to clause subordinator[13]

    a. Speech act verb 'say'
    b. 'Say' as a quotative marker
    c. Complementizer of object clauses headed by speech-act, perception (e.g. 'see', 'hear'), and cognition verbs (e.g. 'know', 'believe')
    d. Complementizer of subject clauses
    e. Subordinator of purpose clauses
    f. Subordinator of cause clauses[14]

---

[13] We are leaving out further specifications of this grammaticalization chain, such as the development of conditional clauses. Note further that there is some evidence that (37d) does not necessarily precede (37e).

[14] One may wonder what the conceptual process is that leads from purpose to cause clause. The answer is complex; in a nutshell this process is of the following kind: Purpose

There is a range of different morphological forms that the verb 'say' may take in this grammaticalization; in one language it is used in some non-finite (infinitival or participial) form, in another it is inflected for person, tense, aspect, etc., or it is used without any morphological trappings. In the course of grammaticalization, the verb form loses verbal properties, gradually being reduced to an invariable conjunction (decategorialization) and losing phonetic substance (erosion).

The following examples from Ewe illustrate the process. Sentence (38a) shows the lexical use as a speech act verb, nowadays highly restricted in the contexts in which it can occur, while (38b) shows the quotative use preceding direct-speech sentences and (38c) the object complementizer use after a cognition verb.

(38) Ewe (Kwa, Niger-Congo)
    a. e   bé?
       2.SG say
       'What are you saying? (Did I understand you correctly?)'

    b. é   gblɔ bé: "ma- á-   vá   etsɔ."
       3.SG say   bé   1.SG- FUT come tomorrow
       'He said: "I'll come tomorrow."'

    c. me-   nyá   bé   e-   li.
       1.SG- know bé   2.SG- exist
       'I know that you are there.'

Similarly, the Austronesian language Buru of Indonesia illustrates this grammaticalization channel (Klamer 2000: 75 ff.). The item *fen(e)* is used as a lexical verb meaning 'think, say, affirm', it also occurs as a quotative marker (39a), and as a complementizer after speech act verbs (39b) and mental and physical perception verbs (39c). After speech act verbs, the use of *fen(e)* is optional. There is no evidence that *fen(e)* has reached the stage of marking subject complement clauses.

---

clauses entail cause while the opposite does not hold. Thus, purpose clauses, such as (a), can be paraphrased roughly by cause clauses, as in (b), but cause clauses (c) can in many cases not be paraphrased by purpose clauses (d).

    (a) John quit the job in order to concentrate on his exams.
    (b) John quit the job because he wanted to concentrate on his exams.
    (c) John caught a cold because he did not wear a coat.
    (d) *John caught a cold in order not to wear a coat.

What appears to happen in this grammaticalization process is that desemanticization has the effect that the purpose meaning is bleached out, with the result that a cause meaning arises.

(39) Buru (Central Malayo-Polynesian, Austronesian; Klamer 2000: 76–8)
  a. Da   prepa **fen,** "Sira   rua   kaduk".
     3.SG say   FEN   3.PL   two   arrive
     'She said, "The two of them came".'
  b. Da   prepa **fen**   sira   rua   kaduk.
     3.SG say   FEN   3.PL   two   arrive
     'She said that the two of them came.'
  c. Ya   tewa   **fen**   ringe   iko   haik.
     1.SG know   FEN   3.SG   go   PFV
     'I know that he has already left.'

This development does not always lead all the way to a complementizer; in a number of languages it has not proceeded beyond the quotative stage. This appears to be the case for example in Georgian (see Harris and Campbell 1995: 168–9 for more details): By the tenth or the eleventh century, the quotative particles *metki* and *tko* (with or without complementizer) were introduced, and these are historically derived from the verb forms *me v-tkv-i* 'I said (it)' and *tkva* 'may you say (it)', respectively. In spite of their fairly old age, the two do not appear to have reached the stage of complementizers. However, they exhibit the salient features of grammaticalization: Their lexical meaning appears to have bleached out (desemanticization), they have undergone decategorialization, having lost the ability to distinguish tense, aspect, and modality, being invariable particles, they can no longer stand alone, and they have also undergone erosion, being reduced from *me v-tkv-i* to *metki* and from *tkva* to *tko*. Finally, they have reached the final stage of context-induced reinterpretation in that they have been conventionalized to the extent that the erstwhile verb can co-occur with the new verb for 'say':

(40) Georgian (South Caucasian; Harris and Campbell 1995: 170)
     me   vambob,   mivdivar-   **metki.**
     I   I.say.it.PRES   I.go-   QUOT
     'I am saying: "I am leaving".'

The more grammaticalized an item is, the more likely it is that it will lose its original meaning and come to be used exclusively as a functional category. This appears to have happened in Tukang Besi, a language closely related to Buru: In Tukang Besi, the same grammaticalization chain is found, with one exception: While the item *kua* of Tukang Besi also functions both as a quotative marker and complementizer, it has no lexical uses as a speech act verb 'say'. That *kua* has lost these uses is suggested by the fact that in the related neigbhoring language Duri, *kua* is used not only as a quotative marker but also as a verb for 'say' (Klamer 2000: 85). In sum, the development of these three Malayo-Polynesian languages can be sketched as in Table 5.4, suggesting that Duri represents the least and Tukang Besi the most strongly grammaticalized stage; there are no data on whether any of the languages has acquired the stage of a subject complementizer.

Similar examples can be found in many other languages. The Khoisan language !Xun of southwestern Africa uses the complementizer *tēē-kōē* when the main verb is a speech-act or cognition verb. This form is historically derived from the phrase *tà kē kwèé* (and PAST say), frequently reduced to *tà-ē kwèé* 'and said'. Thus, example (41a) means historically 'I thought and said: she is at home'. That !Xun speakers may still to some extent be aware of the genesis of the construction is suggested by the fact that they tend to omit the complementizer when the main verb is *kwèé* (or *kòé*) 'to say', as in (41b). Furthermore, the process has not proceeded to the stage where the complementizer marks subject clauses. Thus, the situation is similar to that of Buru (see above).

(41) !Xun, W2 dialect (North Khoisan, Khoisan)
    a. mí   m-   ē   !!ʼŋ́   **tēē-kōē**   gè-   ā   n!āō.
       1.SG TOP- PAST think COMPL COP- T house
       'I thought that she is at home.'
    b. hà   m-   ē   kwèé   (**tēē-kōē**)   hà   ā   g|è.
       N1 TOP- PAST say COMPL N1 FUT come
       'He said that he would come.'

TABLE 5.4. *The development from 'say' to complementizer in three Malayo-Polynesian languages*

Language	Verb 'say'	Quotative	Complementizer
Duri	+	+	−
Buru	+	+	+
Tukang Besi	−	+	+

Source: Based on Klamer (2000).

In Ancient Egyptian, the form *r dd* '(in order) to say' seems to have given rise to a general complementizer (Deutscher 2000: 90; Heine and Kuteva 2002a: 261–2), and Deutscher (2000: 66–90) argues that a similar process happened in the Semitic language Old Accadian (c. 2500–2000 BC) of ancient Mesopotamia, where the expression *enma*, which is hypothesized to be derived from a verb of saying, provided the source for the complementizer *umma* of Neo-Babylonian (c. 1000–500 BC). Another diachronically attested example of the 'say'-to-complementizer channel leading all the way to subject complementizer will be discussed in Chapter 6 (Section 6.4.1.2).

While presumably less common than speech act verbs, similative verbs meaning ('be like', 'be equal', 'resemble') also provide a source for complementizers (Güldemann 2001). The exact nature of this evolution is not entirely clear; from the data available it seems that the process typically leads from the similative via quotative to complementizer uses. The following are a few examples of this grammaticalization. In the Nigerian language Idoma, the verb *bē* 'resemble' has given rise to a complementizer after verbs of thinking, seeing, knowing, and hearing, cf. (42). Similarly, in the Ghanaian language Twi, there is a verb *sɛ* 'resemble', 'be like', 'be equal', that appears to have provided the source for the complementizer *sɛ* (Lord 1993: 160).

(42) Idoma (Volta-Congo, Niger-Congo; Lord 1993: 330, 1993: 200)
    ṇ   je   b̧-   o   ge   wa̧.
    1.SG know   resemble- he   FUT   come
    'I know that he'll come.'

In addition to verbs, it may be similative adverbs and other markers ('like', 'thus') that may be recruited for grammaticalization. The English-based pidgin/creole Tok Pisin provides an example, where the particle *olsem* 'thus, like' became a complementizer (43); cf. also English *like*, which serves as a non-verbatim quotative (e.g. *And I'm like: "Gimme a break, will you!"*; Romaine and Lange 1991).

(43) Tok Pisin (English-based pidgin/creole; Woolford 1979: 116, 118)
    Na   yupela   i   no   save   **olsem**   em   i   matmat?
    and   you:PL   i   NEG   know   that   it   i   cemetery
    'And you did not know that it was a cemetery?'

5.3.2.4 The demonstrative channel
As in the noun channel, the strategy underlying this channel involves a complement of the main clause which is grammaticalized to a complementizer. The complement in this case is a demonstrative pronoun (not a demonstrative

attribute), which is either distal ('that') or proximal ('this'). Accordingly, a structure underlying something like sentence (44a) is grammaticalized into (44b). The main effects of grammaticalization are summarized in (45).

(44) The reinterpretation of demonstrative pronoun as complementizer
    a. [S$_1$ + DEM] [S$_2$]   e.g. *I understand that: [He will come].*
    b. S$_1$ [CPL + S$_2$]   e.g. *I understand [that he will come].*

(45) From bi-sentential structure to clause complementation
    a. S$_2$ is reinterpreted as a complement clause.
    b. The demonstrative object argument (DEM) of S$_1$ is reinterpreted as a complementizer (CPL).
    c. The complementizer moves from S$_1$ to S$_2$.
    d. S$_2$ loses its own intonation contour; there is now only one sentential intonation contour.
    e. The complementizer loses categorial properties that it had as a demonstrative.

This process has been described in some detail for Germanic languages such as English, German, and Faroese (Lockwood 1968: 222–3; Heine, Claudi, and Hünnemeyer 1991: 180; Hopper and Traugott 1993: 185–9; Harris and Campbell 1995: 287–8; Heine and Kuteva 2002a). An example from the Australian language Gunwinggu is discussed by Hopper and Traugott (2003: 185). The following case is taken from the !Xun language of southwestern Africa; cf. (46): The use of the proximal demonstrative pronoun *kā ē* 'this' (referring to nouns of the inanimate noun class 4 [N4]) is exemplified in (46a). In (46b) there are two clauses, where the second clause is interpreted as a complement of the first clause, and the demonstrative pronoun *kā ē* as a complementizer introducing the second clause, even if it could also be interpreted in its non-grammaticalized sense, where (46b) could be translated roughly as 'Ask him this: he'll be at home tomorrow'.

(46) !Xun, W2 dialect (North Khoisan, Khoisan)
    a. mí   má   hŋ́   !ȁhŋ́   kā   ē.
       I   TOP see tree   N4 PR
       'I see this particular tree.'
    b. ȉín   hà   **kā-ē**   hà   ge-   ā khōmē   kē
       ask   N1   COMPL N1 COP- T tomorrow TR
       n!āō   khùyā.
       house at
       'Ask him if he'll be at home tomorrow.'

The development from the demonstrative *kā ē* to the complementizer *kā-ē* involved all of the parameters of grammaticalization (see Chapter 1): Extension had the effect that the form was used in a new kind of context, namely at the beginning of a clause, desemanticization led to loss of spatial deixis, decategorialization meant that the form no longer participates in the variable paradigm of demonstrative categories, and erosion had the effect that the glottal stop and the contour tone (high–low) of the demonstrative [kāʔè] was lost in the complementizer, which is [kāē], or even [kē].

In the early stages of development, the grammaticalization from demonstrative pronoun to complementizer tends to be confined to object complements, and in many languages the process has stopped at this point. But in other languages, the process has proceeded further, in that the use of the complementizer was extended to subject complements. For example, the English distal demonstrative *that* acquired uses as an object complementizer already in Old English, while its use as a subject complementizer in subject position did not evolve prior to the fourteenth century (Hopper and Traugott 2003: 194).[15]

Diachronic evidence for this evolution comes from languages where we have sufficient historical records, such as English, German, etc.: Items such as English *that* were first used as demonstratives before their use was extended to also serve as complementizers. Germanic languages abound with examples showing that the process from demonstrative pronoun in the matrix clause to conjunction introducing complement clauses is documented by means of historical evidence. An example from Faroese is discussed by Lockwood (1968: 222–3; see also Heine, Claudi, and Hünnemeyer 1991: 180); the development of the English complementizer *that* is discussed by Hopper and Traugott (2003: 190–4), and Harris and Campbell (1995: 287f.) analyze the development of German *das/dass* 'that'. Conversely, we are not aware of any data that would suggest that, in the history of English or of any other languages, complementizers developed into demonstratives.

5.3.2.5 The interrogative channel
Question words do not only develop into relative clause markers; they may in the same way give rise to markers introducing complement clauses, that

---

[15] There is an example of *that* from ca. 1000 AD where it appears to be a subject complementizer, though not in subject position (Hopper and Traugott 2003: 192, ex. (41)).

is, complementizers. European languages have plentiful examples of question words (QW) meaning 'what?', 'who?', 'when?', 'where?', 'why?', etc. being grammaticalized to devices for presenting complement clauses. The process appears to have involved clause integration of the following kind: Two clauses, where one is an interrogative one (47a), are conflated into one sentence, where the question word assumes the function of a complementizing pronoun and the interrogative clause into a complement clause, as sketched in (47b).[16]

(47) From question word to complementizer
    a. [S$_1$] [QW + S$_2$?]   e.g. *[I don't know:] [What does he want?]*
    b. [S$_1$ [CPL + S$_2$]]   e.g. *[I don't know [**what** he wants]]*

The Russian interrogative pronoun *čtó* 'what?' has given rise to both a relative pronoun and a complementizer. Example (48a) illustrates the interrogative and (48b) the complementizer use. That the latter is a grammaticalized form of the former is suggested by the fact that it is contextually restricted (decategorialization) and that it has lost the ability to be stressed (erosion), which is a characteristic of *čto* as a pronoun (Noonan 1985: 47).

(48) Russian (Noonan 1985: 47)
    a. Čtó   ty   čital?
       what you read
       'What were you reading?'
    b. Ja ne   znal,   **čto**   ty   čital.
       I   NEG know.PAST COMPL you read
       'I din't know that you were reading.'

While common in Indo-European languages, this channel is also found in some other languages. One example is provided by Georgian, where the complementizer *ray-ta-mca* 'that' is derived from the question word *ray* 'what?' (Harris and Campbell 1995: 298):

(49) Georgian (South Caucasian; Harris and Campbell 1995: 298)
    da   ara   unda,   **raytamca**   icna   vin.
    and not he.want that   he.know someone
    'And he did not want that anyone know.'

Another example comes from Mandarin Chinese, where question words appear to have grammaticalized to pronouns introducing complement

---

[16] So far, no historical data have been found to corroborate this reconstruction.

clauses. For example, in (50a), *shénme* 'what?' is an interrogative pronoun while in (50b) it is an object complementizer.

(50) Mandarin Chinese (Li and Thompson 1981: 522, 554)
   a. nǐnmen   zuò   shénme?
      you   do   what
      'What are you doing?'
   b. nǐ   kàn   tā   xiě   shénme.
      you   see   3.SG   write   what
      'You see what s/he writes.'

Finally, we may cite an example from the North Khoisan language !Xun, where the question word *m̃tcē...-á* 'why?', illustrated in (51a), appears to have been grammaticalized to a reason complementizer; note that the complementizer in (51b) has retained the question marker *á* (Q).

(51) !Xun, W2 dialect
   a. m̃tcē   á   hà̰   kē   g|è-   ā.
      why   Q   N1   PAST   come-   T
      'Why did he come?'
   b. mí   má |ōā   !!èhī   m̥tce̱   á   hà̰   ‖āē-   ā.
      1.SG   TOP   NEG   know   why   Q   N1   die.SG-   T
      'I don't know why he died.'

5.3.3 Adverbial clauses

In the introduction to this chapter we defined adverbial clauses as clauses that have adverbial functions, being modifiers or adjuncts of matrix clauses or predicates. Unlike complement clauses, adverbial clauses are not dependent on valence, that is, on the argument structure of the verb. Adverbial functions typically concern information on where, when, why, if, for what purpose, etc. a given action takes place or a state obtains. And adverbial clauses need not, and frequently do not exhibit the same degree of discourse-pragmatic and phonetic clause cohesion that other subordinate clause types do (Talmy Givón, p.c.).

5.3.3.1 Introduction

Hopper and Traugott (2003: 183) say that "adverbial clauses actually arose out of the reanalysis of adverbial phrases as adverbial clauses". With this statement they identified one important channel of grammaticalization, namely expansion (5.2), but, as we will see below, this does not take care of the entire range of conceptual richness that contributes to the emergence of adverbial clause structures.

5.3.3.2 The noun channel

In a way similar to that for complementizers, general locative and temporal nouns such as 'place', 'interior', 'time' provide a common source for adverbial subordinators. In the Namibian language Khwe (Kxoe), óó is a noun meaning 'inside, place' which appears to have been grammaticalized to a locative postposition ǀo 'at' and to a clause-final subordinator (SUB) of temporal, causal, and modal clauses, for example:

(52) Khwe (Kxoe; Central Khoisan, Khoisan)

tíú	pòo	yaá	xàm̀	ún-	á-	xu-	a-
then	jackal	come	lion	hunt-	JUNC-	TERM-	JUNC-

ta 'ò.
PAST SUB

'Then the jackal came, when the lion had left for hunting.'

In the Ugandan language Ik, the form aɠwéde is composed of the relational noun aɠw^a 'inside' plus the third person possessive suffix -éde, hence its lexical meaning is 'its inside', while its grammaticalized function is that of a reason conjunction 'because', introducing reason clauses which obligatorily follow the main clause, cf. (53).

(53) Ik (Kuliak, Nilo-Saharan; König 2002)

ɗaá	be	ɛɗá	morotoó-	k^e	barats-	⁰
go	ENC	alone	Moroto-	DAT	morning-	ABL
aɠwéde	en-	í	aka	wik-	^a.	
because	see-	1.SG	PERF	children-	NOM	

'He went alone to Moroto in the morning because I have seen the children.'

We observed above (Section 5.3.2.2) that the Ik noun na 'place' was grammaticalized to a locative complementizer, na 'where', which has retained a number of nominal properties: Like its lexical source, it has a locative function, takes the relative clause marker na, even if the use of this marker is optional, and it is case-inflected. As an adverbial clause conjunction it is more strongly grammaticalized: Having either a temporal or conditional function, it has lost the ability to take a relative clause marker and to be inflected for case, cf. (54). Like the complementizer, it can follow or precede the main clause.

(54) Ik (Kuliak, Nilo-Saharan; König 2002)
dét-	an-	a	karáts-	ik-	a	ntí-	kᵉ	**na**
bring-	IPS-	*a*	stool-	PL-	NOM	they-	DAT	when
ats-	áta	dᵉ.						
come-	3.PL	DP						

'Stools were brought for them when they came.'

Temporal subordinate clauses are frequently formed via the grammaticalization of nouns for 'time'. For example, the subordinator most commonly used to explicitly signal a temporal relationship between clauses both in Classical Newari and the Kathmandu dialect of Newari is *belas* (in the modern dialect *bale*), which is formed from a noun meaning 'time' (Genetti 1991: 235). Much the same observation has been made in other languages. In his detailed study of Chadic languages, Frajzyngier (1996: 313) observes: "A source for 'when' expressions in Chadic that [is] one of the easiest to identify is the word for 'time'." This grammaticalization involves either a bare noun for 'time' or 'place', or the noun plus an adposition (55), or a relative clause marker (56).

(55) Giziga (Chadic, Afroasiatic; Frajzyngier 1996: 315)
ta.**pas**	ye	m-	ro	Dlaɓu	'i	pur
with.daytime	1.SG	?-	go	Dogba	1.SG	see
dərleŋge	le.					
leopard	PERF					

'When I went to Dogba, I saw a leopard.'

(56) East Lele (East Chadic, Afroasiatic; Frajzyngier 1996: 313)
kūr	gō	tùwà	ánỳ	déná	sé	kīníiři-	gé.
time	REL	fire	take	then	INCE	run.PL-	3.PL

'When the fire started, they ran away.'

Nouns for 'thing', 'matter' provide a common source for complementizers, as we saw in Section 5.3.2.2, but they may also grammaticalize into adverbial clause subordinators. In the Mande language Susu, the noun *fe* 'matter', 'affair' appears to have given rise to a de-verbal nominalizer and to the purpose clause subordinator *-fe*, or *-fera* (*-ra* = multi-purpose particle) purpose marker, for example:

(57) Susu (Mande, Niger-Congo; Friedländer 1974: 50)
a	nakha	si	sukhu	a	fakha-	**fera**.
(3.SG	TAM	goat	catch	3.SG	kill-	PURP)

'She seized the goat in order to kill it.'

We have been looking at a range of different clause types and nominal sources for adverbial clauses. In concluding, one may mention an example that is unusual with reference to both the lexical source and the grammaticalization channel employed, but which demonstrates the flexibility of the channel from noun to clause subordinator. This example concerns the Chadic language Lele (Frajzyngier 1996: 81–3), and the nominal source is the noun *álà* 'God' (borrowed from Arabic *Allah* 'God'), which appears to have given rise to a marker of adversative clauses ('but'), as illustrated in (58). The reason for adding this example of the noun-to-subordinator channel is that it illustrates the range of cognitive processes that are at work in the rise of clause subordination.

(58) Lele (East Chadic, Afroasiatic; Frajzyngier 1996: 82)

ŋ̄-   jè   tòb   kámyà   **álà**   né-   ŋ   hómyà.
1.SG- HAB   like   milk   but   make- 1.SG sick
'I like milk but it makes me sick.'

Frajzyngier (1996: 83) reconstructs the following process of grammaticalization: The noun *álà* 'God' came to serve as an exclamation marker of surprise, and from there as a marker of an unexpected proposition, cf. (59). In certain contexts, the marker was reinterpreted as an adversative conjunction, as in (58). Such contexts involved situations where in a sequence of two clauses the second clause was interpreted as the unexpected consequence or outcome of the first clause.

(59) Lele (East Chadic, Afroasiatic; Frajzyngier 1996: 83)

álà   sē   hāb-   jè   kùnà-   i   kàyō   dà
CONJ   INCE   find-   VEN   uncle-   3.M   squirrel LOC
kìrè   nī.
road   LOC
'Unexpectedly he encountered his uncle, the squirrel, on the road.'

A crosslinguistically salient grammaticalization process concerns the evolution from noun (or noun phrase) to adposition and further to clause subordinator. Accordingly, a number of instances of the process discussed in this section have an intermediate adposition stage.

There is substantial diachronic evidence to show that there is a unidirectional process from noun structure to adverbial clause subordinator. The German conjunction *weil* 'because' is historically derived from the Old High German noun phrase *al di wila* 'all the while' followed by any of the conjunctional elements *so*, *do*, or *daz*. The phrase was first desemanticized to a temporal and later to a cause/reason subordinator. In addition to decategorialization of the erstwhile noun phrase, the process also involved erosion, in that the phrase *al di wila* + *so/do/daz* was first shortened to *die wile* and eventually to the form *weil* of Modern High German.

A similar process from temporal noun to clause subordinator is provided by English *while*, cognate to the German conjunction *weil* 'because' but being a temporal and concessive rather than a causal conjunction. *While* originated in Old English in an adverbial phrase *þa hwile þe* 'that time that', consisting of the dative distal demonstrative, the dative noun *hwile* 'time', and the relativizer *þe*.[17]

Evidence for a development from noun to adverbial clause conjunction can also be found in signed languages. In German Sign Language (DGS), the noun sign REASON has given rise to a reason subordinator, as in the following example; unlike the noun REASON, the grammaticalized sign REASON shows erosion in that it is linked with the following subordinate clause without any prosodic break:

(60) German Sign Language (DGS; Pfau and Steinbach 2005b: 20)
INDEX$_{1.SG}$ SAD **REASON** POSS$_{1.SG}$ DOG DIE.
'I'm sad because my dog died.'

To conclude, there is clear evidence that adverbial subordinators may go back to noun phrases or adverbial phrases having a noun as their semantic nucleus. This is a unidirectional diachronic process; we are not aware of any clear cases where adverbial subordinate clauses developed into noun phrases or adverbial phrases. And it involved all mechanisms of grammaticalization: The erstwhile noun was exposed to some contexts where the nominal meaning no longer made sense (extension), desemanticization had the effect that the nominal meaning was lost, and a poly-morphemic and poly-syllabic phrase was reduced to an invariable monosyllabic marker (erosion).[18]

5.3.3.3 The verb channel
Depending on the adverbial function concerned, there is a wide range of verbs that supply the conceptual source for adverbial clause subordinators.

---

[17] The phrase had a number of variants:, *ðe hwile ðe, þa hwile þe, þa hwile þa, þa hwila þe, a hwile ðæ, ðe hwile ðæt*, etc. There are differences in detail of interpretation; in other accounts, *hwile* was not a dative but an accusative form.

[18] By Late Old English the noun phrase *þa hwile þe* was already decategorialized and eroded to the simple conjunction *wile* (Traugott and König 1991: 200–1).

The following examples may give an impression of the variety to be found in the languages of the world.

Verbs expressing goal orientation are a not uncommon source for adverbial clauses of purpose, such verbs being, in particular, 'give' (Heine and Kuteva 2002a: 154–5) and 'go to'. In the following example (61) from Thai, the verb *hây* 'give' functions as a purposive marker, while (62) illustrates the grammaticalization of the verb *bang* 'go' to -*bang*, a subordinating conjunction of goal or purpose, in the Central American language Rama.

(61) Thai (Tibeto-Burman; Song 1997: 327)
 khǎw khiǎn còtmǎay **hây** khun tɔ̀ɔp.
 3 write letter give you answer
 'He wrote a letter so that you would answer.'

(62) Rama (Chibchan; Craig 1991: 457)
 tiiskama ni- sung- **bang** taak- i.
 baby 1.SG- see- SUB go- TNS
 'I am going in order to see/look at the baby.'

'Go'-verbs have also been widely used in creole languages to develop purpose subordinators (Bickerton 1981; Rettler 1991; Heine and Kuteva 2002a: 163–5).

Lexical verbs may also develop into temporal clause subordinators. For example, in the Kenyan language Kikuyu, the intransitive verb -*kinya* 'arrive at, come' has given rise to a temporal conjunction *kinya* 'until', for example:

(63) Kikuyu (Bantu, Niger-Congo; Benson 1964: 219–20)
 ikara haha **kinya** nj- ok- e.
 stay.IMP here arrive 1.SG- come- SUBJ
 'Stay here till I get back.'

Verbal sources for adverbial-clause subordinators can also be found in conditional sentences, where the marker of the protasis clause ('if') may historically be derived from an imperative verb form, as in English *suppose*.

Examples for a development from verb to adverbial clause subordinator are also provided by signed languages. For example, in American Sign Language, the verb sign UNDERSTAND has given rise, among others, to a conjunction roughly meaning 'provided that', and in the Sign Language of the Netherlands (NGT), the verb sign FOLLOW can be used as a clause-linking marker introducing temporal and causal relations ('due to'/'after'; Pfau and Steinbach 2005a: 21).

5.3.3.4 The demonstrative channel

Demonstrative pronouns are a salient source of complementizers (see Section 5.3.2.4); less commonly, however, they may also provide the source for adverbial clause subordinators. A striking example is provided by the southwest African Khoisan language !Xun, where the proximal demonstrative *kā ē* 'this (particular one)', usually reduced to *kā-ē*, has turned into a general subordinator of adverbial clauses (alternatively, the second proximal demonstrative *kā ŋ̄ŋ̄* 'this', or simply the personal pronoun *kā* 'it' of noun class 4 (N4), can be used as a subordinator). Depending on the context in which it is used, *kā-ē* may introduce locative, temporal, causal, or conditional adverbial clauses; the following examples illustrate the causal (64a) and conditional (64b) functions of the subordinator.

(64) !Xun, W2 dialect (North Khoisan, Khoisan)
    a. kā-ē   hà   ā   ǁʼàn   hà   má |ōā   g|è.
        SUB   N1   PROG   sick   N1   TOP   NEG   come
        'Since he is sick, he cannot come.'
    b. kā-ē   hà |ōā   tcàʼā   má   mā   nǁàqm   hà.
        if   N1   NEG   hear   TOP   1.SG   beat   N1
        'If he doesn't listen, I'll beat him.'

The same mechanisms that we observed in the grammaticalization of *kā ē* from demonstrative pronoun to complementizer (Section 5.3.2.4) have shaped the clause subordinator *kā-ē*, *kā*, or *kā ŋ̄ŋ̄*.

5.3.3.5 The adverb channel

The channels of adverbial clause subordinators sketched above are strikingly similar to the ones that we observed in the case of complementizers (Section 5.3.2; see also Section 5.4). But there are also channels that appear to be restricted to the former. This applies most of all to the channel leading from adverbs to subordinators. It is spatial and temporal adverbs in particular, such as 'here', 'there', 'now', 'then', etc., that provide a common source for adverbial subordinators. When this channel is used, adverbs grammaticalize into adverbial subordinators (SUB). The kind of construction involved typically takes a form as in (65a) which is reinterpreted as (65b).

(65) From adverb to clause subordinator (SUB)
    a. [main clause] – [adverb + main clause]
    b. [main clause] – [[SUB + subordinate clause]]

Once again we are dealing with a process whereby a lexical item (in this case an adverb) is reinterpreted as a functional category, and a main clause as a subordinate adverbial clause. In the Central African Bantu language Lingala, the locative adverb *áwa* 'here' provided the source for the temporal conjunction 'while', 'when', and further for the causal subordinator *áwa* 'since', 'because', for example:

(66) Lingala (Bantu, Niger-Congo)
 áwa oyɔ́ olingí tɛ́, tokotínda mwána mosúsu.
 (here you you.come NEG we.FUT.look child other)
 'Since you don't come, we'll look for another boy.'

Similarly, the Albanian adverb *ke* 'here' (67a) appears to have provided the source for a conjunction marking a causal clause (67b).

(67) Albanian (Buchholz, Fiedler, and Uhlisch 1993: 221)
 a. ja **ke** erdhi.
 (INTJ here arrive.AOR)
 'Here he is!'
 b. **ke** s'fole ti, [...].
 (here not PTCPL.say.2.SG)
 'Because you did not say anything, [...].'

Finally, one might mention the Latin temporal conjunction *dum* 'while, as long as', which can be traced back to the adverb *dum* 'now' in early Latin, surviving in this meaning as an enclitic of imperative verbs, for example *Accede-dum* 'Come here, now!' (Janson 1979: 104–5).

Of all three main types of clause subordination, adverbial clauses exhibit the largest variety of sources for subordination markers. One may also mention that morphologies used for a mood type frequently called "subjunctive" can also serve to present volitional and counterfactual propositions which may be used for adverbial-clause subordination (Harris and Campbell 1995: 306–7).

5.3.4 *From complementizer or relativizer to adverbial clause subordinator*

In the preceding sections we sketched the main channels leading to clause subordination. We saw that complement and adverbial-clause subordination may follow different channels of grammaticalization, but that they can use the same channel as well. The question then arises how the two kinds of clause subordination differ from one another in their development. The evidence available suggests that both relative and complement clauses can develop into adverbial clauses while a development in the opposite direction is unlikely to happen. The evidence for this hypothesis is as follows. Wherever there is diachronic evidence it indicates that in the grammaticalization of demonstratives to subordinators there was first a relativizer or complementizer stage before the relevant marker turned into an adverbial clause subordinator. For example, in the evolution of demonstrative pronouns, the following stages can be observed:

(68) The main stages in the development from demonstrative pronoun to clause subordinator
    a. Demonstrative pronoun introducing relative or object complement clauses
    b. Demonstrative pronoun introducing subject complement clauses
    c. Demonstrative pronoun introducing adverbial clauses

One piece of evidence is provided by the diachronic criterion: As we will see in Section 6.4.1, there are historical data to show that relativizers and complementizers can develop further into adverbial clause subordinators. A second piece of evidence comes from crosslinguistic observations on grammaticalization. A number of languages have reached stage (68b) but not (68c), that is, a demonstrative has been grammaticalized to a complementizer but not to an adverbial clause subordinator. English is such a language: It uses the item *that* as a demonstrative pronoun (69a), as an object complementizer (69b), and a subject complementizer (69c), but essentially not as an adverbial clause subordinator. In the Khoisan language !Xun, all stages of (68) are represented: In (70a), the form *kā ē* is a proximal demonstrative modifier of noun class 4 ('this'), in (70b) it is an object complementizer, while in (70c) it is an adverbial-clause subordinator—more specifically a reason conjunction, but it has a number of other subordinating functions in addition, such as marking temporal and conditional clauses.

(69) English
    a. I know **that**.
    b. I know **that** she'll come.
    c. **That** he'll not come worries her.

(70) !Xun, W2 dialect (North Khoisan, Khoisan)
a. mí   má   hŋ̀   !ȁhŋ̀   kā   ē.
   I   TOP   see   tree   N4   PR
   'I see this tree.'

b. ǐín   hà̏   kā-ē   hà̏   gè-   ā   khōmē
   ask   N1   CPL   N1   COP-   T   tomorrow
   kē   n!āō   khùyā.
   TR   house   at
   'Ask him if he'll be at home tomorrow.'

c. hà̏   má   ǁꞌàn   kā-ē   hà̏ |ōā   g|è-   ā.
   N1   TOP   sick   SUB   N1   NEG   come-   T
   'She is sick, therefore she cannot come.'

A third piece of evidence comes from verbs for 'say'. As we saw in Section 5.3.2.3 (37), there is a crosslinguistically common channel whereby 'say'-verbs may develop first into complementizers and only thereafter into adverbial-clause subordinators (see e.g. Heine, Claudi, and Hünnemeyer 1991: 158; Ebert 1991: 87; Crass 2002; see below). Languages differ in the extent to which they have proceeded along this channel. For example, whereas the Tibeto-Burman language Chamling has not proceeded beyond the stage of complementizer, the Indo-Aryan language Nepali has gone through all stages, having developed its verb for 'say' into an adverbial clause subordinator (Ebert 1991: 88). We will deal with another example illustrating the extension from complementizer to adverbial-clause subordinator in Section 6.4.1.2.

A final piece of evidence for the development from complement to adverbial clause marking is provided by the adposition channel and the development from allative to purpose markers, involving the following stages, where the main verb denotes goal-directed motion ('go to', 'take to', etc.):

(71) From allative case marker to purpose clause subordinator
   a. [*X goes to place Y*]
   b. [*X goes to (do) activity Y*]
   c. [*X goes in order to do Y*]

In (71a) there is a nominal locative complement marked by an allative adposition or case affix, as in example (72a) from the Papuan language Imonda, where the allative marker consists of the nominal goal suffix *-m* (GL). In (72b), the locative complement is conceived of as an activity, irrespective of whether it is encoded by a noun, as in (72b), or as a nominalized verb (see below). Finally, in (72c) the erstwhile locative complement is now a verb that may take complements and is conceived of as expressing the purpose of the motion.

(72) Imonda (Waris, Trans-New Guinea; Seiler 1985: 161–2)
    a. në-   m   at   uagl-   n.
       bush- GL CPL go-   PAST
       'He has gone to the bush.'
    b. tëta-   m   ai-   fōhō   -n.
       game- GL PL-   go.down- PAST
       'They have gone hunting for game.'
    c. tōbtō soh-   m   ka uagl- f.
       fish   search- GL   I   go-   PRES
       'I am going to search for fish.'

To conclude, there is evidence to establish that whenever one and the same construction gives rise to both relative or complement and adverbial clause subordination, the former types of subordination are likely to have preceded the latter in time. Historical evidence can be found in creole languages, where it is possible on the basis of written documents to establish sequences of historical development; see Section 6.4.1.

5.4 Discussion

As we pointed out in Chapter 1, grammaticalization is unidirectional but there are exceptions, even if they are fairly rare and idiosyncratic—that is, they do not show the crosslinguistic pattern regularity characteristic of grammaticalization processes. With regard to clause combining, one notable exception is provided by Japanese, where there is evidence for the development from hypotaxis to parataxis in adverbial clause constructions (Matsumoto 1998). Newmeyer (1998a: 274–5) takes this case as one out of two examples to claim that there is directionality from hypotactic to paratactic constructions. His second example concerns a case discussed by Harris and Campbell (1995: 284–5), according to which the original English relative marker *ðe*, historically derived from a demonstrative (in accordance with the channel described in Section 5.3.1.1 above), was gradually replaced by markers derived from question pronouns such as *who, which, where*, while there is no indication that the relative clause itself developed from questions. While we agree with Newmeyer on his first example, we do not think that his second example really is in support of his claim. First, it does not concern a change from hypotaxis to parataxis but rather from question word to relative marker. Second, this change—rather than contradicting the unidirectionality principle—is in support of it: It is an instance of a canonical grammaticalization that we described in Section 5.3.1.2.

But this last example concerns an issue that we mentioned in the preceding sections, namely what Harris and Campbell (1995: 284) call "the Marker/Structure Fallacy," according to which the historical sources of subordination markers do not necessarily reflect the sources of the structures of which these markers are a part; accordingly, the etymology of the subordinator does not always reveal the structure of the syntactic source. There are indeed a number of attested cases where, in a situation of language contact, one language just adds an appropriate subordinator replicated on the model of another language to an already existing construction of subordination (see Heine and Kuteva 2005, 2006 for examples). The main argument adduced by Harris and Campbell is that, at least in the history of English, there is no historical evidence for the marker-plus-structure hypothesis:

If relative clauses that make use of relative pronouns that derive from Q-words developed from independent questions, we would expect to find these questions in attested forms of English during this change. But there is no evidence that questions were involved in this change; rather, the facts suggest that the Q-words began to replace the demonstratives and particles found in the existing relative clause structure. (Harris and Campbell 1995: 284–5)

Furthermore, these authors question the hypothesis that subordination commonly develops out of coordinated structures. In fact, as we saw in Section 5.2, there is an alternative development, namely expansion (see (2a)); nevertheless, we side with Hopper and Traugott (2003) in hypothesizing that—in accordance with the integration mechanism of (2b)—a major source of clause subordination can be seen in the development from a structure consisting of two independent main clauses[19] to another structure where one of the main clauses turns into a subordinate

---

[19] We refer to this structure in short as "coordination", even if the structure may take a variety of different forms of clause linkage, not all of which are commonly described as coordination.

clause—rather than assuming, for example, that a new subordination marker was simply grafted onto an old subordinate clause. An indication that this is unlikely to have been the case can be seen in the following diachronic fact. It is as early as Old English, in English translations from Latin—with the latter language believed to have been the triggering factor for the interrogative > relative development in English (Bergs and Stein 2001, among others)—that we come across examples where the erstwhile relative marker *þæt* is not replaced but is *followed* by an interrogative form in the English translation of a Latin (headless) relative clause (which can also be interpreted as a complementizer clause):

>   Latin   non legistis quid fecerit David.
>   Old English   ne rædde ge **þæt hwæt** David dyde.
>   'Have ye not read what David did?' (Lockwood 1968: 245–6)

Now, a usage pattern like this—which in Middle English (Mustanoja 1960: 191; Romaine 1982: 61) may well have been strengthened by French influence, whereby the old *þæt* marker was felt to be more and more unnecessary until it disappeared entirely—is compatible with a scenario of coordinating two independent clauses. In accordance with Mustanoja (1960: 191) as well as Heine and Kuteva (2006: 232), one may hypothesize that the development from interrogative to relative clause marker is the result of a strategy whereby recurrent interpersonal communication structures used for questions become the conceptual templates for expressing relations between clauses. In other words, the starting point of this development could well have been a declarative clause followed by an interrogative clause. At the first stage of the development, an interrogative structure is used as a template to express the relationship between a main clause and a complement clause, where the WH-word expresses an indefinite complement clause. At the following stage, the WH-word introduces a definite complement, which can also be interpreted as a headless relative clause. At the final stage, the erstwhile interrogative has acquired the function of a relative pronoun with a noun phrase as its head.

Another reason for hesitating to adopt the marker-minus-structure hypothesis proposed by Harris and Campbell, and for maintaining that coordination is an important source for subordination is the following. Harris and Campbell show that yes–no questions are like subordinate clauses in not expressing an assertion, and they conclude:

We hypothesize that it is this non-assertion marking that is extended to the function of marking (certain) subordinate clauses. If this is correct, then we might expect that in a language a specific question-marking device would always be extended to marking those subordinate clauses that are non-assertions in that language, before it is extended to marking those subordinate clauses that are assertions. (Harris and Campbell 1995: 303–4)

As the detailed discussion by Harris and Campbell (1995: 298–303) suggests, their primary concern is with the classification of types of English subordinate clauses with reference to the notions of assertion and presupposition: They do not deal with how these clauses relate to the markings used for their expression. The only reasonable conclusion that one may draw from this discussion is that the extension from question to subordination involved either clauses (or propositions) plus their markings or only clauses without markings.[20]

Another observation suggesting that there is evidence for a development from a combination of two independent sentences to clause subordination concerns the rise of complementizers out of verbs for 'say' that we discussed in Section 5.3.2.3. On the basis of crosslinguistic data, some of which are discussed in Section 5.3.2.3, this development is likely to involve the stages distinguished in (73). The source structure is a direct-speech construction (a). In a number of languages, a verb for 'say' is introduced as a quotative marker ('saying')[21] (b), the construction is extended from direct to indirect speech (c), and the quotative marker may gradually develop into a complementizer (d). As commonly happens in the transition from coordination to subordination, the subordinator starts out as a constituent of the matrix clause but is later reanalyzed as introducing the complement clause. This overall process is strongly suggestive of a development that initially involved a combination of two independent sentences but ends up in a main clause—complement clause construction.

(73) From direct speech to clause complementation
    a. *He said:*   "*Mary is pretty.*"
    b. *[He said, saying:]*   *["Mary is pretty"].*
    c. *He said, [saying/CPL*   *Mary will come].*
    d. *He said [CPL*   *Mary will come].*

---

[20] Note also the following generalization by Harris and Campbell (1995: 293): "We suggest here that extension from questions is likewise responsible for the development of question-like structures in subordinate clauses."

[21] The 'say'-verb may take a non-finite form, such as an infinitival or gerundival form, as in our scenario, or a finite form (e.g. 'he says').

A third observation relates to the nature of context-induced inferencing that characterizes grammaticalization. Take the following example discussed by Harris and Campbell:

> A clear example involving extension from a coordinating, rather than subordinating, conjunction is found in Mingrelian. The marker *da*, which forms conditional clauses, comes from the coordinating conjunction "and"; it always occurs in clause-final position, as do several others in the language. (Harris and Campbell 1995: 290)

While the authors present solid diachronic evidence to establish that Mingrelian *da* in fact served coordination before its use was extended to mark subordination, there is no evidence to show whether this process was restricted to this marker or, alternatively, whether *da* was part of a coordinated clause that developed into a conditional clause. But it would seem that there is circumstantial evidence to suggest that the latter hypothesis is to be preferred. There is a crosslinguistically widespread invited inference to the effect that in appropriate contexts a sequence of two consecutive events [X and then Y] is reinterpreted as implying a conditional proposition [If there is X then Y follows]; for example, English *Do it again and I'll take you to court!* may be taken as implying a conditional schema (paraphrasable by *If you do it again I'll take you to court!*). On this account, it is possible that in Mingrelian a coordinating structure was reinterpreted and grammaticalized as a conditional structure.

Such a hypothesis is supported by the following observations. All the cases discussed in Section 5.3 concern the evolution of linguistic substance, that is, morphological expressions that are recruited to encode clause subordination. However, this is not the only way in which clause subordination arises; it can arise without involving any formal linkage; what happens simply is that sentences (or propositions) S_1 and S_2 that are juxtaposed are reinterpreted as one complex sentence (or utterance) where S_1 is the main clause and S_2 a subordinate clause (or vice versa). Basically, this strategy is used in many languages without attaining any degree of grammaticalization; consider the following English examples.

(74) English
    a. Mary is driving to New York. She wants to visit her daughter.
    b. I have a sister. She likes music a lot.

The two examples in (74) each consist of two independent propositions juxtaposed to one another. In specific contexts, however, they can be interpreted as being conceptually linked, in that the second proposition is interpreted as providing information that can be paraphrased by means of a subordinate clause. Such paraphrases for (74a) would be in terms of a purpose clause (*Mary is driving to New York in order to visit her daughter*) or a cause clause (*Mary is driving to New York because she wants to visit her daughter*), and in (74b) in terms of a relative clause (*I have a sister who likes music a lot*). In some languages, such paratactic structures lacking any formal linkage have been grammaticalized to constructions of clause subordination, in accordance with (2b), reprinted here for convenience as (75).

(75) Clause integration[22]
$$S_1 + S_2 > S_1[S_2]$$

The following example from Chinese may illustrate this process. The utterance in (76a) is roughly equivalent in meaning and structure to the English utterance in (74b). In the course of the eighteenth–nineteenth centuries, the structure in (76a) was grammaticalized to a construction where the second proposition is subordinate to the first proposition, as in (76b). This construction is called "The Descriptive Clause" in Chinese grammar, which is similar to a relative clause but contrasts both semantically and morphosyntactically with the relative construction of Chinese (Li 2002: 92).

(76) Modern Chinese (Li 2002: 92)
    a. Wo you yi- ge meimei. Hen xihuan yinyue.
       I have one- CL sister very like music
       'I have a sister. (She) likes music a lot.'
    b. Wo you yi- ge meimei hen xihuan yinyue.
       I have one- CL sister very like music
       'I have a sister who happens to like music a lot.'

As the examples in (76) show, both constructions coexist in Modern Chinese. That (76b) is a grammaticalized form of (76a) is suggested by the following: In (76a), each of the utterances has the full intonation pattern of a declarative sentence, and there is typically a relatively long pause between the two utterances, marking the beginning of a new intonation unit, while in (76b) the boundary between the two propositions is lost and it now forms one intonation unit (Li 2002: 93).

---

[22] The ordering of constituents presented is the one expected in verb-initial (VSO) and verb-medial (SVO) languages, while verb-final (SOV) languages are likely to exhibit the reverse order.

5.5 Conclusions

It is hoped that this chapter has shown that it is possible to find an answer to what Bickerton (2005) calls the "last question" on language evolution (see the introduction to this chapter). Discussion was confined to very few structures of clause combining; however these structures illustrate the two main mechanisms by which patterns of clause subordination may arise: Either via the reinterpretation of thing-like concepts as propositional concepts, leading from nominal to clausal participants (i.e. expansion; Section 5.2) or, alternatively, two juxtaposed sentences $[S_1 + S_2]$ that grow together into one sentence $[S_1 [S_2]]$, where one sentence assumes the function and subsequently also the morphosyntactic structure of a subordinate clause (integration; Section 5.3).

Thus, wherever there is appropriate data it turns out that clause subordination can be traced back to non-subordinated sentences. This does not mean that in the languages concerned there was no earlier subordination. In accordance with the bridge hypothesis that we adopted in Chapter 1, however, we hypothesize that at some stage in the development of human languages there was a situation where the processes sketched in this chapter happened for the first time, that is, where there were no complex sentences but where the means that were available in simple sentences were rearranged in novel ways to create complex sentences.

The findings made in this chapter are at variance with claims that have been made that there was no intermediate stage in the rise of clause combining. For example, Berwick (1998: 338–9) rejects the idea that language could ever have been dramatically different from what it is now, since "there is no possibility of an 'intermediate' *syntax* between a non-combinatorial one and a full natural language" (see also Newmeyer 2002: 363 for discussion). According to the findings on grammaticalization, summarized in Chapter 2, combinatorial syntax does not belong to the earliest layers of language evolution; formal means of clause subordination arise only at a more advanced stage of evolution, namely at layer IV.

And these findings also cast doubts on scenarios proposing a two-stage process, such as those of Bickerton (1990, 1995) or Carstairs-McCarthy (1999). Both assume that there are linguistic correlates to the distinction between *Homo erectus* and *Homo sapiens*, in that the former had no genuine syntax and that it is only the latter that acquired "modern" syntax with its recursive structure and everything else that goes with it. As our findings suggest, there was not one intermediate stage but rather an entire series of stages in the evolution of combinatorial syntax. However, we are restricted here to the analysis of morphosyntactic processes; what they can tell us about neural or cognitive evolution is a question that is beyond the scope of this book.

6 On the rise of recursion

There is an abundant literature on what distinguishes human languages from forms of animal communication, and there is considerable disagreement on what does and what does not (see e.g. Pinker and Jackendoff 2005; Givón 2002a, 2005). But what surfaces in some of the recent literature on language evolution is that if there is one property that is really unique to human languages then that is recursion (Hauser, Chomsky, and Fitch 2002; see also Coolidge and Wynn 2006). It is widely held that recursion belongs to the basic universals of language and that there do not appear to exist analogs in non-human communication—in the wording of Bickerton (2005: 9): Syntax without recursion and movement is about as viable as Hamlet without the Prince of Denmark.

In this chapter we will be concerned with the question of how recursive syntactic structures may arise, using grammaticalization theory as a tool of analysis. We will not be able to deal with all the forms that recursion may take in language structure; rather, we will be restricted to its main manifestations, namely noun modification and clause subordination. This means that we will not be concerned with recursion as a general cognitive phenomenon—an issue that is still largely unclear.

6.1 What is recursion?

Hauser, Chomsky, and Fitch (2002) use what they call "the comparative method",[1] comparing data on a range of biologically diverse species and linguistic vs. non-linguistic cognitive domains to distinguish between a faculty of language in the broad sense (FLB) and a faculty of language in the narrow sense (FLN). According to these authors, most, if not all, of

---

[1] With reference to linguistics, this use of the term "comparative method" is somewhat unfortunate since the term is commonly used for quite a different method, namely one based on the regularity of sound change and sound correspondences.

FLB is based on mechanisms shared with non-human animals, "with differences of quantity rather than kind", while FLN is specific to humans. The following example illustrates their method: They observe that much of phonology is likely to be part of FLB, rather than FLN, either because phonological mechanisms are shared with other cognitive domains (notably music and dance), or because the relevant phenomena appear in other species, particularly bird and whale "song" (Fitch, Hauser, and Chomsky in press: 16). An important hypothesis emanating from this research is that FLN "comprises only the computational mechanisms of recursion as they appear in narrow syntax and the mappings to the interfaces", and that this is the only uniquely human component of the faculty of language, being specific both to language and to humans.[2] On this account, almost everything essential to human language can also be found in other animals, while recursion is believed to distinguish language from the capacities of non-human animals.[3] Recursion is responsible for discrete infinity, that is, for an open-ended and limitless system of communication: It means that sentences are built up of discrete units: There are 6-word sentences and 7-word sentences, but no 6.5-word sentences, and there is no non-arbitrary upper bound to sentence length (Hauser, Chomsky, and Fitch 2002: 1571).

How did recursion evolve in human language? Is it the result of a sudden or a gradual process? One part of the work on language evolution has focused on the question of whether human language evolved by gradual, adaptive extension of pre-existing communication or other systems, or whether it (or important components of it) was exapted away

---

[2] These authors argue that FLN includes the core grammatical computations as they appear in narrow syntax and the mapping to the interfaces, and that these computations are limited to recursion (Hauser, Chomsky, and Fitch 2002: 1570; Fitch, Hauser, and Chomsky in press: 3). Furthermore, they suspect (Hauser, Chomsky, and Fitch 2002: 1574) that many details of language, such as subjacency, Wh-movement, or the existence of garden-path sentences "may represent by-products" of FLN. These authors speculate further that recursion may derive from some computational mechanism that was used by some earlier species for navigation, social cognition, or some other purpose.

[3] The authors are careful to observe that they do not define FLN as recursion by "theoretical fiat"; rather, they wish to offer this as a plausible, falsifiable hypothesis worthy of empirical exploration: "We hypothesize that 'at a minimum, then, FLN includes the capacity of recursion', because this is what virtually all modern approaches to language ... have agreed upon, *at a minimum*. Whatever else might be necessary for human language, the mechanisms underlying discrete infinity are a critical capability of FLB, and quite plausibly of FLN" (Hauser, Chomsky, and Fitch 2002: 1573).

from other capacities, such as spatial, numerical, or other functions. So the question is: Where does recursion come from?

A number of answers have been proposed to this question. One line of answers invokes symbolic reference as a prerequisite for the rise of recursion: Rather than being a necessary ingredient for the emergence of language, recursion (or "non-degrading recursivity") is suggested to be a consequence of symbolic reference and/or symbolic verbal language (Deacon 2003: 126). Hauser, Chomsky, and Fitch (2002: 1569, 1573–4) rightly observe that proponents of the idea that recursion (more precisely, their "FLN") is an adaptation would need to supply additional data or arguments to support this viewpoint, and that existing hypotheses are hard-pressed to explain how "the capacity of language for infinite generativity" would have resulted from a series of gradual modifications.[4] But much the same applies to the idea that recursion is the result of exaptation of any kind, or that the rise of discrete infinity was non-gradual. There is simply no appropriate evidence available to decide in favor of one or the other.

This chapter will not be concerned with how recursion as a cognitive mechanism arose in humans or in human languages; this is an issue that is clearly beyond our methodology. Rather, we wish to show that there are some data applicable to a study of the genesis of recursive structures in human languages, and that these data are in support of the gradualist hypothesis. Accordingly, we will be restricted to recursion as it manifests itself in languages, more precisely in morphosyntax, ignoring the question of whether there may not be additional manifestations of it.

6.1.1 A definition

The linguistic term recursion has been adopted from mathematics and computer sciences, where it stands for the act of defining an object in terms of that object itself—in short: a definition that uses itself as part of itself.[5] As a notion used in some schools of linguistics, it is used to refer to a set of phrase structure rules that allows a category to embed a category of

---

[4] At the same time, these authors note: "However, the available data suggest a much stronger continuity between animals and humans with respect to speech than previously believed. We argue that the continuity hypothesis thus deserves the status of a null hypothesis." (Hauser, Chomsky, and Fitch 2002: 1574).

[5] In mathematics, it is used for example for an expression each term of which is determined by application of a formula to preceding terms, or an expression such that each term is generated by repeating a particular mathematical operation.

the same type. There are structural definitions of recursion, based on phrase structure geometry, and computational definitions, based on processing mechanisms that assign structures (Dougherty 2004).

In its most general form, recursion can be said to be present when some constituent occurs within another constituent of the same type. For instance, a noun phrase can appear within another noun phrase (*John's wife*, or *the book on the table*) or a clause within another clause (*Jane is the woman that John loves*). It can be represented with a rule like (1a), which contains the same symbol on both sides of the arrow and thereby produces an environment for its own reapplication.

Following other linguistic treatments, we will be concerned primarily with only one type of recursion—one that is widely held to be the prototypical one, which we refer to as embedding recursion and, unless otherwise stated, we will henceforth use "recursion" as a shorthand for embedding recursion (see Section 6.1.4). This type can be described thus: [A [B]] is a construction that—in some sense—is structurally derived from another construction [A] which is of the same type, where B is embedded in A, cf. (1b). Recursion can but need not be productive, in that the output (to the right of the arrow) can form the input of another application of the same rule, as in (1c).

(1) a. A → A X (where "X" can be any category)
  b. A → A [B]
  c. A [B] → A [B [C]]

Recursion is not a property of language but rather the product of a given theory designed to describe or account for language structure;[6] given an appropriate theory, one might argue that language is no more recursive than, say, biological reproduction, or some other natural phenomenon. Accordingly, when using this term we are not maintaining that language, or language structure, has recursion but rather that there are certain properties to be observed in language that can appropriately be described in terms of a construct of the form proposed in (1); as we will see below (Section 6.1.5), there are theories that do without "recursion."

We are restricted in this chapter to what may be called direct recursion to distinguish it from other kinds of recursion, such as indirect recursion, where there is not one recursive rule but a recursive "loop", in that two

---

[6] Fritz Newmeyer (p.c.) rightly draws attention to the fact that this applies not only to recursion but to most constructs in linguistics.

(or more) rules jointly produce recursion, as in the set of rules of (2).[7] Indirect recursion raises a number of issues that would be beyond the scope of the present treatment; hence, we will have nothing to say about it and its implications for language evolution.

(2) A → ... B ...
    B → ... A ...

Embedding recursion (1b) entails hierarchy, or phrase structure—be that conceptual or syntactic hierarchy; but not all hierarchy or phrase structure is necessarily recursive: Instances of hierarchy are only recursive when they are in accordance with (1b), that is when the same category occurs both to the left and to the right of the arrow. In dealing with embedding recursion, a number of additional distinctions have been made (see Chomsky 1957; Christiansen and Chater 1999). First, there is a basic distinction between tail and nested (or center-embedded) recursion. In linguistics, tail recursion involves embedding at the edge of a phrase, or invokes another instance of itself as a final step, while nested recursion (or "true" recursion) involves embedding in the center, leaving material on both sides of the embedded component. Second, there is a distinction between counting and mirror recursion. In order to parse strings from left to right it is necessary to count the number of As and note whether it equals the number of Bs, as, for example, in AAABBB. Counting recursion corresponds to sentence constructions such as "*if* S_1, *then* S_2" and "*either* S_1, *or* S_2", which can be nested arbitrarily deeply, for example, "*if if if* S_1, *then* S_2, *then* S_3, *then* S_4". In mirror recursion (which corresponds to center-embedding), the first A is matched with the last A: A[BB]A, as in the English example *The boy girls like ran away*. And crossed embedding (identity recursion, cross-dependency) involves dependencies where A–A are crossed by B–B, that is {A [A B} B].

6.1.2 Manifestations

Recursion tends to be portrayed either as being a characteristic of clause subordination, or more generally as a ubiquitous phenomenon of language, in that the latter has been portrayed as a recursively generated system producing discrete infinity. Neither of these views is entirely

---

[7] In some traditions of generative grammar, a common kind of indirect recursion is found in the following set of rules (where there is an intervening generating operation): S → NP VP; VP → V (S).

appropriate. Clause subordination is not really the domain where recursion manifests itself most frequently in normal language use, nor is recursion found everywhere in language, nor is embedding recursion the only type of rule that is responsible for discrete infinity, nor is discrete infinity something that is in any way characteristic of the way languages are used. Perhaps most commonly, recursion occurs in noun phrase discourse.[8] Outside phrase structure, recursion is less easy to find in language structure; as Pinker and Jackendoff (2005: 216) observe, a case marker may not normally contain another instance of a case marker; an article may not contain an article; a pronoun may not contain a pronoun, and so on for auxiliaries, tense features, etc., and recursion is also absent at the phonological level.[9]

English allows recursive compounds (3a), adjectives (3b), attributive possession (3c), adverbial phrases (3d), and clause embedding (3e), but not really recursive verbs, affixes, or other constituents. Note that the following examples have contrasting branching directions, being left-branching in (a), (b), and (c), but right-branching in (d) and (e).

(3) English
    a. [[[frog]man]team]leader
    b. [[[new] big] red] cars
    c. [[[[Peter's] mother's] brother's] wife's] father
    d. the book [on the table [in the room [behind my office]]]
    e. Judy says [that John claims [that Mary believes [that I want to marry her]]].

However, recursion is not restricted to the internal structure of a sentence—neither in English nor in many other languages; rather, it can also involve direct speech sentences, cf. (4).

(4) English
    Judy said: ["John told me yesterday: ["I want to marry you."]"].

It would be futile to define all the components of grammar that exhibit recursion since this is a theory-related issue. Accordingly, depending on how one defines (embedding) grammatical categories, one will come up with different views on the exact range of recursive structures that exist. For example, if one were to follow those for whom auxiliaries and main

---

[8] This statement is based on impressionistic observations; we are not aware of any quantitative data on the magnitude of recursive structures in language use.
[9] Phonology is hierarchical but not recursive (Fitch, Hauser, and Chomsky in press: 17).

verbs are both verb phrases (VPs), one could argue that there is recursion in verb phrases—that is, something like [VP [VP]], and a theory postulating only one type of grammatical category would be able to produce a grammar with a maximum of recursive structures.

6.1.3 Simple vs. productive recursion

Recursion is a crosslinguistically ubiquitous phenomenon: In most languages it is—or can be—used productively both in attributive possession and in clause subordination. With the term **productive** recursion we refer to the unlimited design of the rule—there are unlimited levels of embedding, that is the rule applies in principle indefinitely many times. But perhaps equally common are cases where a recursive rule applies only once to its own output or is non-productive, we will call such cases **simple** recursion. This distinction is not a trivial one: Simple recursion conforms to the definition of recursion proposed in (1a), but at the same time it has more in common with other, non-recursive, syntactic or semantic properties of human language than with productive recursion.

There are considerable typological differences with respect to this distinction. First, a given language may allow recursion in some specific domain of grammar where another language does not. In German there is simple recursion with modals, cf. (5a), where English has no recursion. On the other hand, German speakers have only simple recursion with inflected genitives, cf. (5b), where English speakers do not have such a constraint, as the English translation of (5b) shows.

(5) German
   a. Peter mag singen können
      Peter may sing   can
      'Peter may be able to sing'.
   b. Peters Bruder   'Peter's brother'
      *Peters Bruders Auto 'Peter's brother's car'

Quite a number of languages, such as Romance languages, have no productive recursion in compounding while others, such as Germanic languages, allow productive noun–noun compounding. Thus, French speakers can use at best simple recursion in compounding, for example *homme grenouille* (man frog) 'frog man'; root compounding is unproductive, limited to frozen and self-conscious coinages (Bauer 1978). For English speakers on the other hand, undersea divers can appear in constructions of productive recursion, cf. (3a). The magnitude that productivity may have can be illustrated with example (6). We have never heard a German speaker uttering this compound, and it probably never will be uttered; however, it is a grammatically correct instance of productive recursive compounding.

(6) German

Auto-	bahn-	rast-	platz-	toiletten-	reinigungs-
car-	line-	rest-	place-	toilet-	cleaning-
personal-	bedarf-	abstell-	schrank-	schlüssel-	dienst-
staff-	need-	storing-	locker-	key-	service-
				telefon-	nummer
				phone-	number

'the phone number of the key service for lockers of the cleaning staff of public conveniences of rest places of highways'

Recursion as it presents itself in, for example, noun combining appears to be rooted in a fairly basic cognitive activity, namely in taxonomy whereby elementary conceptual relations are established among different taxa of the same domain. Simple recursion is a natural product of this activity: Once there is a linguistic expression for relations such as between less inclusive and more inclusive, part and whole, one social role and another, or possessee and possessor, the way is cleared for recursion to enter. And the situation is not dramatically different in the case of productive recursion: What is required is simply that an existing "rule" or convention is re-applied to the same taxonomic entity. Accordingly, the difference between simple and productive recursion appears to be one of degree rather than of kind.

While productive recursion is essentially unlimited, there are generally limits to its use: it is constrained in particular by pragmatic factors, for example by the capacity of a speaker's memory, by what the speaker thinks the hearer can digest, etc. But it is also constrained by syntactic factors; for example, recursive productivity tends to be more limited in center-embedded than in either left- or right-embedded subordination. Some students of connectionist models account for this fact by adding a component of analysis or description that deals specifically with productivity constraints (see Christiansen and Chater 1999), and since Miller and Chomsky (1963) it has been argued that there exists a separate working memory capacity that constrains recursion.[10]

---

[10] We are grateful to Fritz Newmeyer (p.c.) for having drawn our attention to this fact.

6.1.4 Embedding, iteration, and succession

As we noted above, embedding recursion ((7a) = (1b)) is not the only recursive mechanism responsible for discrete infinity in language use; there are two additional mechanisms corresponding to our definition of recursion in (1a), namely iteration (7b) and succession (7c).

(7) Three kinds of recursion
    a. A → A [B]   Embedding
    b. A → A + B   Iteration
    c. For any number n, $s(n)$ is $n + 1$ Succession
       (where "s" is a successor function)

Iteration may take various forms; most commonly it is additive ('A and B') or alternative ('A or B'). Its status vis-à-vis embedding recursion is unclear. The main positions surfacing from the literature are that iteration (a) is an instance of recursion on the same level as embedding recursion; (b) is a special instance of embedding recursion; or else (c) that the two are mutually exclusive mechanisms. The fourth possibility, namely that embedding recursion is included in iteration, does not seem to have found any noticeable support (but see Davidson 2004). In much of the literature on this subject it does not become clear which of these alternatives is implied, and quite a number of authors use the term recursion indiscriminately for both mechanisms.[11] There are reasons for doing so: Both are generative in nature and, hence, both may lead to discrete infinity, and many manifestations of embedding recursion can also be framed in terms of models based on iteration.

At the same time there are also reasons to keep the two apart. First, their effect on language structure is different: Embedding recursion results in conceptual and linguistic subordination or hierarchy, while iteration leads to coordination (or conjunction). Second, there are recursive structures (e.g. center-embedding) that cannot be handled appropriately by iteration models. And third, the two also differ in the range of linguistic phenomena they are associated with. For example, iteration is not restricted to the noun phrase or the clause; it is also a productive mechanism of the verb

---

[11] For Dougherty (2004), for example, recursion includes coordination, subordination, and embedding, and Goldin-Meadow (1982: 54) defines recursion in a way that is not uncommon in linguistics: "Recursion provides a language user with the means for expressing more than one proposition in a single sentence."

phrase (e.g. *He came in, sat down, took the newspaper, and...*), and it is a grammaticalized characteristic of serial verb constructions.

Succession is the kind of recursion that underlies numerosity; thus, (7c) generates units such as the following: $s(4) = 5, s(5) = 6$, for example $1 + 4 = 5$, $1 + 5 = 6$, etc. The ontological status of succession and, more generally, of numerical cognition is far from clear (see below).

The only reason for reducing the term recursion to embedding recursion in this chapter is that this is the use that is commonly implied in discussions on language evolution.

6.1.5 Treatment of recursion in linguistic description

Recursion—more precisely the phenomenon that this term is usually taken to refer to—is not an unproblematical notion. Chomsky himself, who has popularized its use in linguistics (Chomsky 1957), observes that "[T]he *possibility* that languages are nonrecursive, however, is granted by everyone who has seriously discussed the subject, and the question whether this possibility is realized remains an open one", and he concludes that "while languages may be recursive, there is no reason to suppose that this must be so" (Chomsky 1980: 120, 122).

As we noted above, recursion is not a phenomenon of language but rather of a theory that postulates it as a useful device to describe or explain certain properties of language structure. Accordingly, the way it has been treated in linguistics differs greatly from one model to another. First, there are models where its relevance is ignored, or denied. An extreme position is maintained in the paratactic theory of Davidson (2004), where there is no recursion and no embedding; rather, propositions are understood to be paratactically ordered. Accordingly, the English sentence *Pierre believes that snow is white* is said to consist of two distinct utterances linked by parataxis: *Pierre believes that. Snow is white.*

Second, quite a number of linguists acknowledge that there are phenomena that can be described in terms of recursion but that there are also alternative ways of describing such phenomena. And third, there are in fact alternative models that have been proposed to deal with recursive phenomena, perhaps the most noteworthy one being that of endocentricity as used in the early structuralist tradition, which is based on observations of the distributional equivalence of constituents. According to Bloomfield (1933: 194–7), a construction (e.g. *poor John*) is defined as endocentric if it belongs to the same form-class as one (or more) of its constituents. The latter constituent is called the head (*John*) while the other constituent is the attribute (*poor*). Endocentric constructions contrast with exocentric constructions (e.g. *John ran*), which belong to a form-class other than that of any of its constituents. Bloomfield observes that most constructions in any language are endocentric, while exocentric constructions are few. Endocentricity is productive ("there can be several ranks of subordinative position" in his terminology) in that an endocentric construction ([*fresh*] *milk*) can be the head of another endocentric construction ([[*very*] *fresh*] *milk*).

The difference between embedding and iterating recursion (see below) is reflected in Bloomfield's distinction between two kinds of endocentric constructions, namely subordinative or attributive (*poor John*) and co-ordinative or serial constructions (*boys and girls*), respectively. While Bloomfield does not extend the notion of endocentricity to the relation between main and subordinate clauses,[12] later authors did, treating clause subordination in terms of endocentricity. Thus, a noun phrase containing an embedded relative clause, as in *the man who came to tea*, is interpreted as endocentric since it has the same distribution as the noun phrase *the man* (Lyons 1977: 391), and Lehmann (1988: 182) proposes to define clause subordination more generally as an endocentric construction between two clauses where one is the head and the other its dependent. In most formal models, recursion is simply a formal means of capturing (one aspect of) endocentricity (Fritz Newmeyer, p.c.).

6.1.6 Are there languages without recursion?

A number of languages have been claimed to lack recursion, such as pidgins, or the Omotic language Gorze of Ethiopia (for a paradigm example, see Everett 2005). This raises the question of whether embedding recursion is really an indispensable characteristic of human languages. We will not attempt a definite answer to this question (but see Parker 2005), for the following reasons: First, as pointed out above, recursion is a theory-dependent notion; accordingly, no general answer seems possible. And second, the situation is not all that clear in a number of languages because there is not enough information available. Nevertheless, it would seem that all languages on which there is appropriate information do exhibit some kind of recursive

---

[12] Bloomfield (1933: 194) notes that subordinate clauses are exocentric since the resultant phrase (e.g. *if John ran away*) has the function of neither of its constituents (*if* or *John ran away*). But he does not elaborate on how subordinate clauses relate to main clauses.

structures. This applies, for example, to all pidgins that we are familiar with: All African-based pidgins show recursion, at least at the noun phrase level, and the same applies to English-based pidgins such as Tok Pisin, Solomons Pijin, or Nigerian Pidgin English (see Chapter 4).

But there remains the case of the Amazonian language Pirahã: Based on his extensive knowledge of this language, Everett (1986, 2005; Bower 2005) concludes that it does not make use of recursion. A look at the description provided by Everett (1986) suggests, however, that there is an alternative view on this matter. As we noted above, recursion manifests itself in particular in noun modification and clause subordination or, to put it more strongly, if either of these is present, there is recursion. It would seem that both are in fact present in Pirahã. Thus, (8a) seems to be an instance of possessor–possessee modification and (8b) of noun–adjective modification. While Everett notes that noun modification involves paratactic augmentation, the evidence provided suggests that these are cases that in certain schools of linguistics could be described in terms of a recursive rule such as (1b).

(8) Pirahã (Mura, Macro-Chibchan; Everett 1986: 209)
    a. ti   bai   xaagá   giopaí   xahóápatí   giopaí.
       I   fear have   dog   Xahóápatí   dog
      'I am afraid of the dog, Xahóápatí's dog.'
    b. xogaí xogií koíhi hiaba
       field big   small NEG
      '(a) big field, not (a) small (one)'

And there are also examples to suggest that there is some kind of clause subordination in Pirahã; suffice it to quote Everett (1986: 262–3): "Certain types of subordinate clause (nominalized, temporal and conditional) are marked morphologically on the subordinate verb", or "Temporal and conditional... clauses precede the matrix clause, whereas other types of subordinate (adverbial) clauses usually follow the matrix clause."

To conclude, we have so far found no clear evidence for languages that demonstrably lack recursion of any kind; we will return to this issue in Section 7.2.

6.1.7 *Discussion*

The analysis of recursion has triggered considerable debate on the ontological status of this phenomenon. One of the issues raised concerns its status vis-à-vis language, where the following views in particular have been expressed:

(a) Recursion is essentially language-based. Fitch, Hauser, and Chomsky (in press: 19) argue that there are no unambiguous demonstrations of recursion in other human cognitive domains, with the only exceptions (mathematical formulas, computer programming) being clearly dependent upon language. Pinker and Jackendoff (2005: 230) believe that recursive number systems, though not other recursive systems, may have been exapted from the recursive properties of language (see below). But assuming that recursion is language-based, where exactly is it located? (i) Is syntax the locus of recursive structure, as is maintained in much of the generative work, (ii) are syntax and semantics parallel recursive systems (Jackendoff 2002), or (iii) is semantics the source of recursive structure? On the basis of the circumstantial evidence that is available, no definite answer seems possible.

(b) Recursion in language is derivative of other cognitive abilities. Pinker and Jackendoff (2005: 230) argue that the only reason language needs to be recursive is mental: "If there were not any recursive thoughts, the means of expression would not need recursion either."

(c) Recursion occurs independently in different cognitive domains. Hauser, Chomsky, and Fitch (2002: 1569, 1578) consider it possible that their FLN evolved for reasons other than language and that if "recursion evolved to solve other computational problems such as navigation, number, number quantification, or social relationships, then it is possible that other animals have such abilities", and that key computational capacities evolved for reasons other than communication but, after they proved to have utility in communication, were altered. But this view is controversial. For example, as Pinker and Jackendoff (2005: 230) point out, the two principal navigation systems documented in non-human animals do not show the discrete infinity of recursion: Dead reckoning is infinite but not discrete; recognition of landmarks is discrete but not infinite.

There are a number of other issues that are still ill-understood. One concerns the question of whether recursion is innate and/or neurally based (e.g. Calvin 2005: 8) or whether it evolves via cultural transmission (Kirby 2002). Hauser, Chomsky, and Fitch (2002: 1574) observe that recursion is implemented in the same type of neural tissue as the rest of the brain and is thus constrained by biophysical, developmental, and computational factors humans share with other vertebrates. Deacon

(2003: 129) on the other hand points out that an evolved innate neurally based recursive processing faculty is both unnecessary and unsupported by the evidence, in that "the formal recursion that symbolization allows in the external form may nevertheless be processed in non-recursive ways neurologically". And he concludes that the implicit affordance for recursion that symbolic reference provides can be expected in *any* symbolic system.

Another issue that needs to be addressed is whether there are significant stages or degrees in the recursive ability. There is a view that one either has the idea of nested embedding or one does not, and if one does, one can carry it to considerable depths—that is, that recursion is an all-or-nothing phenomenon (Berwick 1998; Calvin 2005: 6). Conversely, we drew attention to the distinction between simple and productive recursion and the relevance it has for linguistic typology. More research is required on this issue.

Finally, there is a problem that we have already alluded to above, namely one relating to the status of the recursive mechanism of succesion and of numerosity in general. Numerical cognition has been analyzed as being closely linked to language—either as being bootstrapped or inferred from language in development or as being derivative of language (Hurford 1975, 1987). Arguing that those human cultures that have developed recursive number systems in their cultural history may have exapted them from the recursive properties of language, Pinker and Jackendoff (2005: 231) conclude that "recursive number cognition is parasitic on language rather than vice-versa". And based on an extensive analysis of number and number systems, Wiese (2003) argues that we do not have to look for an independent development of recursivity outside the language system.

On the other hand, numerical cognition and language are also claimed to belong to different and independent mental faculties. The following observations in particular suggest that succession is independent of other recursive mechanisms, and of language in general (see Gelman and Butterworth 2005; Grinstead *et al.* n.d. for details):

(a) Considering their different nature, it would be hard to establish for example that rules underlying embedding recursion move from syntax to numerosity.
(b) Some mentally retarded individuals who demonstrate intact grammatical development nevertheless possess limited or no ability to calculate and have little grasp of basic counting principles.

(c) Some hearing-impaired adults who have developed no grammar in the sense of natural human languages nevertheless are able to use number.
(d) There is crosslinguistic evidence to the effect that speakers of languages having no recursive number system, lacking the ability to count, such as the Pirahã of Amazonia, !Xun of Namibia, or Warlpiri of Central Australia, are nevertheless able to learn and handle such a mechanism in arithmetic tasks (see e.g. Gelman and Gallistel 2004: 441).

6.2 Animal cognition

In Chapter 3 we gave an overview of the cognitive language-related abilities that have been found in animals. We saw, for example, that a non-primate, the grey parrot (*Psittacus erithacus*) Alex can identify about 50 different objects using English labels. He can also label seven colors, five shapes, and quantities up to and including six, and he has functional use of phrases like 'I want X' and 'I want to go Y', where X and Y, respectively, are object or location labels. He combines these labels to identify, refuse, request, and categorize more than a hundred different items. He is said to have concepts of sameness and difference, of bigger and smaller, of absence of information, and of number. Thus, even in non-primates there are a number of properties having analogs—conceivably also homologs—in human languages.

But there is one clear difference. In their survey of animal communication systems, Hauser, Chomsky, and Fitch (2002) did not find any evidence for recursive embedding. And, working with cotton-top tamarin monkeys (*Saguinus oedipus*), Fitch and Hauser (2004) arrive at the same conclusion. They tested the animals on their abilities vis-à-vis two grammatical models, namely finite state grammar, taking the form $(AB)^n$, and phrase structure grammar ($A^n B^n$). Whereas the latter can embed strings within other strings, thus producing center-embedded hierarchical structures ("phrase structures": AAA–BBB...) and long-distance dependencies, the former model cannot (producing AB–AB–AB...). These authors conclude that, unlike humans, tamarins cannot handle phrase structure grammar, suffering "from a specific and fundamental computational limitation on their ability to spontaneously recognize or remember hierarchically organized acoustic structures," and they speculate that the acquisition of hierarchical processing ability may have represented a critical juncture in the evolution of the human language faculty (Fitch and Hauser 2004: 380).

This claim has not gone unchallenged. Kochanski (2004) suggests that the conclusion reached by these authors needs to be taken with care. He argues that if humans can learn patterns of (meaningless) syllable sequences of the form AABB, or AAABBB, whereas tamarin monkeys can not, then this does not tell much about *how* these two groups process the sequences, that is, whether both use the same cognitive technique in tackling this task—in other words, Kochanski doubts whether this is really a meaningful comparison to establish a difference between humans and animals. For example, rather than relying on recursive mental structures, the human subjects could have determined the "grammaticality" of the stimuli by simply counting and comparing the number of syllables of type A and type B and checking that the numbers match.

Still, our observations on animal behavior in Chapter 3 confirm the conclusion reached by Hauser, Chomsky, and Fitch (2002): In spite of all the intense training that some animals received, none of them showed the ability to understand or produce structures that could in any sense be called recursive (but see Pepperberg 1992; 1999b for a different view). Some animals arguably have recursion in the form of iteration (see (7b)); for example, the songs of male humpback whales (*Megaptera novaeangliae*), commonly described as courtship signals, are usually sung in repetition, often for half an hour or more, and repetition appears to be communicatively significant: The more repetitions, the greater the desire of the male to attract a female, and the more it demonstrates the male's physical fitness. But irrespective of how this situation is to be accounted for, our concern here is not with iteration but rather with embedding recursion.

It would seem, however, that there is another way to approach this issue. One prerequisite for recursion—we argue—consists of the ability to form abstract concepts. That at least some animals are capable of abstraction has been pointed out independently in a number of studies; Hauser, Chomsky, and Fitch (2002: 1575), for example, note that animals use a wide range of abstract concepts, including tool, color, geometric relations, food, and number.

But abstraction on its own is not sufficient. We noted in Chapter 3 that what is required for noun phrase recursion as a syntactic mechanism to arise is, first, the cognitive ability to understand asymmetric relations of dominance between objects, as they hold, for example, between inclusive (*tree*) and less inclusive items (*apple tree*), between wholes (*finger*) and their parts (*fingernail*), between non-possessed (*car*) and possessive concepts (*Anne's car*), between simple objects (*book*) and spatially defined objects (*the book on the table*), etc. All languages that we are aware of have ways of expressing such relations in some way or other, be that in the form of compounding, of attributive possessive constructions, or of noun–modifier constructions—hence, in all these languages there is recursion expressed linguistically (see below). But what about animals?

Hurford (2004) describes the problem appropriately thus:

> If non-human animals know in some sense that things have parts that have subparts which have subparts, then again their mental representations, independent of language, have a recursive structure. It is not known whether animals are capable of such mental representations.
>
> If one entertains the hypothesis that recursion evolved to solve other computational problems, such as navigation, number quantification or social relationships, then it is possible that other animals have such abilities. (Hurford 2004: n.p.)

Certainly the way Alex, the grey parrot, combined "attributes" with "nouns" as in *rose paper* or *rock corn* might point to the presence of a modifier–head construction, hence of a simple recursive structure [[B] A] in accordance with rule (1b). However, the data provided by Pepperberg (e.g. 1999a) are not sufficient to compellingly demonstrate that such a construction really exists. Nevertheless, there is more convincing evidence from some non-human primates. We argued in Chapter 3 that some chimpanzees can be said to have acquired some basics of conceptual inclusion and perhaps also of part–whole relations. Remember the ability of the chimpanzees Sherman and Austin to label the hypernyms 'food' and 'tool' and to assign, respectively, food items and tool items to these two categories (Savage-Rumbaugh *et al.* 1980: 924), or of the chimpanzees Peony and Elizabeth, who could sort the plant parts leaves, stems, seeds, and flowers to a plant category, and the animal parts fur, teeth, hair, and bones to an animal category (Premack 1976: 217–18), thereby showing some taxonomic understanding of a part–whole relationship. One may also wish to draw attention to the following observation, which might be suggestive of a hierarchical social relationship:

> Baboons classify themselves and their conspecifics both in a linear hierarchy of dominance, and in matrilineal kin groups. In other words, they are capable of forming conceptual structures such as [X is mother of Y [who is mother of Z

[who is mother of me]]] or [X is more dominant than Y [who is more dominant than Z [who is more dominant than me]]]—tail-recursively embedded associations, which (unlike the iterative counterparts) cannot be re-ordered while maintaining the correct relations. (Parker 2005: 5)

Accordingly, there is reason to assume that while these animals do not clearly have (embedding) recursion of any kind, they appear to dispose of one important prerequisite for the rise of recursion in the noun phrase, namely the cognitive ability to establish a hierarchical relationship between inclusive and included objects or between social roles.

In fact, a number of traits in animal behavior have been reported that—given appropriate experimental testing—might turn out to be conducive to an interpretation in terms of simple recursion. Possible cases are provided, for example, by the food preparation techniques of mountain gorillas, which could be suggestive of hierarchical reasoning (Parker 2005).

6.3 The noun phrase

We now turn to recursion in human languages. As we observed above, recursion in its paradigm manifestations presupposes either noun modification or clause subordination. Whenever either of these is present then there is at least simple recursion. Accordingly, in order to reconstruct the genesis of recursion it suffices to show how noun modification or clause subordination arise.

Within the noun phrase, recursion can be described on the one hand in the form of a taxonomic relation hypernym–hyponym, on the other hand it can—in accordance with (1b)—be represented by means of a formula as in (9), where the category to the left of the arrow is the inclusive taxon and the one to the right of the arrow is the included taxon. Accordingly, whenever there is a rule like (9) there is at least simple recursion, and our concern in this chapter is with showing how such rules evolve.

(9) $N \rightarrow N$ [modifier]

As we argued in Section 3.2.9, recursion within the noun phrase is contingent on certain conceptual structures which we referred to as hierarchical taxonomic relations. Each of these relations tends to be associated with specific linguistic encoding structures; Table 6.1 lists the ones that are perhaps, crosslinguistically, most common.

TABLE 6.1. *Hierarchical taxonomic relations and typical linguistic expressions for them*

Type of relation	Description	Canonical linguistic expression
Inclusion	A is a kind/type of B	Attributive possession, modifying compounding
Property relationship	B has property A	Adjectival modification, modifying compounding
Partonomy (or meronymy)	A is a part of B	Attributive possession, modifying compounding
Social relationship	A is a relative of B	Attributive possession
Possession	A possesses B	Attributive possession
Location	B is located at A	Nominal post-modification

The kinds of linguistic constructions that we will be confined to are attributive possession (or genitive constructions) (Section 6.3.1), modifying compounding (Section 6.3.2), and adjectival modification (Section 6.3.3). Finally, there is another common construction conforming to (9), namely relative clauses, which we will return to in Section 6.4. While there are considerable crosslinguistic differences in the way these relationships are encoded, all languages that we are familiar with have some recursive form of grammatical expressions at least for some of these relationships.

6.3.1 *Attributive possession*

Possessive constructions and the recursive structure associated with them can frequently be traced back to non-recursive structures. These constructions are frequently encoded by case markers, in particular genitive inflections. Most of these case markers are etymologically opaque, that is, their genesis is beyond the scope of the methodology of historical linguistics. But there are case markers and the possessive constructions associated with them for which there exists sufficient diachronic evidence to allow for generalizations on their genesis. The diachronic sources of these constructions are discussed in Heine (1997a; see also Heine and Kuteva 2002a: 34–5) and the reader is referred to these works for more details. The main conceptual schemas serving as sources for the rise of attributive possession are listed in Table 6.2.

TABLE 6.2. *The main source schemas used crosslinguistically for the expression of attributive possession*

	Formula	Label of event schema
a	Y at X	Location
b	Y from X	Source
c	Y for/to X	Goal
d	X with Y	Companion
e	(As for) X, X's Y	Topic

Source: Adapted from Heine (1997a: 144).

Three of the schemas listed there (location, source, and goal) can be described in a broad sense as being spatial in nature, whereby the possessor is conceptualized as a spatially described participant. In European languages, the schema predominantly recruited is the source schema, where the possessor is presented by means of an ablative preposition ('(away) from', 'out of'), for example English *of*, German *von*, Dutch *van*, Frisian *fan*, Catalan *de*, Macedonian *od*, Upper Sorbian *wot*, etc. Thus, there was a historical development whereby a locative prepositional construction of the form ['from' NP] developed into a possessive/genitive construction [GEN possessor]. For example, the Latin prepositional construction [*dē* X 'from X'] is the historical source of attributive possessive constructions to be found in the modern Romance languages, as well as of the productive recursive structure exhibited by these constructions, for example French *le chien de l'ami de mon père*, in much the same way as the corresponding recursion in the English translation 'the dog of the friend of my father' is the result of a process from prepositional phrase to genitive/possessive construction.

This process can be illustrated with an English example. In Old English, *of* was a low stress variant of an Old English adverb and preposition meaning 'away', 'away from', as in (10a). In Late Old English and Early Middle English, *of* came to acquire uses occurring between two noun phrases, where its meaning gradually shifted from the older 'springing or coming from, belonging by origin to' to 'belonging to a place, as a native or resident', and in the eleventh century to 'belonging to as inhabitants or occupants', 'living in', and 'things situated in or at'; an example is provided in (10b).

(10) a. Old English (Anglo-Saxon Chronicle (Parker), 658 AD; OED)
Þis wæs gefohten siþþan he *of* Eastenglum com.
b. Early Middle English (*ca.* 1160, Anglo-Saxon Chronicle (Laud), 1132 AD; OED)
Was it noht suithe lang þer efter þatte king .. dide him gyuen up ðet abbotrice *of* Burch.

The change from Early Middle English to Modern English was marked by extension and desemanticization: *of* was extended to more and more contexts, spreading from one noun to another, increasingly encroaching on the domain reserved for the inflectional genitive. And from Middle English onwards it became the primary means for expressing attributive possession—turning into a fully productive marker of linking nouns. In other words, what used to be an adverb and preposition turned into a new case marker of noun phrase syntax in a process leading from a structure like (11a) to one like (11b). The result is that English acquired a new pattern of recursive syntax, illustrated in (11c), whose structure can be described as in (11d).

(11) Change from Old to Modern English
    a. Verb   *of* + NP   Preposition
    b. Noun   [*of* + NP]   Possessive/genitive
    c. the dog   [*of* the friend   [*of* my father]]
    d. [NP   [*of* NP   [*of* NP]]] ...

Much the same process happened in other European languages. In a number of northern European languages it was not the source schema of Table 6.2 but the location schema that was used to create new possessive constructions along the same lines, where a locative preposition 'at' was grammaticalized to a possessive/genitive case marker, such as Faroese *hjá* 'at', Scottish Gaelic *aig* 'at', or Irish *ag* 'at'; cf. (12a). In the case of the goal schema, attributive possession can be traced back to a directional, allative or benefactive adverbial phrase, where the adposition turned into a possessive case marker, as appears to have happened with Norwegian *til* 'to', illustrated in (12b).

(12) a. Faroese
    hestur- in   hjá   Jógvan- i
    horse- DEF.SG.M.NOM at   John-   SG.DAT
    'John's horse'

b. Norwegian
hatt- en til Per
hat- DEF to Peter
'Peter's hat'

To conclude, one way in which recursion in possessive constructions arises is via the reinterpretation and subsequent grammaticalization of locative or comitative adverbial phrases as possessive/genitival modifiers (Heine 1997a). Grammaticalization has the effect that the erstwhile preposition undergoes desemanticization: It loses its spatial meaning (of location, source, goal, etc.) and turns into a marker whose function is restricted essentially to marking a case relation. Decategorialization can be seen in the fact that the erstwhile preposition loses its ability to introduce participants of the clause, becoming restricted to one kind of context, namely that of linking a modifier to its head constituents within the noun phrase. And frequently there is also erosion, in that the preposition tends to lose its ability to receive stress or may be phonetically reduced on the way to becoming a case marker.

6.3.2 *Modifying compounding*

Of the four common types of noun–noun compounding,[13] only one, namely modifying compounding, clearly exhibits recursion in accordance with (1b), and we will be restricted here to this type of compounding.

Evidence for the hypothesis that modifying compounding has its origin in the combination of independent, non-recursive, nouns is of the following kind. First, this hypothesis is in accordance with the parameters of grammaticalization. The following processes can be observed: When a new noun–noun compound is formed, the modifying noun loses in referential and semantic properties: In the compound *apple tree*, the constituent *apple* no longer refers to a referential entity or to the semantics of an apple; an apple tree may be a tree that no longer has or—maybe— never had any apples on it (desemanticization). Furthermore, the modifying noun loses the ability to be itself modified, or to receive affixes such as plural markers (?*apples tree*), thus turning into an invariable form (decategorialization), and the modifying noun also tends to lose its

---

[13] The four types are modifying (e.g. *apple tree*), additive (*whisky-soda*), appositive (*poet-doctor*), and alternative (*egg head*) compounding.

individual stress or intonation pattern in favor of a suprasegmental pattern that is characteristic of the compound as a whole (erosion). Thus, compared to its use as an independent noun, the modifying noun lacks the salient morphosyntactic and phonological properties of a noun.

A second piece of evidence comes from historical observations. Noun–noun compounding typically arises via a process whereby free, referential nouns are combined in accordance with established conjoining patterns of the language concerned (condensation), which starts with loose combinations that are gradually transformed into tighter ones. And it is by and large a unidirectional process: The development from morphosyntactically loose combinings of nouns to tight compounds forming phonologically and semantically one single word is ubiquitous, while a development in the opposite direction is fairly rare.

Another strategy of forming modifying compounds is provided by attributive possession ("genitive constructions"), where specific attributive modifier combinations of free nouns (e.g. [[B's] A]) turn into regularly used noun–noun compounds (e.g. [[B-]A]). This is the process that can be held responsible crosslinguistically for many instances of modifying compounds. Evidence for this process can be seen in the fact that in a number of languages, compounds exhibit morphosyntactic relics of possessive constructions, one such relic being genitive case markings that survive the process from possessive construction to compound. For example, German compounds frequently contain the genitive case suffix -s which bears witness to their origin as possessive constructions, for example *Kalb.s.braten* (veal.GEN.roast) 'roast veal'. However, since possessive constructions are already recursive (see Section 6.3.1), we are simply dealing with a process from one recursive structure to another recursive structure; hence, this process is not immediately relevant to the present discussion, which is concerned with the rise of recursion.

Another kind of evidence for the hypothesis that noun–noun compounds are historically derived from the combination of self-standing nouns comes from languages that already have a construction for modifying compounding and develop new instances of this construction: In a number of instances it is possible to establish that a given compound cannot be traced back to earlier phases of the history of the languages concerned while the nouns making up the compound can. Thus, the English alternative compound *skyscraper* presumably did not exist prior to the occurrence of the relevant buildings whereas its constituents were there earlier as independent words; a more recent area of such new coinages involving previously existing nouns involves the area of computing, for example *website, laptop*. Even in languages for which we have no historical records it is possible to show that independent nouns were combined into compounds, while a process in the opposite direction does not seem to occur. For example, in the West African language Ewe there is a wide range of compounds which must have arisen after Ewe speakers came into contact with Western civilization, e.g., *ga-ŋkúí* (metal-eye.is.it) 'spectacles', *ga-só* (metal-horse) 'bicycle', or *ga-mɔ́* (metal-way) 'railway', while the constituents of these compounds, *ga* 'metal', *ŋkú* 'eye', *só* 'horse', and *mɔ́* 'way', already existed as independent nouns prior to this contact situation.

A final piece of evidence comes from synchrony: In many languages, compounding forms a productive process, where independent nouns can be combined creatively into new nouns expressing new meanings. Accordingly, we witness how new compounds arise and evolve. This process can be observed in actual language use, in that novel compounds are constantly emerging. Conversely, a process whereby compounds regularly develop into simple nouns is uncommon. To be sure, it may happen that in the course of time some specific compound may be lexicalized to the extent that it is no longer conceived as a compound and is reinterpreted as a simple noun. But even in such cases, the earlier development was one where independent nouns were combined into compounds.

That the directionality is generally from noun to noun–noun compound is also in accordance with findings made in language acquisition studies: Children learn first simple nouns before they proceed to acquire compounding, and hence recursion. Conversely, we are not aware of any convincing evidence to suggest that young children start out with compound nouns before they learn nouns in isolation. It is only between age 2;0 and 2;6 that English-speaking children are able to produce compounds (*birthday cake*), while they comprehend and use simple nouns (*cake, birthday*) clearly earlier, and Clark (2003: 298) observes: "The child coinages, *Dalmatian-dog* and *boxer-dog*, pick out subtypes of dogs; and, in each case, the modifier noun adds critical information for distinguishing which subtype is intended. The same goes for compounds like *car-smoke* vs. *house-smoke* (for car exhaust vs. smoke from a chimney)."

6.3.3 Adjectival modification

Adjective–noun constructions exhibit a number of restrictions. First, a number of languages do not have a distinct category of adjectives—hence there is no construction of this kind. And second, such constructions also show constraints on recursion in some languages, allowing simple but no productive recursion.

One way in which adjectival notions may be expressed is via the relativization of verbs; thus, in the Ik language of northeastern Uganda, qualities such as 'big', 'red', or 'good' need to be encoded by means of relativizing verbs of state, for example *dɛ-Ika ní ze-ík^a* (foot-PL REL.PL be.big-PL) 'big feet'. Since relative clauses are already recursive, we will ignore this kind of structure.

Instead, of central interest here is a process to be observed in quite a number of languages where two nouns are juxtaposed—with the effect that one of them is grammaticalized to an adjectival modifier while the other is the head of the construction—accordingly, a structure such as (13a) is grammaticalized to a recursive structure (13b) (see Section 2.2.1.1).

(13) a. N–N
b. N [Adj]

In this way, nouns for 'man' and 'woman' tend to acquire characteristics of adjectives in many languages, taking the position of adjectives and denoting 'male' and 'female', respectively. Semantic domains that are crosslinguistically most likely to undergo grammaticalization from noun to adjective are those of plants, animals, and metals, cf. English *a pink dress, a silver pot*. Grammaticalization involves (a) extension, in that a given noun is moved from the canonical position it occupies as a referential noun to a position next to another noun, (b) desemanticization, in that the nominal meaning is reduced to some semantic property of color, sex, size, etc., and there is also (c) decategorialization since that noun loses salient morphosyntactic properties of nouns, such as taking its own modifiers, determiners, and inflections.

That the directionality is generally one from noun to noun modification is also in accordance with evidence from child language development: Children learn nouns before they proceed to acquire noun modification. Conversely, we are not aware of any convincing evidence to suggest that young children start out producing modified nouns before they learn nouns in isolation. English-speaking children acquire simple nouns typically before age 2, while modifier–noun structures, such as possessive–noun (*your hand*) and adjective–noun phrases (*a brown car*) are produced only between age 2 and 2;6 (Valian 1986; Tomasello 2003a: 208). The fact that these children consistently place adjectives before, rather than after, nouns might suggest that they do have some knowledge of the taxonomic relationship between the two entities—that is, that already at this age they use this as a hierarchical structure ([A] B) rather than as a shallow, juxtapositional "island" structure (A–B).

6.3.4 Conclusion

In the preceding paragraphs we dealt only with a limited array of processes leading from nominal constituents showing no recursion to recursive noun phrases. These processes involved the grammaticalization of one constituent as a modifier of another constituent. Not all processes were of the same kind: On the one hand, they concerned the creation of new forms of recursive structures where previously there was none; on the other hand, they simply meant that new instances of recursion were added to an already existing pattern of recursion. As the data discussed suggest, the rise of recursive structures of the kind examined in this section are a predictable product of grammaticalization within the noun phrase—at least in grammatical models that assume that there is a rule of the kind proposed in (9).

6.4 Clause subordination

We saw above how recursive forms of language use may arise out of the grammaticalization of combinations of nouns or noun phrases. The problem that we are concerned with in this section is recursion in clause combining. In order to show how recursion in clause subordination arises, one simply needs to show how clause subordination arises. As we argued in Chapter 5, there are crosslinguistically only two ways essentially in which this may happen: Either via the expansion of a noun phrase into a clausal structure (expansion), or by conflating two independent sentences within one new sentence (integration). These two processes are sketched in (14) (= (2) of Chapter 5). Once either of these two processes has taken place, there is recursion, in that there is one clause (S_2) embedded in another clause, and the outcome of the diachronic process sketched in (14) can be captured by means of a synchronic rule of the kind proposed in (15).

(14) The rise of recursion in clause combining
    a. S [NP] > S_1 [S_2]   Expansion
    b. $S_1 + S_2$ > S_1 [S_2]   Integration

(15) S → S_1 [S_2]

Accordingly, to demonstrate that there is sentence recursion means simply to show that there is clause subordination. In Chapter 5 we surveyed how clause subordination arises. In the next section we will look at this process in more detail, illustrating with two historically documented cases the emergence and development of clause subordination, both involving clause integration (14b).

6.4.1 Case studies

The two cases concern situations where at some earlier historical stage there demonstrably was no recursive structure but where at some later stage recursion arose. Both cases are taken from a creole language; one concerns the emergence of restrictive relative clauses (Section 6.4.1.1) and the other concerns complement and adverbial clauses (Section 6.4.1.2). Creoles that exist today are fairly young languages; they all arose in the course of the last 400 years, and for some of them there are sufficient historical records to reconstruct the rise of recursive structures.

6.4.1.1 The rise of a relative clause construction

As we observed in Chapter 5 (Section 5.3.1.1), the most frequent development in the languages of the world is to create a new relative clause structure via the grammaticalization of demonstrative pronouns (not attributes), and creoles have also made use of this mechanism, as we will now demonstrate with an example from the English-based creole Sranan of Suriname.[14] Sranan arose in the late seventeenth century on the plantations of Suriname, which was made a British Colony in 1651; in 1668, Dutch became the colonial language. As in pidgins and newly emerged creoles, there were essentially no formally marked relative clauses in early

---

[14] For the data on this process, see Bruyn (1995a, 1995b, 1998).

Sranan: The English relativizers were lost, and most of the text examples that were seemingly unmarked relative clauses might also receive alternative interpretations, and there is no evidence that there was a recursive structure of relativization. Zero marking of relative clauses continued throughout the nineteenth and twentieth centuries.

For most of the seventeenth century, the demonstrative *disi* 'this' (< English *this*) was functioning both as a nominal modifier and a pronoun (we will refer to this as stage 0). Probably around the end of the seventeenth century, *disi* developed from demonstrative to relativizer within the short period of half a century, thereby giving rise to an overtly recursive structure (stage 1). In the course of the eighteenth century, *disi* underwent extension as an explicit relativizer. However, while *disi* could be understood as a relative clause marker determining the preceding noun, an interpretation of *disi* as a demonstrative in an independent clause could not be ruled out in most cases. However, there are eighteenth century examples such as (16), where *disi* was already unambiguously a relativizer.

(16) Sranan (Bruyn 1995a: 168)
    Hoe   fa   mi   zel   fom   wan   zomma
    Q   manner 1.SG FUT   beat INDEF.SG person
    [**diesi** no   doe   ogeri].
    REL   NEG   do   harm
    'How would I beat someone who didn't do any harm.'

But *disi* underwent a further grammaticalization, namely one from relative marker to adverbial clause subordinator (CONJ): In the second half of the eighteenth century, if not earlier, its use was extended to introduce temporal, causal, and concessive clauses; cf. (17a), which illustrates its temporal, and (17b) its causal uses.

(17) Sranan (Bruyn 1998: 29)
    a. ary   fadom   trange   **disi**   mi   de   na   gron.
       rain   fall.down strong   CONJ 1.SG be   at   ground
       'Rain fell heavily when I was in the fields.' (Source from 1765)
    b. **di**   ju   brokko   mi   nefi,   ju   musse
       CONJ 2.SG   break   1.SG   knife   2.SG must
       gi   mi   wan   so   srefiwan.
       give   1.SG   a   so   same.one
       'Since you've broken my knife, you must give me a similar one.'
       (Source from 1783)

290 *The genesis of grammar*

TABLE 6.3. *The grammaticalization of the Sranan demonstrative* disi *'this' to a relative clause marker*

Stage	Time	Function
0	seventeenth century	*disi* is a demonstrative pronoun meaning 'this'. Relative clauses are zero marked
1	From the end of seventeenth century	*disi* specializes as a relativizer. There is frequently ambiguity between its demonstrative and relative uses in specific contexts
2	Mid eighteenth century	*disi* is extended to also introduce temporal, causal, and concessive adverbial clauses
3	nineteenth century	*disi* is shortened to *di* as a relativizer, though not as a demonstrative modifier or pronoun

In eighteenth-century sources, the relativizer had the form *disi*; in the course of the nineteenth century, the new relativizer *disi* underwent erosion, that is, it was shortened to *di*, leading to the separation of demonstrative and relative marker. As example (17b) shows however, instances of erosion of *disi* can be found much earlier.

The whole process of the rise and development of new forms of clause subordination, however, did not mean that the original zero-marked structure was abandoned; rather, the latter survived to some extent and is still accessible today. The major stages in the development of Sranan *disi* are summarized in Table 6.3.

To conclude, Sranan speakers appear to have created subordinate clause structures in accordance with universal parameters of grammaticalization as sketched in Chapter 5. A strikingly similar example was presented in Section 4.3, involving the development of the Kenya Pidgin Swahili demonstrative *ile* 'that'. As grammaticalization theory would predict, in both cases there was a sequence of stages as sketched in Table 6.4.

Accordingly, in both cases there was a demonstrative pronoun used non-recursively that was grammaticalized to a new syntactic structure, thereby giving rise to a recursive structure of clause subordination.

TABLE 6.4. *Stages in the development from demonstrative to relative clause marker*

Stage	Function
0	There are no formally marked relative clauses
1	A demonstrative is introduced as an optional relative clause marker
2	The demonstrative tends to be ambiguous between its earlier and its new uses
3	The demonstrative is exclusively a relative clause marker in specific contexts

6.4.1.2 The rise of complement and adverbial clauses
Our second case concerns another crosslinguistically widely attested pathway from non-recursive to recursive construction of clause subordination, namely one leading from verbs meaning 'say' to complementizers as well as to adverbial clause subordinators. The main stages of this process are summarized in Table 6.5 (see Section 5.3.2.3 for general discussion). The following is a summary of the detailed analysis by Plag (1993, 1994b, 1995), describing the genesis of clause subordination in Sranan, where the verb *taki* 'say, talk' (< English *talk*) has been grammaticalized to a complementizer and eventually to an adverbial clause subordinator.

The initial stage of this development in Sranan falls into a period probably dating from 1780 to 1850. In (18a), *taki* is suggestive of a quotative marker (stage 1), although it can still be understood in its literal sense (stage 0). Sentence (18b) shows the use of *taki* as an object complementizer (stage 2), while (18c) is an example of the subsequent stage of a marker of subject complement clauses (stage 4). Finally, in (18d), *taki* has acquired the function of an adverbial clause subordinator.

(18) Sranan (Plag 1994b: 41, 1995)

  a. Na   Papa piki   hem,   a   **taki** "Luku,
     (the   father answer him/her 3.SG say   look
     sowan bigi   gro   mi   habi."
     such   big   field   I   have)
     'The Papa answered her, he said "Look, I have so large a field."'

b. Anansi si **taki** tok te a bori, a mu
   (Anansi saw that ? he cooked he must
   gi   eng uma.
   give   his woman
   'Anansi saw that when he cooked he must give (something) to his wife.'

c. **Taki** Kofi no kiri Amba meki wi breyti.
   (that Kofi NEG kill Amba make we happy)
   'That Kofi didn't kill Amba made us happy.'

d. Den de so don **taki** yu musu [...].
   (they be so dumb that you must)
   'They are so dumb that you have to [...].'

The historical development from main clause combining to clause subordination is summarized in Table 6.6. The data provided by Plag suggest that up until the middle of the nineteenth century, *taki* was largely confined to introducing arguments, that is quotes and complement clauses, and it is only after 1850 that *taki* came to introduce adjuncts such as purpose clauses (see especially Plag 1995: 130, 134).

Like the preceding example of Sranan *disi/di*, the present case shows how two independent sentences merge into one sentence via integration (cf. (14b)), where one of the sentences is grammaticalized to a clause that is subordinate to the other sentence, thereby giving rise to a morphosyntactic structure of embedding recursion, first of complement clauses

TABLE 6.5. *Stages in the evolution from 'say' to complementizer*

Stage	Function	Context
0	Unconstrained verb 'say'	Free
1	Quotative marker introducing direct and indirect speech	Headed by speech act verbs
2	Restricted complementizer	Headed by verbs of perception and cognition
3	General marker of object complement clauses	Headed by verbs of volition, intention, etc.
4	General marker of subject complement clauses	Headed e.g. by verbs of state
5	Marker of adverbial clauses	Headed by matrix clause

TABLE 6.6. *The grammaticalization of Sranan* taki *'say' to a complementizer and adverbial clause subordinator*

Stage	Time	Function
1	?	*taki* is a general speech act verb meaning 'say'
2	Late eighteenth century	*taki* turns into a quotative marker after other speech act verbs like *piki* 'answer' or *bari* 'cry', introducing direct speech
3	Nineteenth century (if not earlier)	*taki* is established as a complementizer after speech act, perception and cognition verbs (e.g. *si* 'see')
4	Mid nineteenth century	*taki* also serves as a complementizer of noun complement clauses
5	Late nineteenth century	*taki* now serves as a conjunction of consecutive adverbial clauses.
6	Twentieth century	*taki* is now a conjunction of purpose adverbial clauses as well as of subject complement clauses

and subsequently of adverbial clauses—in other words, a non-recursive combination of two sentences [$S_1 + S_2$] turned into a recursive combination [$S_1[S_2]$] in accordance with (14b) and (15).

6.5 Loss of recursion

As the preceding discussions suggest, recursion is a product of the rules that one wishes to set up in order to describe and/or account for structural properties of a given language. Once one's description contains a rule of the form (1b) then there is recursion. The position we adopted in this chapter is that there are pragmatic reasons in favor of such a rule, and our main concern was with showing how recursive morphosyntax emerges out of non-recursive structures. But there is also evidence for the opposite development, whereby recursive morphosyntax ceases to exist.

One kind of decline is provided by the development of adpositions. Crosslinguistically the most common pathway in which adpositions arise is via the reinterpretation of nouns for body parts or environmental landmarks, forming the head of possessive constructions, as prepositions or postpositions (Heine, Claudi, and Hünnemeyer 1991; Heine 1997b). For example, the phrase (19) is historically a possessive construction consisting of the modifier noun g‖ú 'water' and the head noun !x՚ā 'heart', but the latter has been grammaticalized to a postposition meaning 'inside'. The earlier syntactic structure of (19) can thus be represented as in (20a) while its present structure takes the form (20b). Note that this is by no means an uncommon example; similar examples can be found in other languages across the world.

(19) !Xun, W2 dialect (North Khoisan)
g‖ú   !x՚ā
water heart
'inside the water' (lit.: 'heart of water')

(20) From noun to adposition
 a. [NP_1] NP_2
 b. [NP]   adposition

In accordance with the rule convention adopted for this example, such cases can be described as ones where a recursive possessive construction of the form (20a) has been grammaticalized to a non-recursive adpositional construction (20b)—that is, as a construction that no longer corresponds to our definition of recursion in (1b).

The second kind of decline is of quite a different nature; it concerns loss of productivity in grammatical constructions: Once a recursive construction is lexicalized to an unanalyzable unit, it loses its recursive nature. For example, English has a recursive [[adjective] noun]-rule. But in adjective–noun compounds, such as *darkroom, hothouse,* or *blackbird,* recursion is lost.

To conclude, the languages of the world offer many examples where new recursive constructions arise, but also examples where existing recursive structures can disappear. Loss may only concern individual, lexically determined use patterns, as in the English examples, but it may as well affect more inclusive phrase structures.

6.6 Conclusions

Hauser *et al.* (2002) argue that the language faculty in the narrow sense (LFN) consists essentially only of recursion. The present chapter was not meant to challenge this view; rather, its main purpose was to show how recursive constructions may arise. We argued that there is a regular process leading from non-recursive to recursive noun phrase and sentence constructions, and that this process is a by-product of grammaticalization and, hence, can be accounted for by means of grammaticalization theory. Our concern was not with structure in general but rather with specific constructions. For example, we observed in Section 6.3.1 that English speakers created the recursive *of*-construction, but there was already a recursive construction of attributive possession, marked by genitive -*s*. Thus, what we showed was simply how a non-recursive construction turned into a recursive one. In most cases this process involved integration, whereby constituents that were juxtaposed—be that in the form of asyndetic, paratactic or any other kind of combining—gradually turned into hierarchical structures of modification or subordination.

In Section 6.1.1 we distinguished between two basic types of recursion: tail recursion and nested (or center-embedded) recursion. Our main concern was with the former, and the question arises whether grammaticalization theory also accounts for nested recursion, which for some constitutes the "true" form of recursion: It entails not only that category X is nested in another category Y, where some parts of Y precede while others follow X, but also that these parts of Y, although forming one syntactic unit, are separated by X. Grammaticalization theory in fact accounts in much the same way for both kinds of recursion.

An example of nested recursion was provided from Kenya Pidgin Swahili in Chapter 4 (example (17e)). We described in that chapter how this pidgin arose in the early twentieth century without any relative clause construction, and how it acquired such a construction by grammaticalizing the distal demonstrative *ile* 'that' to a relative clause marker—in much the same way as had happened earlier in the history of English. Now, whenever the newly acquired relative construction determines the subject noun phrase, it immediately follows the latter, with the effect that the subject is detached from the verb phrase. In this context, the predictable outcome of grammaticalization is therefore center-embedding and nested recursion. Note that nested recursion is productive, as example (21) shows, even if such double nestings are rarely found in normal discourse.

(21) Kenya Pidgin Swahili

Ile	mtu	[ile	iko	watoto	tatu	[ile	na-
DEM	person	REL	have	children	three	REL	NF-
kosa	chakula]]	kwisha	fung-	wa.			
lack	food	PERF	shut-	PASS			

'That man, who has three children, who lack food, has been jailed.'

Our main concern in this chapter was with simple recursion, and this raises the question of how simple recursion may lead to productive recursion, that is, to a characteristic of human languages that tends to be described as discrete infinity. The development from simple recursion to productive recursion is a fairly basic one: It simply means that an existing rule is applied more than once. However, as we noted in Section 6.1.3, there are considerable differences with regard to which structures of a language allow productive recursion. In clause subordination it tends to be fully productive—that is, there are in principle no limits to its application. And much the same applies to attributive possession: In many languages there are essentially no constraints on its productivity. But other noun phrase structures frequently exhibit constraints of one kind or another, and in the case of modifying compounding, there exists crosslinguistically a broad range of variation: While some languages, such as German or Ewe, have virtually no limits in productivity, other languages, such as French or Swahili, allow only simple recursion, if at all. Overall one may say that recursion tends to be more productive with structures that are syntactically relatively free, such as clauses or possessive constructions, and less productive with bound structures such as compounds; but, to our knowledge, there are no general typological studies on the degree to which recursion is productive, or on why there are such differences.

Our discussion may have shown the following: First, on the basis of the grammaticalization evidence that we found, recursion does not appear to be created for its own sake but rather arises as a by-product of other cognitive or communicative intentions, such as modifying some argument or proposition—hence, it is a predictable outcome of certain grammaticalization processes. However, while grammaticalization is a necessary requirement for a recursive structure to arise, it is not a sufficient one; as we saw in Section 6.2, what is required in addition is an understanding of taxonomic hierarchy, as it manifests itself in particular in the conceptual relationship between inclusive and included taxa.

Second, what recursion achieves grammatically can essentially be achieved as well by other grammatical means, such as concatenation of phrases or sentences, or by pragmatically determined strategies, such as appositional or afterthought specification. Accordingly, there are languages that have been argued to do well with little use of recursion. And second, there are in fact forms of linguistic communication that lack recursion; as we saw in Section 4.7, no form of recursive structures has been found in restricted linguistic systems such as those of homesigners, twins' language speakers, and isolated children—however, these speakers are able to interact linguistically fairly well, both among themselves and with others.

Third, the manifestation of recursion that is perhaps most frequently mentioned in the relevant literature, namely its ability to introduce discrete infinity into language structure, is one that turns out to be fairly peripheral if one looks at how languages are actually used, as can be determined, for example, by studying larger samples of texts: Most instances of recursion in language concern simple embedding, while instances of three or more embeddings are rare. And finally, we wish to reiterate that recursion is a theory-dependent notion—it is a property of the system of rules proposed rather than of the phenomenon to be studied. As such, the notion can be applied to many kinds of different phenomena, including phenomena that are unrelated to human cognition; for example, biological reproduction can be described efficiently in terms of this notion. We do not see any intrinsic reason why a linguistic theory that ignores this notion should be less appropriate than one that does not.

7   Early language

The goal that we want to achieve in this book is a fairly ambitious one: Considering that there are 4,000–6,000 languages spoken in the world, exhibiting a wealth of contrasting structures and socio-cultural environments, reconstructing all this diversity back to one earlier form of language is a task that is near to impossible. The reason we are nevertheless tackling this task is that we are restricting analysis to one specific methodology and to a limited spectrum of phenomena. The procedure adopted in the preceding chapters was to concentrate on grammaticalization in reconstructing some properties that are hypothesized to have characterized language at an earlier stage of human evolution. In the present chapter we will recapitulate the main findings presented and deal with the questions that we listed in Chapter 1 (example (1)). While discussion in the preceding chapters was concerned with observations made within the framework of grammaticalization theory, in this chapter we will also look at findings made by other authors using alternative approaches and/or drawing on wider ranges of relevant phenomena.

7.1   Grammatical evolution

The reconstructions presented in Chapter 2 were meant to provide a general outline of grammatical evolution, and to this end we discussed a network of pathways of grammaticalization. In this section we wish to explore what this network can tell us about language evolution.

7.1.1 *Layers*

The pathways that were the subject of Chapter 2 were described in terms of a sequence of six layers of evolution, which are reproduced in Table 7.1. In this table, we also propose a scenario of grammatical innovations on the basis of the structural properties characterizing the various layers.

TABLE 7.1. *Layers of grammatical evolution (based on Figure 2.1)*

Layer	New categories introduced	Hypothesized main grammatical innovations
I	Nouns	One-word utterances[a]
II	Verbs	Mono-clausal propositions
III	Adjectives, adverbs	Head-dependent structures
IV	Demonstratives, adpositions, aspect markers, negation	Elaboration of phrase structures
V	Pronouns, definite (and indefinite) markers, relative clause markers, complementizers, case markers, tense markers	Temporal and spatial displacement, the beginning of clause subordination
VI	Agreement markers, passive markers, adverbial clause subordinators	Obligatory expressions, elaborated clause subordination

[a] That we hypothesize that at layer I there were one-word utterances can by no means be taken to lend support to the "holistic hypothesis" that we mentioned in Section 1.1.3: There is neither evidence to suggest that such utterances would have been semantically complex, nor that they were of a kind that would have allowed for a segmentation process whereby complex but unanalyzable signals were broken down into words and syntactic structures.

The only parameter we used for distinguishing these layers was the relative degree of grammaticalization that a given category exhibits vis-à-vis less or more strongly grammaticalized categories of the same overall pathway. To determine the exact diachronic significance of these layers is a matter of future research. One problem such research will face is that grammaticalization proceeds at different paces in different domains of grammar and/or in different languages. This may mean, for example, that a given category can be more strongly grammaticalized in domain D_1 or in language L_1 than another category in domain D_2 or language L_2 although it belongs to a less grammaticalized layer than the latter category. The classification that we present in Table 7.1 therefore has to be taken with care.

Another problem concerns the internal nature of layers, as can be illustrated by the following example. The categories discussed under layer V include definite markers and case markers. More recent research (König forthc.) has shown that there is a pathway from definite markers (sometimes in combination with some other marker) to case markers, most of all to subject or ergative case markers. This pathway appears to be unidirectional; there is no evidence suggesting a development from case morpheme to definite article. Furthermore, as we showed in Section 5.2, there is a unidirectional pathway from case markers to complementizers, even though both are hypothesized to belong to layer V. Such examples suggest that the internal structure of the various layers is more complex than portrayed here, in that not all pathways making up a layer necessarily exhibit the same relative degree of grammaticalization. However, the layers that have been distinguished are impressively supported by some body of crosslinguistic data and we will now use them to formulate some tentative hypotheses on the development from early language to modern languages.

Layer I On the basis of the reconstructions proposed in Section 2.2.1 we assume that at layer I there was only one kind of category, namely "nouns", that is, time-stable, referential units expressing primarily thing-like concepts.[1] Assuming that this reflects a significant stage in the evolution of language this implies that language structure at this stage was associated with one-word utterances (see below), conceivably also with sequences of such utterances.[2]

Bickerton (2005: 7) asserts that "a verbless protolanguage seems intrinsically implausible." One may wonder whether this is a credible position. That it is possible to establish communication by means of one-word utterances without verbs is suggested, for example by observations on agrammatic patients and children at the one-word stage of language acquisition. And there are a number of modern communication systems, all of them restricted linguistic systems, that may be used to a considerable extent without verbs. One example is volunteered by Bickerton himself (1990: 118–22). In his discussion of pidgins, he describes two varieties,

---

[1] Maggie Tallerman and Jim Hurford (p.c.) rightly point out that a category "noun" does not make much sense unless contrasted with other kinds of word categories. We are restricted here to the application of grammaticalization theory, which is concerned with the reconstruction of basic linguistic categories, and "noun", that is, a category referring to thing-like entities, is one of these categories. What exactly the reconstruction of a category "noun" means with reference to the pragmatic situation that may have characterized linguistic behavior at layer I is a question that is open to further research (see below).

[2] An anonymous referee of this work rightly points out that if in fact it was possible to combine utterances then there was already word order at layer I.

namely Hawaiian pidgin English and Russonorsk (which was used in trading contacts between Russian and Scandinavian sailors). He observes on the former that "verbs may be missing altogether", and on the latter that "[T]he longest utterance, *Big expensive flour on Russia this year*, contains no verb." Another example can be found in untutored late second language acquisition by adult immigrant workers who did not receive explicit instruction in the host language: The first phase of development, referred to as the pre-basic variety, is characterized by noun-based utterance organization consisting essentially of nominal structures without verbs, where the meaning of words is highly context-dependent (Klein and Perdue 1992; Perdue 1996; Benazzo 2006). While our methodology does not allow for any elaboration on this issue, it would seem plausible to assume that the meaning of nominal utterances at layer I was propositional in nature, and that it was highly context-dependent. Thus, a nominal utterance *N* may have been used in appropriate contexts to express propositional contents, such as *(I see) N, (Give me) N,* or *(There is) N*; however, this is no more than a possibility—one that remains conjectural without any further empirical support.[3]

Another question that arises is whether the entities making up layer I were exclusively common nouns or whether they may also have included proper nouns (or names). This is an issue that has been the subject of controversy: Whereas Hurford (2003) argues that there were no proper names in earlier forms of human language, Bickerton (2005) maintains that his "protolanguage" already had both common nouns and proper nouns. Grammaticalization theory does not provide any clues on this issue, except for the following observation: As we saw in Chapter 2, common nouns are one of the main sources of functional categories, while proper nouns do not normally undergo processes of grammaticalization.[4] However, not all common nouns undergo grammaticalization, only a small portion of them do; accordingly, this observation is of little help in deciding on this issue.

---

[3] An anonymous referee of this work makes the important observation that lexical nouns are useful for referring to entities that are not immediately accessible, whereas pronouns, demonstratives, and other pointing devices are characteristic for talking about the here-and-now. Conceivably early language speakers made extensive use of gestures for deictic and other communicative purposes, but our methodology does not allow for any insights on this matter.

[4] There is one notable exception that we pointed out in Section 5.3.3.2: In the Chadic language Lele, the noun *álà* 'God' was grammaticalized to an adversative conjunction 'but' (Frajzyngier 1996: 81–3).

Aitchison (1998: 24–5) suggests that language very likely began "messy", and only gradually neatened itself up. Structure began to emerge when there were preferences on how to arrange existing items, which became habits, which again may have turned into rules. This characterization is in accordance with findings on grammaticalization: There is abundant crosslinguistic evidence to show that the emergence of new categories is somehow "messy" in its early stages: There is variation but no new structure, there are use patterns but no categories. It is out of this variation that new use patterns evolve, that is something that comes close to Aitchison's habits, and some of these use patterns may develop further into stable, conventionalized categories (see Heine and Kuteva 2005, ch. 2). Accordingly, the emergence of noun-like units was presumably no more than the peak of the iceberg surfacing from the cognitive and communicative activities making up the earliest forms of human linguistic communication.

Layer II Layer II was also an essentially lexical stage of language evolution, characterized by the presence of two kinds of categories, namely nouns and verbs.[5] This implies, first, that there were time-stable concepts for thing-like phenomena, and conversely non-time-stable concepts describing actions or events (cf. Aitchison 1996); but there is no way of telling how many members each of these categories included, that is, what the size of the lexicon at that layer was. Second, this also implies that there were means of combining the two, that is, that it was possible to form noun–verb combinations, let us say, mono-clausal verb-argument constructions. There is no evidence to conclude that at layer II it was already possible to use a verb with more than one argument; but the presence of at least two constituents suggests that there may have been some principle of linear arrangement of these constituents (see Section 7.1.4).

Thus, at layer II there must have been some rudimentary form of grammar, but there were no items whose primary function it was to express relations among words, that is, there was no morphology or—in the words of Bernard Comrie (1992: 209; see also 2002)—there was at best

[5] We noted in Chapter 2 that the evidence for distinguishing layers I and II is of a different nature than the one that we are using in the case of all other stages: While grammaticalization deals with grammatical forms, the distinction between I and II concerns the structure of how these forms are presented, rather than the forms themselves—hence, the evidence is circumstantial rather than direct; we therefore marked the evolution from noun to verb in Figure 2.1 with a dotted line.

"a language with isolating morphological structure", where the only productive means of syntax must have been word order. Whether the introduction of word order marked the beginning of grammaticalization in language evolution is an issue that requires further research.[6] While it is possible that there were notions for conceptual relations such as spatial orientation or possession, there were no grammaticalized means to express them.

Casey and Kluender (1998: 74–6) argue that both deaf children with impoverished language input and some trained non-human primates may represent an intermediate form of communication in the evolution of language. In fact, layer II appears to be suggestive of what we tentatively referred to in Section 4.8 as an elementary linguistic communication system, reflected in the behavior of homesigners, twins' language speakers, and isolated children and in the abilities to be found in some non-human animals (see Chapters 3 and 4), having structural characteristics such as the ones listed in (42) of Section 4.8 (see also Johansson 2002: 122, 2005). But, as we also saw in Chapter 4, layer II is unlike anything that one commonly finds in stable or expanded pidgins. Whether such an elementary linguistic system may have constituted a distinct stage—one that stands out in the evolution of language and could be meaningfully related to notions such as "protolanguage", "proto-grammar", or "pre-grammar" (e.g. Calvin and Bickerton 2000: 137; Hurford 2003: 53; Bickerton 1990, 1995, 2005; Givón 1995, 2002a, 2005)—is an issue that cannot be answered satisfactorily on the basis of findings of grammaticalization.

Layer III Once there is a noun–verb distinction, many other design features can collect around it (Jackendoff 2002: 259), and grammaticalization now becomes the driving force of grammatical evolution. The third layer of development was marked by the emergence of new word categories via the grammaticalization of nouns and verbs, namely adjectives and adverbs. Since these categories are functionally dependent on the categories of layers I and II, we hypothesize that this was the stage where the first phrasal structures arose, namely noun–adjective and verb–adverb constructions.[7]

---

[6] To the extent that word order is a grammatical device that is based for example on a conceptual transfer from temporal and spatial relations to the linear arrangement of utterances and their parts it can be said to be a manifestation of grammaticalization.

[7] This hypothesis is in accordance with Jackendoff's (2002: 252) observation that word order on the clausal level precedes that of the phrasal level: "The provision of headed phrases in grammar allows principles of *word order* to be elaborated into principles of phrase order."

Such a process can be observed quite commonly in modern languages. What is needed for a new grammatical category such as adjectives to evolve is, for example, simply to have a combination of two lexical nouns: Once this combination is used frequently enough over an extended period of time, one of the nouns may gradually assume an auxiliary function, turning into a grammatical modifier of the other. For example, as we noted in Section 2.2.1.1, in a number of languages, combinations of two nouns, $N_1 + N_2$, where one of the nouns denotes a plant, a plant part, or a metal, have given rise to adjectival (color) modifiers, as in English, where plant parts such as *orange* or *pink*, or metal names such as *bronze*, *brass*, or *silver*, have acquired properties of color adjectives, as in *an orange cup*, *a pink dress*.

Our reconstruction thus implies that sentence structure preceded phrase structure in time. And this innovation would mark the beginning of head-dependent structures,[8] that is, of hierarchical syntactic organization. And since hierarchical head-dependent structures can give rise to recursive structures, where one category is embedded in another category of the same type—in accordance with rule (1b) of Chapter 6, it is possible that this innovation led to the introduction of recursion in language structure. This does not necessarily mean, however, that recursion as a *cognitive* principle was not already in place before layer III; our concern here, as elsewhere in this book, is exclusively with morphosyntactic exponents of recursion.

Recursion in language structure is, as we argue, most of all a product of noun modification on the one hand and clause subordination on the other. In accordance with our scenario of grammatical evolution that we proposed in Chapter 2 (Figure 2.1) we therefore hypothesize that it emerged at the earliest at layer III, when modifying categories such as adjectives and adverbs made their appearance. In clause combining this must have happened much later, namely at layer V (see below).

Layer IV Layer IV does not seem to have been characterized by dramatic syntactic innovations: It led essentially to an elaboration of phrase structure. The introduction of demonstratives and adpositions meant an

---

[8] That a head-dependent structure in the form of part–whole relationships was conceptually present at the latest some 10,000 to 15,000 years ago might be suggested by rock engravings made by hunter-gatherers of the Upper Paleolithic depicting animal bodies without head, or body parts without body.

elaboration of the already existing head-dependent phrase structure centering around the noun. Demonstrative categories are believed to belong to the earliest materials to evolve in language (see, e.g., Diessel 2005: 24) and, in fact, they are among the first functional categories to appear; the evidence that we were able to access suggests, however, that their appearance cannot reasonably have preceded layer IV. And in the same way as the noun phrase, layer IV also affected the verb phrase: The rise of verbal aspect marking and negation made it possible to manipulate verbal predications. However, discourse structure remained restricted to non-embedded mono-clausal predications.

Layer V The hypothetically set up layer V is associated with the most dramatic changes in language structure. First, it introduced formal means for presenting multi-propositional contents, more particularly clause subordination, as is suggested by the availability of relative and complement clause subordinators. Second, the introduction of pronouns and definite markers allowed for displaced reference: Rather than being restricted to referents accessible within the immediate speech situation, it was now possible to refer to temporally and spatially displaced participants (Givón 2002b: 32). And finally, with the introduction of tense markers, events and states could now be represented as detached from the here-and-now.

Clause subordination, which was discussed in detail in Chapter 5, must therefore have arisen only at a more advanced stage of evolution, that is, well after phrase structure was already in place. Note, however, that we are restricted here to the emergence of the morphosyntactic means used for subordination; this does not necessarily mean that at some earlier layer it had not already been possible to express subordination by juxtaposing sentences.

Thus, layer V could have brought about another significant innovation in the development of human language. Evidence from primatology and ontogeny has been taken to suggest that in the earliest forms of human language, communication was overwhelmingly about here-and-now, I-and-you, and this-and-that reference,[9] that is, about the immediate

---

[9] An anonymous reviewer of an earlier version of this book draws attention to the following issue that needs more attention in future research: Modern everyday communication is also overwhelmingly about the here-and-now and the linguistic categories employed are more likely to be pronominals and other pointing devices rather than, for example, lexical nouns. This would seem to suggest that pronominals might have been a fairly early acquisition of early language speakers; but see also Section 2.5 on this issue.

speech situation and that there were no operations for perspective shifting (Givón 2002b: 32–3). While a number of non-human animals have been found to show object permanence in the sense that they are able, for example, to refer to objects in their absence (Section 3.2.2), it is safe to assume that expressions for displaced reference, that is, reference to participants, states, and events that are outside of the immediate speech situation, were not characteristic of the earliest forms of language that are accessible via grammaticalization theory; it is only at layer V that categories that may be taken to be suggestive of concepts relating to perspective shifting, such as tense markers, pronouns, definiteness markers, relative clause markers, and complementizers are attested.

And the introduction of recursion in the structure of clause combining must also have taken place at layer V, when early language users developed the ability to form relative clauses and complement clauses. At the earliest stages in the rise of recursive structures there may have been only simple recursion, but the information that is available is not sufficient to determine whether this was so and at what stage human language acquired productive recursion.

Layer VI The final layer VI led most of all to an obligatory marking of functional categories. It meant on the one hand that optional means of expressing agreement or distinctions of argument structure that may already have existed became obligatorily encoded, and that clause combining became increasingly more complex. Subordination of adverbial clauses was possible to some extent already at layer V, but it now became a fully-fledged means of embedding. And with the emergence of formal markings for passive constructions, there were now grammaticalized means for manipulating discourse participants in utterances. Thus, layer VI represents a stage in evolution that marks the transition from early language to the modern languages as we find them today.

Discussion For a better understanding of this scenario, three points need to be emphasized. First, our methodology provides access only to linguistic expressions, while the question of what kinds of concepts may have been distinguished by early language speakers is quite a different issue.[10] For example, it is possible that speakers as early as layer II had functional concepts for, say, negation or tense and aspect, but that these concepts

---

[10] We are grateful to Fritz Newmeyer (p.c.) for having drawn our attention to this problem.

were expressed by lexical means, as is still possible today: There are verbs such as 'fail', 'reject', 'stop (doing)', 'ignore', etc. that can be used to render the notion of negation, or 'finish', 'stay', etc. to express aspect distinctions. Our scenario is able to show how such concepts over time turned into regularly used grammatical expressions but not when they first appeared in human evolution.

Second, the hypotheses presented are based on regularities in grammatical evolution rather than on typological comparisons. For example, whereas most languages distinguish the categories of layers I through IV,[11] there are quite a number of languages that lack a number of the categories figuring in layers V and VI—a situation nicely summarized by Li thus:

> Many languages in the world have little or no grammatical agreement, derivation, inflection, declension, tense, number, gender, case, and many languages with long histories of both written and spoken traditions rely primarily on some word order conventions and grammatical particles with little or no morphology as their grammar. The Chinese language family is but one of numerous examples. (Li 2002: 92)

This situation may be due to a number of factors (see Section 1.2.2 on Riau Indonesian). One possibility is that such morphology did exist at some earlier stage in the development of these languages but was subsequently lost; for example, it is possible that the Chinese language family had most or all of the categories of layers V and VI at some earlier stage of its development. Alternatively, it is equally possible that such categories never existed in the Chinese language family or in some other languages. To decide which of these alternatives applies is beyond the scope of linguistic methodology, including that of grammaticalization theory. Accordingly, our reconstruction of the categories of layers V and VI is based on the analysis of languages that distinguish the relevant categories. Chinese would therefore not be among the languages that are of use to reconstruct categories of agreement, gender, or case.

And third, we are concerned here with the development of individual functional categories rather than with the evolution of languages as wholes. Accordingly, it would be futile to try to locate a given language along the scale of layers distinguished: In the same way as grammaticalization produces new functional categories it may lead to the decline and eventual loss of such categories. The English language is a case in point: In

---

[11] There are a few exceptions; for example, not all languages distinguish adjectives as a morphosyntactic category.

the course of its history over the last millennium, it lost much of its strongly grammaticalized morphology that it had inherited from early Indo-European times, including its case system and most of its agreement structures.

The scenario presented above is generally in agreement with that proposed by Givón (2002a, 2005), which is also based on observations on grammaticalization even if he includes a wide range of other phenomena in addition. Givón hypothesizes that grammatical morphology arises overwhelmingly from lexical words, hence this presupposes the prior existence of a well-coded lexicon. He goes on to propose the following major lines of language evolution (Givón 2002a: 158–9), which are substantiated by the reconstructions discussed above:

(a) iconic syntax before arbitrary syntax;
(b) manipulative before declarative speech-acts;
(c) one-word before two-word before longer clauses;
(d) simple clauses before complex clauses;
(e) mono-propositional before multi-propositional discourse.

Our scenario also shows similarities to alternative hypotheses that have been proposed on language evolution (e.g. Smith 2006). Thus, layer I can be likened to Jackendoff's (2002: 238 ff.) stage I of single-word utterances consisting of "palaeo-lexical" or "defective" lexical items. And layer II may be likened to Calvin and Bickerton's (2000: 136–7) stage 2 situation, where there were categories of agent, theme, and goal which "were exapted to produce a basis for sentence structures," or to Jackendoff's (2002: 238 ff.) hypothesized stage II of language evolution, characterized by the concatenation of symbols into larger utterances and the innovation of a large lexicon. Note also that in Jackendoff's hypothesized protolanguage there was "no grammatical differentiation of parts of speech, only Object words versus Action words. There is no inflection for case and agreement, and no use of pronouns and other proforms", but there were semantically defined notions like agent and patient (Jackendoff 2002: 255).

No attempt is made here to relate our scenario to these alternative hypotheses, however, especially to Bickerton's notion of a protolanguage,[12] for the following reasons. First, our goal is a more limited one,

[12] What Bickerton (1990, 2005) calls protolanguage is a structure that he claims to have characterized the speech of *Homo erectus* (or *Homo ergaster*) about 1.6 million years ago. Protolanguage, he argues, had vocal labels corresponding to concepts, but it had no true syntax, and no, or hardly any, grammatical items, and the arrangement of elements in an utterance was determined by pragmatic rather than syntactic mechanisms.

it is not, for example, concerned with human biological evolution. Second, our hypothesis is based on grammaticalization theory, that is, on regularities of grammatical change—hence it can be falsified with reference to this theory. Falsification is more difficult in the case of the former two hypotheses, which rest on observations on a range of different phenomena (see Bickerton 1990, 1995; Calvin and Bickerton 2000; Jackendoff 2002: 240), but do not take diachronic generalizations into account.

And third, our reconstruction suggests that language evolved gradually in an incremental fashion, constantly acquiring new grammatical use patterns and categories, and becoming increasingly more complex (see also Pinker 1994; Jackendoff 2002). There is no convincing evidence for something like "protolanguage" as standing out as a distinct stage in this evolution. If in fact there was a more salient stage then more likely it would have been located somewhere around layer V, when some more dramatic innovations were made, as we saw above. And even this stage falls out naturally from the preceding stages, being an elaboration of previous structures via parameters of grammaticalization—parameters that we hypothesize to have been essentially the same from the beginning of human language up to the modern languages as we find them today.

But there is an alternative hypothesis that strikingly resembles the one proposed here, even though it is based on a different perspective. Noting that human languages possess four salient features that are relatively independent of one another, Johansson (2002, 2005, 2006) argues that these features must have arisen successively in language evolution. In accordance with these features, languages are:

(a) Structured: there are some rules ordering sequences of items;
(b) Hierarchical: there are levels of structures within structures;
(c) Flexible: words can be moved around, sentences can be restructured;
(d) Recursive: structures may contain substructures, where the latter are "incarnations" of the former.

Confronting this classification with a number of different communication systems, Johansson maintains that these features are suggestive of differential levels of linguistic elaboration, and he finds support for an implicational scale where (d) presupposes (c), (c) presupposes (b), and (b) presupposes (a).[13]

---

[13] Concerning a similar procedure of reconstruction, based on logical dependence among hypothesized stages underlying language origin, see Bierwisch (2001).

On the basis of these features he proposes a five-stage model of successive grammatical elaboration, as depicted in Table 7.2. This table shows that there are significant correspondences with the scenario of grammaticalization of Table 7.1. While the status of feature (c) in this model does not become entirely clear, it would seem that, overall, the sequence postulated by Johansson captures a major line of progression in the grammaticalization of language structure, namely one leading from elementary utterances without internal structure to structural relations among word units, subsequently to complex, hierarchical structures, and finally to clause subordination and recursive syntax.[14]

And Johansson's model is also in agreement with another reconstruction that we proposed above, namely that the language-like abilities of non-human animals can be located roughly at layer II of language evolution: Johansson (2002: 122; 2005: 233–40) suggests that stage 2 of Table 7.2

TABLE 7.2. *Sequence of successive grammatical elaboration according to Johansson*

Step of grammatical elaboration	Layer of grammaticalization roughly corresponding to Johansson's steps of elaboration
1 One-word stage: basic semantics with no syntax (= unstructured)	I One-word utterances
2 Two-word stage (= structured)	II Mono-clausal propositions
3 Hierarchical structure, much like a basic phrase structure grammar (= hierarchical)	III Head-dependent structures
4 Recursive syntax (= recursive)	IV Elaboration of phrase structures
5 Full modern human grammar	V Temporal and spatial displacement, the beginning of clause subordination
	VI Obligatory expressions, elaborated clause subordination

Source: Johansson (2005, 2006).

---

[14] Fritz Newmeyer (p.c.) suggests that the linguistic dividing line between *Homo erectus* and *H. sapiens* may have been between step 2 and step 3.

very likely is within the reach of chimpanzees, involving nothing but activating already existing capabilities.

7.1.2 From non-language to language

One question that we asked in the introductory chapter ((1g)) was whether language genesis can be linked to the behavior of non-human animals, especially that of apes: How does the hypothesis on early language proposed above (Section 7.1.1) relate to what we know about cognitive abilities and communication systems to be found in non-human animals? An answer to this question is faced with a number of problems; we discussed them in Chapter 3. Ignoring these problems, it would seem that at least a partial answer is possible. Research findings that we discussed in Section 3.3 suggest, first, that language should not be viewed as contingent upon speech and, second, that the prerequisites to language are presumably not uniquely human. We saw that some non-human animal species have been found to display a respectable catalog of language-related abilities, such as the ones listed in Section 3.3.2 ((2), repeated here for convenience):

(a) to understand salient characteristics of concepts;
(b) to distinguish form–meaning units ("words");
(c) to acquire form–meaning units of more than 100 items, including items denoting objects, actions, and some numbers;
(d) to handle functional items for negation and questions;
(e) to have an elementary understanding of the notion of deixis;
(f) to use an elementary argument structure;
(g) to acquire some understanding of linear arrangement between form–meaning units;
(h) to conjoin propositions and/or form–meaning units;
(i) to acquire some basics of taxonomic hierarchy as it manifests itself in inclusion and part–whole relations.

Taken together, these abilities provide a sufficient basis for developing an elementary linguistic communication system (see Section 4.8), consisting at least of some word units that can be combined, where combining is compositional, a small lexicon, a basic sentence structure, some principle of linear arrangement of words, some mechanism of conjoining propositions, and a few essential functional categories for questions, negation, and deixis. What such a system would lack most of all are the following structural properties: an elaborated noun phrase and verb phrase structure, derivational and inflectional morphology including agreement, pronominalization, subordination and, more generally, any marked degree of grammaticalization—in other words, properties arising from layer IV onwards.

Keeping in mind that this is not an actual system but rather one that we construct on the basis of behavior found in non-human animals, how does such a system relate to our scenario of grammaticalization in Table 7.1? There can be little doubt that it corresponds fully to layer II, in some respects—that is with regard to properties such as functional categories and the conjoining mechanism—even to more advanced layers. This suggests that the gap between human languages and the abilities exhibited by other animals is by no means unbridgeable, in that there is an overlap area between the two. And this would also suggest that the transition from non-language to language fell squarely within the period covered by our grammaticalization scenario of early language.[15]

This was a period prior to the rise of recursion as a morphosyntactic phenomenon, as we observed above. We noted in Chapter 6 that after explicit training some non-human primates developed a conceptual ability that we consider to be a prerequisite for an understanding of recursion, namely to comprehend taxonomic relations between more and less inclusive objects such as wholes and their parts. Such relations, we argue, are suggestive at least of implicit conceptual recursion. Accordingly, the transition from non-language to early language was presumably not nearly as dramatic as has been suggested by some authors.

To conclude, the question raised above can be answered in the affirmative: There is a possible link between language genesis and language-related capacities of non-human animals, and—as the evidence discussed in Chapter 3 suggests—this link is not restricted to apes. How the transition from non-language to language exactly proceeded is a subject of further research; most likely it was not animal communication but rather cognitive abilities already in place in animals that provided the bridge to an increasingly open communication system (Rumbaugh *et al.* 1978: 137). But once the transition was concluded, the subsequent evolution must

---

[15] Fritz Newmeyer (p.c.) considers it possible that layers I and II happened in the period of *H. erectus*, and that our scenario is not incompatible with Bickerton's (1990, 1995, 2005) "protolanguage" notion.

have continued along the lines sketched in Section 7.1.1, proceeding from layer to layer right into what we find in modern languages.

7.1.3 *Lexicon before syntax*

A much discussed question in works on language evolution is the following: Which of the two, syntax or lexicon, was there first? Or did both arise simultaneously? Our work on grammaticalization leads to an unambiguous conclusion: The reconstructions proposed in Chapter 2 suggest compellingly that lexical distinctions, in particular the distinction between nouns and verbs, must have been in place before functional categories and syntax appeared in human language—hence, there must have been some kind of lexicon before there was grammar.

This hypothesis is in accordance with what most other authors have argued for on the basis of circumstantial evidence. Thus, Givón (2002a: 130, 2002b: 9) finds a coherent body of suggestive evidence in support of this reconstruction, in particular the following:

(a) Ontogenetically, both hearing and signing children acquire the lexicon first, using pre-grammatical communication before acquiring grammar.
(b) Natural second language acquisition often stops short of grammaticalization.
(c) Birds, dogs, horses, primates, and other pre-human species can easily be taught auditory, visual, or gestural lexical code-labels for nouns, verbs and adjectives.

The relative ease with which the teaching of a well-coded lexicon takes place in many pre-human species strongly suggests, according to Givón, that the underlying neuro-cognitive structure is already in place.

Additional evidence for the lexicon-before-grammar hypothesis is provided by Comrie (2000, 2002). After reviewing a wide range of situations where new forms of linguistic communication came into being, using evidence from the analysis of feral children, creoles, deaf sign languages, and twin languages, Comrie suggests:

In the few cases where we can be reasonably certain that a normal child has been exposed to no input, language has not developed.... If a lexicon is provided, then it seems that, at least in the presence of a community of potential speakers, language will develop, and will develop rapidly.... Thus, perhaps somewhat surprisingly, the main task in creating language seems to be providing the lexicon. (Comrie 2000: 1000)

There are in fact some kinds of modern linguistic communication systems, such as pidgins, second-language varieties of the Basic-Variety type, or homesigns that rely overwhelmingly or exclusively on lexical items, while we are not aware of any types of linguistic systems that consist largely or entirely of functional categories only. And if there are non-lexical categories then they normally arise later than the lexicon (e.g. Morford 2002). Note further that in "fully-fledged" modern languages it is at least theoretically possible to make oneself understood by using lexical items only, while a propositional message containing grammar only without a referential lexicon would be contentless (Givón 2002b: 31). Jackendoff and Pinker (2005: 5) therefore bluntly observe that "it would make little sense for syntax to evolve before words, since there would be nothing for it to combine into utterances."

To conclude, we take it to be fairly uncontroversial that the lexicon preceded syntax in language genesis. But there is one caveat:[16] At layer II, but perhaps already at layer I, it was possible to combine units—hence, there may have been some elementary syntax, that is to say some form of grammaticalization of the linear concatenation of information flow.

There are further questions that one might wish to raise, such as the following: What exactly was the nature of the lexicon in early language? An answer to this question is essentially not within the scope of our methodology; still, it seems possible to narrow down the range of concepts that might have been lexically distinguished. The main driving force underlying grammaticalization is a strategy according to which existing means of expression are recruited to express novel meanings and grammatical functions. In this way, expressions for concrete, physically perceptual contents are employed for less concrete and perceptually less easily accessible contents (see Heine, Claudi, and Hünnemeyer 1991). We hypothesize that the same strategy was operative in the development of the early lexicon—in the words of Givón (2002b: 28), "the early human lexicon must have been equally concrete, confining itself to, primarily, sensory-motor spatial-visual objects, states and actions." In support of this hypothesis, Givón relies, first, on what we know about the conceptual capacity and lexical learning of non-human primates, second, on early child vocabulary, and, third, on the fact

---

[16] We are grateful to an anonymous referee for having drawn our attention to this issue.

that "even in the lexicon of extant human languages, the bulk of abstract vocabulary is derived by analogy and metaphoric extension from concrete core vocabulary" (Givón 2002b: 28).

Another question is whether there was any significant correlation between the expansion of the lexicon and that of functional categories— considering the claim made by Li (2002: 90) that an important landmark in the evolutionary process leading toward the crystallization of language was reached when the size of the lexicon attained a critical mass of a few hundred items. Linguistic communications systems that have a small lexical inventory, such as pidgins and other restricted linguistic systems, also tend to dispose of only a limited range of functional categories, suggesting that there is some correlation between the expansion of both lexicon and functional categories. Nevertheless, an answer to this question, or an assessment of Li's claim, is essentially beyond the scope of our methodology, and we tend to side with Jackendoff (2002: 242), who maintains that the capacity for an open vocabulary is independent of that for grammatical elaboration.

7.1.4 *Word order*

What was the order of linear arrangement of meaningful elements in early language?[17] The main hypotheses that have been proposed on this issue are listed in (1).

(1) Hypotheses on linear word arrangement in early language
   a. Words were simply juxtaposed without any ordering principle.
   b. Linear arrangement was determined by pragmatic principles, involving an arrangement topic–comment (or given–new), or non-focus–focus constituent.
   c. Word order was based on semantic principles, where e.g. an agent precedes a patient/undergoer.
   d. Word order was based on the syntactic notions subject, object, and verb.

Hypothesis (1a) has occasionally been mentioned but, to our knowledge, not been elaborated in any detail. It is implied, for instance, in Calvin and Bickerton's (2000: 137) understanding of early language, according to

---

[17] Since it is only at layer II that a differentiation of meaningful elements arose, this question does not concern layer I.

which language began without any formal structure—being just handfuls of words or gestures strung together.

A number of students who have worked on this issue have come up with proposals in terms of (1b). What Bickerton (1990) proposed to call "protolanguage" is said to have had an arrangement of elements determined by pragmatic rather than syntactic mechanisms. This view can be reconciled with that of Jackendoff (1999, 2002), although his position is more difficult to locate, being a mixture of (1b) and (1c): On the one hand he invokes semantic principles, maintaining that in what he calls protolanguage there was no notion of the subject and object of a sentence, only semantically defined notions like agent and patient. On the other hand, he proposes principles that are suggestive of pragmatic notions, arguing that in the discourse coding of given and new information there were the principles "Agent First" and "Focus Last", where the agent was expressed in subject position and the informationally focal element resided in the last position. Claiming that these are "fossil principles from protolanguage, which modern languages often observe and frequently elaborate," he finds evidence for this hypothesis not only in the Basic Variety of late second language acquisition, but also in pidgin languages and agrammatic aphasics.

A hypothesis in favor of (1c) is proposed by Newmeyer (2000): He argues that syntactic categories presuppose semantic categories, and that the latter preceded the former in time:

In particular, since there is a rough correlation between the semantic notions 'predicate', 'argument', and 'proposition' and the syntactic categories 'V', 'NP' and 'S', respectively, it seems reasonable to hypothesize that, as language evolved, the latter were grammaticalizations of the former. (Newmeyer 2000: 388)

There is in fact support from grammaticalization for (1c). For example, semantic head–dependent patterns may give rise to syntactic head–modifier constructions, and in Chapter 5 we discussed cases where a sequence of two clauses S_1-S_2, having a semantic relation where S_1 expresses the main proposition while S_2 provides qualifying or modifying information on S_1, may grammaticalize into syntactic main clause–subordinate clause constructions. Conversely, we are not aware of regular processes leading from syntactic constructions to semantic configurations. Accordingly, (1d) would seem to be a less plausible hypothesis.

But Newmeyer (2000, 2003) also proposes, at least for a stage following that of a "protolanguage", a hypothesis in terms of (1d), according to which word order was syntactically organized—in that the human language had verb-final (OV) word order. Unlike the other authors mentioned, he provides an impressive list of arguments in support of this hypothesis, in particular the following: (a) OV predominates among the languages spoken today; (b) the earliest languages were OV; (c) the historical change OV > VO (= verb before object) is both more common than the change VO > OV and more "natural"; and (d) OV languages are more likely to have alternative orderings of S, V, and O than do VO languages. Nevertheless, this hypothesis has to be taken with care: With the exception of (a), for which there is sufficient evidence, none of the arguments listed is really uncontroversial, and we still lack appropriate typological and diachronic information to verify them.

Findings on grammaticalization suggest in fact that there is a fairly regular development leading from pragmatically motivated structures to syntactic structures—a process referred to as syntacticization (Givón 1979b, 1979c). On the basis of our bridge hypothesis (Section 1.1.3), this would lead to the conclusion that (1b) preceded (1d) in time—in other words, that (1b) would be the most plausible hypothesis. While many languages provide some kind of evidence that pragmatically motivated structures tend to acquire syntactic properties, such as those characterizing clausal subjects or objects, we are not aware of any language undergoing a process in the opposite direction, where a subject or object developed into a topic or focus constituent.

But there is an alternative perspective to this issue. Pragmatic functions are most commonly encoded by means of variation in word order, suprasegmental forms (e.g. stress), morphological material, or periphrastic constructions that are non-pragmatic in nature. It would therefore seem that the unidirectionality sketched above describes only one aspect of the development of pragmatic structures such as topic and focus constructions, in that the latter themselves can be derived from non-pragmatic structures. These observations suggest an overall development of the kind sketched in (2). Note that while (2) depicts a common pathway of grammaticalization, this is not the only one leading to the rise of pragmatic and syntactic functions.

(2) Complex syntactic structure > pragmatic function > syntactic function

With reference to the reconstruction of early language, this means that grammaticalization theory does not allow for any general hypothesis that would be supported by substantial evidence. There are, however, two conclusions to be drawn from grammaticalization evidence. One is that, in accordance with Newmeyer's (2000) hypothesis that semantic arrangements can give rise to syntactic ones while the reverse is unlikely to happen, hypothesis (1d) can be ruled out. The second conclusion is that the remaining hypotheses (1b) and (1c) are mutually compatible in the following: Since topics generally precede comments and agents precede predicates, the two hypotheses converge on the ordering topic and agent before comment and predicate, respectively. This convergence is also in accordance with the preferred argument structure in the management of information flow in discourse, where speakers tend to place given information in the agent slot of transitive clauses (Du Bois 1987).

Accordingly, irrespective of whether early language from layer II onward was shaped by pragmatic or semantic principles, the order must have been topic/agent before comment/predicate—as hypothesized by Givón (2002a), Jackendoff (1999, 2002), and others.

7.1.5 *Functions of early language*

In a recent discussion on the evolution of the language faculty, Fitch, Hauser, and Chomsky conclude:

Thus, from an empirical perspective, there are not and probably never will be data capable of discriminating among the many plausible speculations that have been offered about the original function(s) of languages. (Fitch, Hauser, and Chomsky in press: 5)

Given the fact that language genesis is not immediately accessible to empirical analysis, there is no reason to question this conclusion. Nevertheless, from a grammaticalization perspective there are a few observations that might shed some light on this issue. The question that we wish to address in this section is the following: What induces people to design new grammatical structures? Thus, we will be concerned with the motivations underlying grammatical change.

7.1.5.1 *Cognition or communication?*

A survey of works in contemporary linguistics suggests that there are two main contrasting hypotheses on the functions of human languages. These hypotheses are:

(3) Hypotheses on the main function of languages
   a. To express thought,
   b. To communicate.

Let us refer to (3a) in brief as the cognition hypothesis and to (3b) as the communication hypothesis. According to (3a), language is for internal knowledge representation; it is designed for cognition or the computation underlying reasoning, that is, for thought or inner speech, rather than for social interaction. This position is held by Chomsky (2002: 76–7), who observes that "language use is largely to oneself: 'inner speech' for adults, monologue for children," and communication may even be of no unique significance for understanding the functions and nature of language.

That inner speech and monologues are part of language use is uncontroversial; but that language use is largely to oneself rather than to others is a claim that contrasts with what other students of language structure have written on this issue over the last centuries, many of whom find (3b) to be a more attractive hypothesis: The way languages are used and structured, it is argued, can be accounted for appropriately only with reference to communicative intentions of the people who create and use these systems, rather than with reference to cognitive abilities (see e.g. Givón 1979c, 1995).

There are different views on (3b), depending on whether it is meant to refer to communication of information or for establishing or maintaining social relationships (Fritz Newmeyer, p.c.); we will ignore this distinction in the paragraphs to follow. That language structure is based on a speaker–hearer setting—and hence lends support to the communication hypothesis—is suggested in particular by two different kinds of observations. The first kind relates to the genesis of new linguistic systems, such as pidgins, homesigns, Nicaraguan Sign Language, and basic varieties: All these systems appear to have arisen in situations where communities were seeking communication (Senghas and Coppola 2001; Pinker and Jackendoff 2005: 19). Conversely, we are not aware of language systems that evolved in situations where a person developed a language system without intending to communicate with other persons.

The second kind of observations relates to structural properties characterizing human languages.[18] One property can be seen in the virtually

---

[18] For additional observations, see Pinker and Bloom (1990: 714 ff.), Hurford (2002), and Newmeyer (2004: 2–4).

universal presence of personal pronouns for second-person deixis ('you') in the languages of the world, which suggests that a speaker–hearer dichotomy is a central component of both language use and language structure. Note further that second-person pronouns are among the most frequently used linguistic forms in many languages. And the structure of personal deixis offers perhaps even more plausible evidence in favor of the communication hypothesis (3b): In many languages there is a functional distinction between inclusive ('I/we including you') and exclusive personal pronouns ('we excluding you') (see e.g. Nichols 1992). The presence of such functional categories can be interpreted meaningfully only if one assumes that communication based on a speaker–hearer dichotomy is a central function of human language, while we are not aware of any convincing structural evidence supporting the cognition hypothesis.

A second property can be seen in morphosyntactic categories having an interpersonal-manipulative function, such as imperatives and other categories of deontic modality on the one hand, and interrogative structures on the other. Both the virtually universal presence of such categories and their frequent use in linguistic discourse suggest that social interaction is a paramount function served by human languages.

Another piece of evidence in favor of the communication hypothesis is the following: That language use presupposes a speaker–hearer setting can also be seen in the occurrence of definite and indefinite articles. One of the main uses of indefinite markers concerns speech acts where the referent of a noun phrase is identifiable for the speaker, and where this referent is presented by the speaker in such a way that it is left unidentified for the hearer. Definite articles on the other hand are nominal determiners whose functions include that of marking definite reference, where the referent is uniquely identifiable for both the speaker and the hearer. Accordingly, the grammaticalization of a distinction between indefinite and definite articles, to be found in quite a number of languages across the world, implies that for the speakers concerned a dichotomy between speaker and hearer is crucial for using language.

Other examples suggesting that a speaker–hearer dichotomy is a central component of both language use and language structure are not hard to come by. For example, many languages have a functional distinction proximal ('this') vs. distal ('that') in demonstrative deictic categories, where the former means 'near to speaker (and hearer)' and the latter 'at some distance from speaker (and hearer)'—hence, a grammatical distinction that also presupposes a communicative setting. But in a number of languages there is in addition another grammaticalized category, namely a spatial-deictic demonstrative meaning 'near to hearer but not to speaker'. That such a category exists crosslinguistically also suggests that the cognition hypothesis is not sufficient to understand the nature of language structure.

Nevertheless, the communication hypothesis is also not without problems. Chomsky argues, for example, that language design appears to be in many respects dysfunctional, "yielding properties that are not well adapted to the function language is called upon to perform". But even if one does not subscribe to this view (see Pinker and Jackendoff 2005: 224 for a contrasting position), there remains another problem that the communication hypothesis shares with the cognition hypothesis, namely the question of how they can be tested. Both communication and cognition are complex phenomena having many different manifestations—accordingly, depending on which of these manifestations one has in mind, different testing methods are required, and it is unlikely that each of them will lead to the same conclusions.

The situation may be different once the notion "communication" is narrowed down to some specific manifestation. This is what Pinker and Jackendoff (2005: 231) attempt to do when they argue that "language is an adaptation for the communication of knowledge and intentions". But even this hypothesis is hard to verify or falsify as long as the relevant terms are not properly defined and justified;[19] as it stands, it remains unclear what exactly "adaptation" stands for—especially what the selection pressures were that led to adaptation, and the terms "knowledge" and "intention" are clearly contrasting phenomena which each require different tools of analysis.

To conclude, there are structural properties across languages as we know them today that can be taken to lend support to the communication hypothesis; still, there does not appear to be conclusive evidence in support of one hypothesis against the other. But what about early language? Newmeyer (2003: 74, 2004) proposes a perspective that combines both hypotheses. He argues that the origin and evolution of grammar cannot be reduced to one motivation only, rather, that there was a conjunction of two factors, namely cognition-aiding (knowledge representation) factors

---

[19] Fitch, Hauser, and Chomsky (2005) note that communication "is far too vague to constitute such a hypothesis.... Consider the analogous question: 'What is the brain for?' No one would question the assertion that the brain is an adaptation (in some broad and not particularly helpful sense), but it would seem senseless to demand that neuroscientists agree upon an answer before studying neural function and computation."

and vocal interaction-aiding ("functional") factors, roughly corresponding to our distinction cognition vs. communication—in other words, according to him there is room for both hypotheses. He goes on to hypothesize that the former preceded the latter in time, that is, that "cognition left its mark on language before communication":

> In other words, cognitive factors were the first to shape grammars. But with the passage of time, the exigencies of communication came to play an ever-more important role in grammar. (Newmeyer 2004: 7)

He presents in particular the following arguments in favor of the cognition hypothesis:[20]

(a) Grammars are "propositional", and one's cognitive representation embodies all the arguments of the sentence. This contrasts with actual language use, where arguments tend to be either reduced to pronominal or affix status or omitted entirely.

(b) Recursion is presumably not necessary for communication, but human thought has recursive properties.

(c) Considering the amount of ambiguity that they allow, human languages "are horribly designed for communication". Since different meanings are represented differently from the cognitive standpoint, grammars seem well adapted to cognition.

(d) We are able to utter pure nonsense sentences: the presence of communicatively useless sentences suggests that language is "over-designed" for communication.

(e) Grammatical categories tend to have a closer relation to cognitive categories than to communicative ones: Parts of speech and units of word formation are almost always definable semantically. Compared to that, communicative categories such as topic and focus are less likely to be marked by a special category.

(f) Those aspects of language structure that seem designed to better aid communication are historical in nature, while those suggestive of cognition are not "learnable", which suggests "that they were there from the dawn of human language itself" (Newmeyer 2004: 4–6).

We will return to this issue in Section 7.1.5.3.

[20] We are ignoring his sixth argument, which is theory-dependent and controversial, as he admits (Newmeyer 2004: 6).

7.1.5.2 Motivations underlying grammaticalization
We will now turn to grammaticalization theory with a view to reconstructing possible motivations underlying grammatical change. More specifically, the question we wish to look into is the following: What induces people to design new forms of grammatical expression? The literature on grammaticalization is rich with hypotheses that have been volunteered to answer this question. Looking at a wider range of processes of grammatical change, it would seem that it is in particular the catalogue of motivations proposed in (4) that can be reconstructed. These motivations are by no means mutually exclusive; as we will see below, more than one of them can be involved in a given instance of grammatical change and it remains frequently unclear what the relative contribution of each of them is in the rise of new functional categories.

(4) Motivations for grammaticalization
    a. To express abstract concepts.
    b. To express complex concepts.
    c. To be social.
    d. To be "extravagant".
    e. To speak like people using other languages do.

We will now look at each of these motivations in turn.

To express abstract concepts One strategy for finding expressions for abstract concepts consists of extending the use of forms for concrete (e.g. physically defined) entities to denote abstract concepts. For example, terms for body parts (e.g. English *front, back, head*) are a constant source of grammaticalization for expressions of spatial orientation, as in the case of the English prepositions *in front of, in back of, ahead of,* and these may further develop into temporal markers (*ahead of time*). In a similar fashion, verbs serving the expression of physical actions, such as English *go, keep,* or *use* are grammaticalized to fairly abstract markers for tense and aspect, as in *He's going to come, He keeps complaining,* or *He used to wear ties* (see Section 1.1.3) while we will not expect a tense or aspect marker to develop into a verb denoting physically defined actions.

A survey of the data that have become available suggests that this is the primary motivation of grammaticalization—on a rough estimate, it accounts for more than half of all instances of the processes that have become known.

To express complex concepts In a similar fashion, forms for less complex concepts tend to be used to also express concepts that are more complex in content. For example, demonstrative pronouns, such as English *this* or *that*, typically refer to concrete concepts such as a person or an object, for example *This is an apple.* But in many languages, including English, they have been grammaticalized to refer also to complex contents, such as propositional information, for example *This* (= *what you say*) *is not true.* Another example is provided by the grammaticalization from adposition to conjunction: Adpositions, such as the English prepositions *after, before, for,* etc., first served as heads of noun phrases (e.g. *after dinner*) before their use can be extended to express more complex, propositional contents, that is to introduce subordinate clauses (e.g. *After he had mailed the letter*). Furthermore, cognitive, speech-act, and various other verbs tend to grammaticalize their complements from noun phrases (e.g. *I know that person*) to clauses (e.g. *I know that he did not tell the truth*); see Section 2.2.6 and Chapter 5 for more details.

Expressing complex concepts is a motivation that Pinker and Jackendoff view as being decisive for language genesis: These authors claim that "the language faculty evolved in the human lineage for the communication of complex propositions" (Pinker and Jackendoff 2005: 204). In fact, this motivation is fairly common, second only to (8a), accounting for a considerable portion of cases of grammaticalization.

To be social That social interaction, and social bonding, was a major function of early language has been claimed in particular by Dunbar (1998, 2004). In fact, another motivation that can be reconstructed on the basis of findings on grammatical change, even if it appears to be less common than either (8a) or (8b), is to look for linguistic expressions that are taken to be most appropriate in a given context of social interaction, for example to act in accordance with social norms or to impose social norms on others. Obviously, this motivation is reflected most of all in dialogue situations, where speakers and hearers search for suitable forms for addressing one another.

A few examples may illustrate this point. One goal that surfaces from cases of grammaticalization concerns the expression of social distance. Well known examples are provided by deferential noun phrases in some

Romance languages that were grammaticalized to personal pronouns for 'you': Spanish *Vuestra Merced* and Portuguese *Vossa Mercê*, both meaning 'Your Grace', or Italian *La Vostra Signoria* 'Your Lordship' turned into second person pronouns (Spanish *Usted*, Portuguese *Você*, Italian *Lei*; see Head 1978: 185 ff.). Another goal concerns the avoidance of direct reference, where the speaker uses concepts from other domains of experience for addressing the hearer or to refer to himself (Heine 2002b). Perhaps the most common strategy to introduce what is felt to be socially appropriate forms of address is pluralization, whereby the use of second- or third-person plural pronouns is extended to second-person singular address. Paradigm examples are found in highly stratified societies; thus, the last emperor of Ethiopia, Haile Sellasie, spoke of himself as əɲɲa 'we' and expected to be addressed in the plural (Zelealem Leyew, p.c.), and a similar situation obtained in the late Roman empire, where the emperor spoke of himself as *nos* 'we' and he was addressed by means of the plural pronoun *vos* (you.PL), which thereby acquired a new, additional meaning, namely that of denoting a singular referent. In medieval Europe, generally, the nobility used the second-person singular form to the common people but received the plural form. Later on, this distinction was extended from social rank to social relation, leading to a situation whereby in many western European languages there arose a grammatical distinction between two forms of second-person singular reference.

The creation of such new forms of address and pronouns is not commonly considered to be a case of grammaticalization; still, it is in accordance with two major parameters of grammaticalization: Extension has the effect that existing forms are extended to new contexts, for example plural pronouns are used in contexts involving singular referents; desemanticization means that in these new contexts part of the old meaning (e.g. plural reference or spatial deixis) is bleached out. It is not unusual for erosion to be involved as well, in that the new category may lose in phonetic substance, as has happened perhaps most dramatically in the case of address forms such as Spanish *Vuestra Merced*, being reduced to *Usted* 'you'.

Motivation (4c) is clearly less common than the preceding two motivations, and it differs from them in being a more peripheral strategy of grammaticalization, in that it does not lead to the rise of more abstract meanings, nor does it always involve decategorialization. But in the same way as the other motivations, its application involves unidirectionality, leading from plural to singular forms, from third-person to second-person reference, from spatial to personal deixis, etc., that is, hardly ever in the opposite direction.

To be "extravagant" Extravagance, that is, "to talk in such a way that you are noticed" (Haspelmath 1999a) is the factor that Haspelmath considers to be not only the main motivation underlying grammaticalization but also the one that he holds responsible for the unidirectionality principle of grammatical development.[21] This factor has been pointed out by a number of other authors, usually referring to it as "expressivity", or as the expressive function of language. Hurford (2003: 46) draws attention to the human capacity for social manipulation: When a human speaks, she or he does so with some estimation of how her hearers will react. This is an important factor in grammaticalization: Speakers constantly propose novel, and sometimes "ungrammatical" expressions. Most of these novel uses will not be accepted, that is, re-used by the hearer, and will fall into oblivion. But in some rare cases, such novel uses may be accepted and thereby turn into new grammatical use patterns or even into new functional categories.

Haspelmath's reason for proposing what he calls the maxim of extravagance is the following:

> The crucial point is that speakers not only want to be clear or "expressive", sometimes they also want their utterance to be imaginative and vivid—they want to be little "extravagant poets" in order to be noticed, at least occasionally. (Haspelmath 1999a: 1057)

Compelling evidence that extravagance is instrumental in triggering certain kinds of language change is provided by the development of intensifiers, especially intensifiers on adjectives expressing the notion 'very'.[22] Intensifiers are coming and going at all times, and they arise via a canonical process of grammaticalization; we will return to this process in Section 7.1.6. While this example is a fairly uncontroversial manifestation of extravagance, in many other cases it is hard to separate it from other motivations since it tends to be involved in many processes which also involve (4a) or (4b). Depending on whether one adopts a broad or a narrow

---

[21] Haspelmath (1999: 1058) proposes the following scale of grammatical development: extravagance > increased frequency > routinization > obligatoriness > rule.

[22] An alternative label proposed is degree adverb. The term "intensifier" has been used in a wide range of different functions; more recently, it has come to be widely used for what is traditionally called "emphatic reflexive".

definition of this notion, extravagance will either be seen as a more general motivation, as Haspelmath sees it, or as a more specific one, as we suggest in this work.

To speak like people using other languages do In much of the work on grammatical change there is an implicit or explicit assumption to the effect that grammaticalization is a language-internal process while contact-induced change is externally motivated. As is suggested by more recent observations (Heine and Kuteva 2003, 2005, 2006), this assumption is in need of revision. It would seem, in fact, that one further motivation of grammaticalization consists of creating new usage patterns and functional categories by replicating categories from other languages. Replication means, as a rule, that speakers use parameters of grammaticalization to design new grammatical structures in language A on the model of language B, with which they are in contact, and in doing so they draw on linguistic forms available in language A. For example, in the earlier history of the Basque language there was no indefinite article, while the surrounding Romance languages Spanish, French, and Gascon had indefinite articles. As a result of centuries of close contact with these Romance languages, speakers of Basque grammaticalized their numeral for 'one', *bat*, to an indefinite article. This was by no means an isolated case of contact-induced grammaticalization; as Haase (1992) demonstrates, it was only one out of a large number of instances of grammatical replication that Basque speakers introduced on the model of their dominant Romance neighbor languages, and it is also not the only case where a European language grammaticalized its numeral for 'one' to an indefinite article as a result of language contact; see Heine and Kuteva (2006) for more examples.

Contact-induced grammaticalization is a young field of study and at the present stage of research it is hard to assess generally what its contribution to grammatical change is. Since it follows the same principles as grammaticalization that does not involve language contact, many instances of it can simultaneously be interpreted with reference to the other motivations mentioned above. For example, numerals for 'one' are fairly concrete referential entities whereas specific reference, which tends to be the main function of indefinite articles, is a more abstract concept. Accordingly, when Basque speakers grammaticalized *bat* from numeral to indefinite article on the model of Romance languages, they did so also in accordance with motivation (4a). Since our concern is with language as it arose for the first time, language contact is not an issue that we need to be concerned with here; hence, we will ignore it in the remainder of this section.

Other possible motivations In addition, a number of other factors have been proposed. One of them is structural simplification, in that grammaticalization is argued to simplify syntactic structure by changing the nature and/or number of movement operations (Roberts and Roussou 2003; van Gelderen 2004). That such a motivation exists can in fact be maintained if one subscribes to linguistic theories based on assumptions of parsimony or economy of linguistic description, where movement or other syntactic operations are proposed to be central components of linguistic analysis. If one does not adhere to such theories then there does not appear to be any need to assume a motivation of this kind; on the contrary, one could argue that the creation of new functional categories can make syntactic structure more complex rather than simplifying it. We may illustrate this with the following example. One common strategy for designing new functional categories is by means of periphrastic constructions. The German present tense expresses both present and future, for example *Er kommt morgen* (he comes tomorrow) 'He'll come tomorrow'. Nevertheless, German speakers have developed a new future tense category by means of the verb *werden* 'become' plus the infinitival main verb, for example *Er wird morgen kommen* (he becomes tomorrow to.come) 'He'll come tomorrow'. Rather than simplifying the grammar of German, this new tense made it more complex: Instead of the simple verb-medial (SVO) syntax of the present tense, it requires a more complex verb-final (SOV) syntax, where the main verb is separated from the tense auxiliary by object and/or adverbial constituents. In a similar fashion, introducing the locative preposition *in front of* in English involved a fairly complex process, whereby the morphosyntax of a possessive construction was required to express a simple spatial concept.

And much the same applies to another factor that has been proposed, according to which grammaticalization makes language production easier. Newmeyer (1998a: 276) argues that lexical categories require more production effort than functional categories, hence a change from the former to the latter is far more common than vice versa, and he concludes: "All other things being equal, a child confronted with the option of reanalyzing a verb as an auxiliary or reanalyzing an auxiliary as a verb will choose the former." While this may be so, Newmeyer's hypothesis does not account for the fact mentioned above, namely that the introduction of new functional categories frequently involves fairly complex grammatical constructions. It is hardly plausible to argue that German speakers created the periphrastic future category to make the production effort easier, especially since there was no really pressing need to create such a category: The expression of future was—and still is—very well taken care of by the existing present tense; in fact, in spoken German the periphrastic future is hardly used. And much the same applies to the creation of the English preposition *in front of*: Drawing on an adverbial phrase which acts as the head of a possessive construction would seem to require quite some production effort for the simple purpose of expressing a schematic grammatical function.

The list of motivations distinguished above is by no means exhaustive, it is confined to factors that surface most commonly, or for which there is appropriate crosslinguistic evidence. One may mention that there are additional motivations, such as euphemism, or playful language use (see also Heine 2003). What this discussion suggests is that grammaticalization cannot easily be reduced to one particular motivation; rather, there seems to be a cluster of different goals contributing to it.

7.1.5.3 Discussion
Taking the different motivations distinguished in (4) as the basis of evaluation, there is seemingly no clear evidence in favor of either the cognition or the communication hypothesis. To be sure, two of the factors would seem to lend support to the latter hypothesis: To act in accordance with or to impose social norms (4c) requires some kind of communicative context, and this also applies to (4d): To talk in such a way that you are noticed implies that talking relates to people other than the speaker. On the other hand, (4a) and (4b), which are the ones most frequently observed in grammaticalization, can be reconciled with both hypotheses.

But there is another perspective on this issue. Newmeyer (2004) observes that parts of speech and units of word formation are almost always definable semantically, and he therefore concludes that grammatical categories tend to have a closer relation to cognitive categories than to communicative ones. While this is an important observation, it is also possible to argue the other way round, namely that parts of speech and units of word formation arose because of their relevance to communication and that cognition was simply an aiding factor. Human linguistic communication is mostly about actions, events, and states, it involves people acting or experiencing actions, objects acted upon, places, time, circumstances, etc. These are entities evolving naturally in human interaction, hence there is reason to assume that the categories used to encode these entities, such as the semantic roles agent and patient (or undergoer), or the corresponding parts of speech, are motivated by communicative needs. On this view, the cognitive machinery that is used for representing these entities in linguistic discourse is derivative of their communicative functions (Givón 1979a, 1979b, 1979c, 1984, 1995).

How does this perspective relate to the motivations underlying grammaticalization that we listed in (4)? As we noted above, (4a) and (4b) are compatible with both hypotheses. There is abundant evidence in the literature on semantic and grammatical change to show that describing abstract and less easily accessible concepts in terms of concrete and easily accessible concepts—in accordance with (4a)—constitutes an important strategy of linguistic communication, and so does (4b): Describing complex or less clearly delineated contents with reference to less complex, more readily intelligible concepts is also a salient strategy to be regularly observed in human day-to-day interaction (see, e.g. Lakoff and Johnson 1980; Lakoff 1987). Accordingly, an interpretation of (4a) and (4b) with reference to the communication hypothesis is at least as plausible as one with reference to the cognition hypothesis.

This leaves us with the remaining motivations that we identified in (4): All of them are incompatible with the cognition hypothesis. We are thus left with the following situation: Since all motivations can be reconciled with the communication hypothesis, and none is exclusively in support of the cognition hypothesis, the only reasonable conclusion is that the communication hypothesis is the one that has to be adopted, while cognition may be defined as an auxiliary function in structuring early linguistic communication.

This conclusion is in accordance with the circumstantial evidence that has been adduced in favor of the communication hypothesis: A number of authors, most of all Givón (2002a, 2005), argue that communication in early language was characterized predominantly by manipulative speech-acts and that the shift towards declarative speech-acts may constitute a later development in the evolution of human language. The evidence for his hypothesis is taken from both primate communication and from early child communication: Both are said to be predominated by manipulative speech-acts, while the bulk of the grammatical machinery of modern languages is invested in the coding of declarative speech-acts (Givón 2002b: 32).

Accordingly, the most plausible scenario proposed so far on this issue is that of Pinker and Jackendoff (2005: 223), according to whom the language faculty evolved gradually in response to the adaptive value of more precise and efficient communication in a knowledge-using, socially interdependent lifestyle, where "later stages had to build on earlier ones in the contingent fashion characteristic of natural selection" (see also Haspelmath 1999c; Croft 2000), even if it is still largely unclear what exactly the nature of the selection pressure leading to adaptation in language evolution was.

7.1.6 *Who were the creators of early language?*

The question of who exactly it is that creates new functional categories and modes of grammatical organization has aroused a remarkable interest in various domains of linguistic analysis. Discussion on this issue has centered especially around demographic variables, and most conspicuously on age. It has been claimed in studies of creole languages, for example, that it is young children rather than adults who produced new forms of grammar. Thus, Bickerton (1981, 1984) argued that in the polyglot slave and servant populations the only lingua franca among adults was a pidgin, a makeshift system with little in the way of grammar. The children in those plantations did not passively have the pidgin culturally transmitted to them, but quickly developed creole languages, which have all the basic features of established human languages.

That (pre-school age) children are the agents who are responsible for language change has been claimed by a number of scholars. According to Kiparsky (1968: 194–5), one of the early proponents of this view within formal linguistics, children construct oversimplified intermediate grammars, and some feature of these grammars may survive into adulthood and be adopted by the speech community, resulting in a new linguistic norm. More generally, a major line of generative diachronic linguistics is that first-language acquisition is the main engine of grammatical change: Faced with such mixed data, it is argued by some, young learners can acquire grammars that are distinct from those of the previous generation. One major thread of reasoning in this tradition is that children are portrayed as simplifiers of grammars while adults are elaborators. Perhaps most vociferously, Lightfoot (1979, 1991, 1999a: 77 f., 1999b) claims within the framework of a principles-and-parameters theory that changes in grammar result from resettings of parameters in language acquisition. And children have also been claimed to play a crucial role in grammaticalization (Newmeyer 1998a: 276). Evidence in support of this claim has been provided by the development in Nicaragua from homesign to signed language roughly between 1977 and 1985 (Senghas 1995, 2000; Kegl and McWhorter 1997; Kegl, Senghas, and Coppola 1999; Senghas and Coppola 2001; Morford 2002; Senghas, Kita, and Özyürek 2004). Goldin-Meadow summarizes the findings made by herself and others in the following way:

> For example, many children who are congenitally deaf have hearing losses so severe that they are unable to acquire spoken language, even with intensive instruction. If these deaf children are not exposed to sign language input until adolescence, they will be for all intents and purposes deprived of a usable model for language during early childhood—although, importantly, they are not deprived of other aspects of human social interaction. Despite their lack of linguistic input, deaf children in this situation use gestures to communicate. These gestures, called "homesigns," assume the form of a rudimentary linguistic system, one that displays structure at both word and sentence levels and is used for many of the functions served by conventional language. (Goldin-Meadow 2002: 344)

But there is an alternative view on what actually happened in Nicaragua and elsewhere. Slobin observes that all of the grammatical innovations that have been studied were already present in the first cohort of deaf people, that is in the old signers, and "what seems to have happened was that younger signers—that is, those who entered a community that already had a developing communication system—used the existing grammatical elements more frequently and fluently" (Slobin 2002: 388). He notes further that Senghas and Coppola (2001) report that children who acquired Nicaraguan Sign Language before the age of 10 sign at a faster rate and are more skilled in comprehending grammatical forms. What this suggests is that the claim that young children invented a new sign (or gestural) system has to be taken with care. While children were in fact involved in the emergence of homesigns in general and Nicaraguan Sign Language in particular, there is no evidence to suggest that these were actually young children in their early years of language acquisition.

The question of whether the creators of early language were children or adults is an issue that falls squarely within the scope of grammaticalization theory; but unfortunately, an answer must remain unsatisfactory at the present stage of research because so far there is hardly any information on the sociolinguistic conditions of grammaticalization: For most grammaticalization processes there is essentially no empirically sound knowledge on who exactly did what in instigating and propagating the process, or on whether there were any specific socio-psychological requirements for it to take place. Still, there is some information at least.

The pidgin/creole Sango, national language of the Central African Republic, has undergone a number of structural changes since its genesis at the end of the nineteenth century. One of these changes involved the rise of a new functional category via the grammaticalization of the second-person plural pronoun *álà* to a deferential second-person singular pronoun (Samarin 2002). What makes this a particularly interesting case is that it allows us to study such a change in its *status nascendi*. The rise of the category is a recent one, it occurred after the Central African Republic attained its independency; there is no evidence for its existence prior to the 1960s. Essentially any Sango speaker can give *álà* to anyone else instead of the traditional second-person singular pronoun *mò*. Nevertheless, there are degrees of probability in the use of the strategy, based on the demographic variables of sex and age: Female addressees are generally more likely to receive *álà* than males, and the probability that grandparents and parents receive *álà* is the highest, being in the range between 77 and 89 percent of Samarin's data.[23] The social categories least affected by this grammaticalization process are younger siblings and friends, with whom *álà* is used with less than 25 percent probability.

As the description by Samarin (2002) suggests, it was adolescents who were crucially involved in the rise of the new category: They are the ones in the modern Sango-speaking world that are most sensitive to social identity, status, style, and linguistic change, and it is young male adults and adolescents who make the most pronounced use of deferential *álà*; very likely therefore they were the initiators of the process. Children below 13 years of age do not seem to have been involved in the creation of the new category, and they hardly use it.

---

[23] Unfortunately, Samarin (2002) does not make it clear what the corpus is on which these figures are based.

In cases such as the present one it is possible to narrow down the range of possible agents involved in the genesis of new grammatical forms. What is fairly obvious on the basis of the evidence available is that it was not preschool-age children who instigated the process. In creating the new category, Sango speakers drew on a universal strategy, namely pluralization, whereby expressions used to refer to plural referents are extended to singular referents, giving rise to (deferential) second-person singular pronouns (see Heine 2002a). That it is not young children who can be held responsible for the genesis of new functional categories is also suggested by other examples of the pluralization strategy. We observed in Section 7.1.5.2 that the Roman emperor sometimes spoke of himself as *nos* 'we', and that he was addressed by means of the plural pronoun *vos* instead of the singular form *tu* 'you (SG)'. *vos* thereby acquired a new, additional meaning, namely that of a polite form denoting a singular referent. Subsequently this usage turned into a grammatical change: It was extended to a range of new situations of social interaction, with the effect that in medieval Europe, generally, the nobility said *tu* to the common people but received *vos*—a usage that spread to a number of European languages and is retained, for example, in the French distinction between *vous* and *tu*. It appears that the agents who initiated this grammatical change were of a different kind than those that created the new pronoun *álà* in Sango: Rather than adolescents, very likely they were adults of the Roman nobility. In the further development of the *vos*-form there were other agents shaping the meaning of the category, possibly also including adolescents. But what this case has in common with the Sango one is the fact that on account of the social environment in which the process took place it is unlikely that young children played any significant role in it.

Another domain where there is demographic information on the age of speakers innovating grammatical change is that of intensifiers, especially intensifiers on adjectives expressing the notion 'very'.[24] An analysis by Claudi (1998) on 108 intensifiers of contemporary German suggests that this development represents a canonical process of grammaticalization having the following properties:

(a) One of the main conceptual sources is provided by adjectives expressing negatively evaluated qualities, such as 'terrible', 'mad',

---

[24] An alternative label proposed is degree adverb. The term "intensifier" has been used in a wide range of different functions; more recently, it has come to be widely used for what is traditionally called "emphatic reflexive".

'crazy', 'sinful', 'frightful', 'insane'; their number is three times as high as that of adjectives having positive meanings.

(b) Extension has the effect that the use of these items is extended to contexts that are incompatible with the meaning of the erstwhile intensifiers, such as use as modifiers of positively evaluated adjectives, for example English *wicked good, crazy good, mad good*, etc.

(c) The effect is desemanticization, in that the intensifier loses its negative semantics and is reduced to something like an emphatic equivalent of 'very'.

(d) The intensifiers are decategorialized, losing their categorical properties as adjectives, no longer being members of the paradigm of adjectives, and they tend to be restricted to the context where they qualify adjectives.

As the study by Claudi (1998) suggests, it is young people below the age of 30, in "constant search for new emotional kicks", that are the main instigators and propagators of the process, taking innovative and hyperbolic language use as a means of efficient communication; there is no indication that pre-school age children are involved in the process.

Further support for the hypothesis that young children are not crucially involved in grammatical change in general and the development of functional categories in particular is provided by Bybee and Slobin (1982) after a study of innovations in English past-tense verb forms such as *builded, hitted*, and *brang* in three different age groups. The authors found that it is only the forms produced by school-age and adult speakers that mirrored ongoing changes in the English verb system, while many of the "errors" produced by pre-school age children were transient, showing no chance of becoming part of English. Accordingly, the authors conclude that earlier learners are not innovators in this part of grammar.

That children can be involved in grammatical change is suggested by the following examples. A number of languages in many parts of the world have grammatical systems of numeral or noun classification, and the primary source of nominal classifiers is provided by nouns via grammaticalization (Aikhenvald 2000). Senft (1996: 235) found that in the Austronesian language Kilivila of the Trobriand Islands there is a change in the structure of classifiers involving the extension parameter, whereby specific classifiers, for example, tend to be replaced by shape-based classifiers. The greatest number of innovations has been found among school children between the ages of 8 and 14 "because of their readiness for a playful exploration of the possibilities the CP [classifier particles; a.n.] system offers and because of their increased linguistic awareness". Innovations and language change patterns in classifier usage have also been observed among adults between 21 and 35. Once again, as in all the preceding cases, it was not pre-school age children that were involved.

Studying the data that are available on the rise and development of linguistic systems such as pidgins, creoles, homesigns and signed languages, Slobin (2002: 386–9) argues that in situations where people who have only limited linguistic resources at their disposal are put together and begin to communicate about a range of topics, it is not young children who are the creators of new languages, nor are they the innovators of new grammatical categories (see also Section 4.4). His discussion suggests, however, that the situation is more complex: Young children also seem to make some contribution in both processes.

The first piece of evidence comes from situations where pidgin languages have acquired, or are acquiring, native speakers. One example is provided by the English-based pidgin/creole Tok Pisin when it began to acquire first language speakers (Sankoff and Laberge 1973, 1974). There is no evidence that the intervention of young children was required in the transition from pidgin to creole, nor in the innovation of new forms. Still, in the wording of Slobin (2002: 387), "children learners apparently did what children are good at: making a system more regular and automatic". Speaking with much greater speed and fluency, children appear to have been responsible for the decategorialization and erosion of Tok Pisin structures by turning optional constituents into obligatory ones and reducing their phonetic substance.

Similar examples have been reported from Nigeria (Shnukal and Marchese 1983) and the Solomon Islands. In Southern Nigeria it is adults rather than first language speakers of Nigerian Pidgin English who were responsible for most structural changes, while use of this pidgin/creole among younger speakers is characterized by an increase in tempo and fluency, and by some amount of erosion (phonological reduction). In a similar fashion, among speakers of Solomons Pijin English it is adults who have a creative impact on the pidgin (or creole) by expanding the syntactic resources and the lexicon, whereas children have a regularizing impact, particularly as they streamline and condense phonology and generalize grammatical patterns.

These observations are all consonant with the findings made by other students of grammaticalization (e.g. Traugott and Dasher 2002: 41–2). Accordingly, on the basis of a review of data on child development, diachronic linguistics (including grammaticalization), homesigns, signed languages, and pidgins and creoles, Slobin finds that (pre-school) age children are neither the agents of language change nor do they create new languages. And this is confirmed by Haspelmath's (1999b: 589) analysis: Using observations on a wide range of grammatical changes, he concludes that "they all involve adult innovations and diffusion to other adult grammars through language use."

Additional evidence in support of this view can be seen in the conceptual and pragmatic nature of grammaticalization processes: A number of these processes presuppose a fairly sophisticated knowledge and concepts about referential relations across sentences and social situations, about distinctions in epistemic modality and evidentiality, or about relative time spaces. Research on language acquisition suggests that children in their early years of language learning have problems understanding and distinguishing such concepts. In a similar fashion, grammaticalization entails mechanisms of inferencing, for example on causal, conditional, concessive, and other concepts, and it is hard to imagine how pre-school age children would be in a position to use such mechanisms in a productive way:

New meanings of grammatical forms arise in adult language use on the basis of pragmatic inferences drawn from existing referential and propositional meanings. Preschool-age children are not yet able to draw most of such inferences, and are limited to core semantic concepts and pragmatic functions. (Slobin 1994: 384)

Furthermore, it is hard to see how small children would be able to perceive the complex network of text relations characterizing narrative discourse, for example, and to contribute substantially to the grammaticalization of discourse structure (Traugott and Dasher 2002: 41).

To conclude, there is a fairly clear answer to the question of who were the creators of early language: On account of all the observations made above they must have been adolescents and adults, and it seems unlikely that young children were the driving force in creating early language; rather, the observations on the rise of new grammatical forms and constructions are all in support of Aitchison's (2003: 739) categorical statement that "babies do not initiate changes. Groups of interacting speakers do, especially adolescents."

However, these observations also suggest that children as well may have contributed to the evolution of grammatical categories. While not being creators, proposing new meanings in new contexts (extension), they can be important agents in the second part of grammaticalization where existing grammatical forms undergo decategorialization and erosion, becoming more regular, automated, and phonetically reduced due to increased frequency of use, fluency, and fastness of production.

7.1.7 Did language arise abruptly?

A question that has divided opinion perhaps more than many others is whether the emergence of human language was continuous, taking place in a gradual process over an extended period of time, or whether it happened in one abrupt leap from non-language to language. We have discussed this distinction in Section 1.1.1, where we referred to the former as the gradualist hypothesis and to the latter as the leap hypothesis. The gradualist hypothesis tends to be based on the Darwinian notions of natural selection and adaptation and assumes a continuous evolution, where there may have been a number of intermediate stages (Pinker and Bloom 1990; Newmeyer 1998b; Bichakjian 1999; Corballis 2002b: 173; Jackendoff 1999, 2002; Givón 2002a, 2002b; Heine and Kuteva 2002b). Supporters of the leap hypothesis tend to draw on alternative concepts of evolutionary biology such as exaptation, co-optation, or mutation, and/or invoke a modification of the brain structure to account for language evolution (Chomsky 1988; Piattelli-Palmarini 1989; Davidson and Noble 1993; Bickerton 1990, 1998, 2005; Gould 1997).

That these contrasting positions are to some extent due to differences in theoretical orientation, rather than to the facts examined, is suggested by the following example. Both Newmeyer (1998b) and Berwick (1998) invoke findings on syntax in support of their position; but whereas the former proposes a gradualist hypothesis, the latter insists on the leap hypothesis. Newmeyer argues that syntax is made up of various subsystems each of which is governed by a distinct set of principles which cannot be derived from a more fundamental principle or property. He concludes that these principles must have evolved in an incremental way, thus suggesting a gradualist scenario. In contradistinction, Berwick assumes that there cannot be a "partial syntax":[25] Many syntactic relations and constraints can be

---

[25] That there are good reasons to argue that there is something like "partial syntax" is suggested by the structure of restricted linguistic systems (see Section 4.7). Also, Pinker and Bloom (1990) provide examples that can be interpreted as exemplifying "partial syntax" (Fritz Newmeyer, p.c.).

derived from a combinatorial operation used in the derivation of sentences, called "Merge", and this operation cannot have evolved incrementally from non-combinatorial syntax but rather must be the result of a discontinuous process from a situation without syntax to one with syntax in all its glory (but see Burling 2002 for an alternative view).

Findings on grammaticalization suggest that the leap hypothesis is not really a plausible one and that it is essentially only the gradualist hypothesis that can account for grammatical evolution. We observed in Chapter 1 that grammaticalization is a cognitively motivated mechanism. But this mechanism is only one component of the process, the second component being regular human interaction and communication, in the course of which novel, and frequently more abstract concepts are proposed in specific contexts and are expressed in terms of old and less abstract concepts—with the effect that existing linguistic forms and constructions are extended to new contexts and new meanings. This process starts out with fluid discourse patterns acquiring a higher frequency of use, gradually crystallizing into new functional categories. Not all of the newly proposed concepts really end up in this way; in fact, it is only a tiny portion that, over time, are propagated and accepted as new grammatical structures. Accordingly, it takes, as a rule, centuries before a new functional category evolves (but see Section 4.6), and the time until such a category runs its full course, turning into an inflectional affix may be in the range of one millennium or more.

As the scenario of grammatical evolution discussed in Chapter 2 and Section 7.1.1 suggests, the evolution of grammar did not involve just one kind of grammatical categories but rather a sequence of layers of different kinds of categories, with each layer expanding and refining the categories of the preceding layer. It is hard to estimate the timespan required for the evolution from layer I to VI to take place; judging on the basis of chronological observations from modern languages, it is reasonable to assume that several thousand years are required for a language to acquire a new grammatical makeup, but certainly much longer to go through all of the layers from I to VI.[26]

---

[26] On the other hand, Fritz Newmeyer (p.c.) suggests that early language did not have fully-fledged modern languages spoken around it, which could possibly act as a conservativizing, braking force on the speed of language change.

Our hypothesis of a gradual evolution is in agreement with that of other students of language evolution (e.g. Pinker and Bloom 1990: 721; Jackendoff 2002; Givón 2002a, 2005). Agreement concerns the following points: First, language evolution was gradual and continuous rather than abrupt or catastrophic. Second, there was a series of intermediate steps or stages on the way from early language to the modern languages. Third, evolution led incrementally from less complex to increasingly more complex linguistic structures. And fourth, increase in complexity led from smaller to larger utterances and from non-hierarchical to hierarchical discourse organization.

In defense of the leap hypothesis, Crow (2002: 94, 107) observes that "[l]anguage is an embarrassment for gradualist evolutionary theory because, according to some authors, it requires a sa[e]ltation, that is, a discontinuous 'speciation event'," and he asserts that there are two problems with a Darwinian gradualist hypothesis. First, there is no evidence of a gradual accumulation of linguistic capabilities over a long period. And second, he asserts that the dispersal and geographical isolation of modern humans must have occurred after the appearance of language, with the propensity for language being present in all humans despite that dispersal and subsequent isolation.

In light of the findings presented here, neither of the problems discussed by Crow (2002) are necessarily an embarrassment for gradualists. With reference to his first assertion, we saw earlier that findings on grammaticalization suggest that there must have been a gradual accumulation of grammatical distinctions in the evolution of human languages (see also Newmeyer 2002). Crow's second assertion—supposing that there is evidence to substantiate it—opens a number of possible interpretations but does not lead to any clear conclusion. Irrespective of whether the appearance of language occurred prior to or after the dispersal and geographical isolation of modern humans, that appearance could have been discontinuous or continuous. If it had occurred after the dispersal, this would have meant that the sequence of grammaticalization processes described in Chapter 2 took place more than once; as we will argue in Section 7.3, observations on grammaticalization can be reconciled with both a mono- and a poly-genetic hypothesis (for a similar conclusion using a different basis of argumentation, see Bickerton 2005). To conclude, neither of Crow's arguments can be taken to be a challenge to the gradualist hypothesis.

But there is one observation suggesting that the two hypotheses sketched above are not necessarily mutually incompatible. As we observed, grammaticalization is a process that needs time to take place. Still, as we also observed, it is reasonable to assume that within a few thousand years a language can acquire a new grammatical makeup. For example, two millennia were enough in the development from Latin to modern Romance languages to introduce new modes of marking case, tense, aspect, modality, and subordination, and even less time was required for a similar development from Old English to present-day English. There are in fact scholars who argue that, while the transition to modern language was not abrupt, it nevertheless proceeded rapidly. Li (2002) is one of them:

> Once words are sequenced to form larger linguistic units, grammar emerges naturally and rapidly within a few generations. The emergence of grammar in the first generation of speakers of a creole attests to the speed of the process. The speed of the emergence of the first grammar at the inception of language is astronomical in comparison to the speed of Darwinian evolution. (Li 2002: 90)

Note that Li's concern is not with the transition from early language to modern languages but rather with the transition to a grammar consisting of a few word order principles and grammatical markers. While we agree overall with his scenario, we are hesitant to accept that a few generations were really sufficient to produce grammatical structure, or that observations on creole development provide an appropriate basis for reconstruction. Ignoring the fact that the neurological, cognitive, and socio-cultural conditions characterizing the situation of early language creators cannot have been the same as that of early creole speakers, there remains the fact that the former had no linguistic models to rely on whereas creole speakers were able to draw on multiple models provided by other language varieties, be that a "lexifier" language or "substrate" languages, or any other languages that presumably were available in the multilingual setting characterizing early creole development; the possibility that these contrasting conditions affected the relative pace at which grammars developed cannot be entirely ruled out.

Nevertheless, the evolution of language from the lexical structure as hypothesized for layers I and II to a well-grammaticalized morphosyntax of layer VI may have happened within a relatively short period of, say, ten to twenty thousand years: Assuming that human evolution from *Homo erectus* to *Homo sapiens sapiens* extended over a timespan of around 1.5 million years, the impression of a kind of leap or catastrophic shift can in fact arise.

However, this is but one possibility. An alternative scenario would be that the evolution of grammatical complexity from layer I to VI extended over a large period of time, possibly as much as 100,000 years. Considering the fact that the speakers of early language did not have any models to draw on in order to develop, for example, new structures of combining clauses, of establishing reference relations, of expressing abstract grammatical concepts of reference and modality, etc., such a scenario appears to be more plausible. But so far there is no appropriate evidence to decide in favor of one scenario over the other.

This means on the one hand that the evidence from grammaticalization does not allow for a firm conclusion on the gradualist vs. leap issue. On the other hand, this evidence is clearly in favor of the former hypothesis: It suggests that the evolution from concrete lexical to abstract inflectional structures, or from basic sentences to complex structures of clause subordination required a number of intermediate stages and must have taken millennia to be accomplished, and since such processes probably took considerably more time when they happened for the first time in human history, the timedepth of this evolution was presumably much larger. We therefore maintain that a hypothesis in terms of an abrupt, catastrophic transition from non-language to language, at least of the kind described above, does not provide an appropriate basis for reconstructing the evolution of grammar.

7.2 Grammaticalization—a human faculty?

We argued in Section 7.1.1 that without grammaticalization, early language would not have proceeded beyond the lexical layers I and II. And grammaticalization is also responsible for the rise of the recursive morphosyntactic structures that we discussed in Chapter 6:[27] Wherever there is appropriate diachronic evidence it turns out that head–modifier constructions arise via grammaticalization, and so does clause subordination. These two kinds of structures, hierarchical noun modification and clause subordination, are paradigm manifestations of recursive structures—hence the latter can ultimately be interpreted as a by-product of grammaticalization. Grammaticalization is thus suggestive of an elementary human activity or "faculty".

---

[27] As we observed in Section 6.2, grammaticalization is a necessary requirement for recursive structure to arise, but it is not a sufficient one; what is required in addition is an understanding of conceptual taxonomic hierarchy, as it manifests itself in particular in the relationship between inclusive and included phenomena.

But does such a "faculty" also surface in types of languages that have been described as restricted linguistic systems, such as pidgins, basic varieties, Nicaraguan Sign Language, twins' languages, or homesigns?[28] That grammaticalization is in fact instrumental in shaping pidgins has been demonstrated in Chapter 4: Those pidgins that develop further after having passed beyond what we described as the "stripping" process tend to acquire new grammatical categories on the basis of the parameters that we defined in Section 1.2.1. Speakers of Kenya Pidgin Swahili were found to have created an array of new functional categories, and much the same applies to speakers of extended pidgins such as Tok Pisin, Solomons Pijin, or Ghanaian and Nigerian Pidgin English.

There is at least some evidence that grammaticalization parameters are also made use of by the creators of basic varieties (i.e. the Basic Variety), that is, linguistic systems arising in specific situations of untutored late second language acquisition. For example, verbs for 'finish' are commonly used as boundary markers, as in *work finish* 'after work is/was/will be over', and demonstratives may function as definiteness markers (Klein and Perdue 1992, 1997: 318, 321; Perdue 1996). Both the development from verbs meaning 'finish' to completive markers and from demonstrative attributes to markers of definiteness are crosslinguistically common pathways of grammaticalization, as we saw in Chapter 2 (Sections 2.2.1.2, 2.2.3); we are, however, not aware of cases where grammaticalization was carried to completion in basic varieties.

There are also a few indications of grammaticalization in Nicaraguan Sign Language, which appeared after the Sandinista government was elected in 1979 (Kegl and McWhorter 1997; Kegl, Senghas, and Coppola 1999; Senghas 1995, 2000; Senghas and Coppola 2001; Senghas *et al.* 1997; Senghas, Kita, and Özyürek 2004; Morford 2002). While there is no evidence for it in the rudimentary gestural system of the first cohort, the elaborated sign language of the second cohort that evolved after 1985 is suggestive of some kind of grammaticalization. Morford (2002: 333, 337) argues that there is both what she calls innovation and "grammaticization" in Nicaraguan Sign language. With the former term she refers to the process whereby individuals who had no exposure to language create within a single grammar "system-internal grammatical properties", whereas "grammaticization" follows the automation of language processing in subsequent

---

[28] See Botha (2003a, 2003b, 2005, 2005/6) for a definition and discussion of restricted linguistic systems.

generations of users, leading, for example, to the rise of classifiers, serial verb constructions, and compounds as "instances of processing-dependent properties". Such properties, especially the rise of classifiers, might be taken to reflect grammaticalization, although the data provided are not really sufficient to establish that they are.

This situation contrasts with that found in other restricted linguistic systems: As we saw in Section 4.7, there is no clear indication of grammaticalization in homesign systems, twins' languages, isolated children (that is, children raised intentionally in social isolation), or in non-human animals. While some animals acquired form–meaning pairings that can be interpreted as equivalents of functional categories in human languages, such as markers for negation, interrogation, or deixis, the acquisition of such items was not based on parameters of grammaticalization; rather, these items appear to have been learned in the interaction between animal and human trainer in much the same way as lexical form–meaning pairings.

An answer to why grammaticalization is essentially absent in homesigns, twins' languages, isolated children, and non-human animals has been volunteered in Section 4.8: Grammaticalization requires a linguistic system that (a) is used regularly and frequently within a community of speakers and (b) is passed on from one group of speakers to another (or from one generation to the next). Clearly, this applies to none of these forms of communication, with the possible exception of homesigns spoken in village communities. Accordingly, when Nicaraguan Sign Language developed from a kind of homesign system of the first cohort to an elaborated sign language within the second cohort, this appears to have triggered grammaticalization. And this also accounts for why neither homesigners, twins' language speakers, isolated children, nor non-human animals acquired recursive structures:[29] There were no appropriate pragmatic (or sociolinguistic) conditions for grammaticalization to arise, and since grammaticalization is a prerequisite for recursive struc-

---

[29] Note that we are restricted in this book to linguistic manifestations of embedding recursion. Goldin-Meadow and Mylander (1990: 348) found that all ten children of their study could form "complex sentences" by combining gestures into at least two propositions, mostly involving a temporal sequence of events, and these authors therefore conclude that the deaf children exhibit the property of recursion in their gesture systems. This ability is suggestive of iteration, while we have found no clues that there is embedding recursion in homesigns; see Section 6.1.4.

tures to arise, there is also no recursion—at least not as a morphosyntactic phenomenon.

We argued above that grammaticalization might be suggestive of a human "faculty". But this suggestion is in need of qualification. First, we found no evidence for it in some forms of human linguistic communication, viz. in certain restricted linguistic systems. Second, it may well be that apes and other animals are able to grammaticalize and that it is only the lack of an appropriate pragmatic environment, as defined above, that has so far prevented them from doing so. Accordingly, future research might establish that grammaticalization is not a distinctly *human* faculty. And third, our analysis—here as elsewhere in this book—is restricted to some linguistic manifestations of cognitive abilities. It is quite possible that, given a wider range of knowledge about both human and non-human cognition, the ontological status of grammaticalization needs to be re-defined.

7.3 Looking for answers

In the preceding sections we dealt in some way or other with questions that were raised in (1) of Chapter 1, and in the present section we will look for answers—at least as far as this is possible and feasible on the basis of the narrow framework that was used in this book. To this end we will now deal with each of those questions in turn.

(a) Why did human language evolve, and what purpose did it serve? This question has occupied students of language evolution for centuries, and we do not attempt to summarize all the answers that have been volunteered, especially since none of them is really satisfactory; see Johansson (2005: 193–218) for a convenient summary (see also Section 1.1.1). In search for plausible reasons for why language evolved, a large variety of factors have been invoked, including economic activities such as co-operative hunting or new techniques of foraging (e.g. Bickerton 2005), cultural achievements such as tool-making (e.g. Wildgen 2004), new forms of social interaction such as the transition from grooming to gossiping (Dunbar 1996), human niche construction (Odling-Smee *et al.* 2003), or cognitive and neural factors, such as the growth and/or a modification of the human brain.

From the point of view of grammaticalization, the most promising hypothesis is one in terms of Darwinian adaptation and natural selection (Pinker and Bloom 1990); however, so far it has not been possible to develop a plausible hypothesis on what the nature of selection pressures was that led to adaptation. Bickerton (2005) observes that no other species has language and if any adaptation is unique to a species, the selection pressure that drove it must also be unique to that species. This may well be so, but presumably the situation characterizing language genesis was more complex. Findings on grammaticalization suggest that the rise of new functional categories is the result of a complex interaction of cognitive, pragmatic, and morphosyntactic variables. Very likely therefore there was not just one kind of selection pressure that was responsible for language genesis; rather, we suspect that there was a number of different forces that jointly contributed, or "conspired", to create human language.

But in this book, discussion was limited to one particular issue, namely to what purpose early language may have served. The conclusion that we reached in Section 7.1.5 was that communication must have been the primary motivation for creating early language, while cognition ("inner speech", monologue, etc.) is likely to have served an important auxiliary function in structuring early linguistic communication.

(b) When and where did early language evolve? There is as yet no empirically sound answer to (b), which raises an issue that cannot be decided purely on the basis of linguistic evidence. Estimates on when humans began to talk range from between 2,000,000 and 9,000 years ago and—unfortunately—none of these estimates can really be falsified. Is the appearance of stone tools some 2.4 million years ago an indication that language was on the way in? Can the fact that our ancestors became anatomically modern humans and were able to comprehend and apply the concept of part–whole relationships to tool-making some 300,000 years ago (Wynn 1979; Lock 1988) offer clues on language origin, or does the explosion in culture and the size of population, the creation of art, and burying the dead roughly 40,000 to 60,000 years ago provide the proof required that there was language (see, e.g., Li and Hombert 2002)? Claims that core syntax must have been in place not later than 90,000 years ago (Bickerton 2005), or arose only some 50,000 years ago (Krantz 1980; Klein and Edgar 2002), are as good as any other of the many guesses that have been made on this issue. And whether there was already some form of

rudimentary language ("protolanguage") spoken by *Homo erectus* some 1.6 million years ago (Bickerton 1990) is a question that at the present stage of research is largely a matter of belief, that is, it is not testable in principle (see Section 1.1.3). And finally, even the questions of where the ultimate home area of our ancestors and of human language are to be located have so far not received an entirely satisfactory answer. There is good evidence that Africa was the cradle of the human species, and that human languages must also have originated in Africa; but even these assumptions are not entirely uncontroversial (Dennell and Roebroeks 2005).

Recent research on mitochondrial DNA lineage patterns, on behavioral innovations such as the emergence of hunting missiles, perforated shell beads, bone tools, and abstract geometrical designs that were found in southern African sites (Blombos Cave, Klasies River, and Diepkloof in particular) might be taken as providing converging evidence that the period between 85,000 and 50,000 BP was crucial for the rise of modern human languages (e.g. d'Errico *et al.* 2005; Mellars 2006; Henshilwood 2006). But it seems that there is still a long way until all this evidence can be turned into a viable hypothesis on the time and the location of early language evolution.

(c) Who were the creators of early language? With regard to this question, our discussion was restricted to one particular issue, namely to whether the creators of early language were young children, as has been claimed or implied in a number of works (e.g. Piattelli-Palmarini 1989). Our observations in Section 7.1.6 suggest that it must have been adults and adolescents that created early language, while young children presumably played an important role in the second part of grammaticalization, making existing grammatical forms more regular, and being responsible for an increase of frequency of use, fluency, and fastness of production.

(d) Was its origin mono-genetic or poly-genetic, that is, do the modern languages derive from one ancestral language or from more than one? It has been argued that it makes no difference whether language began in one place or several (Bickerton 2005). We find this claim, which is based on specific theoretical assumptions, problematic. Observations on grammaticalization can be reconciled with both a mono- and a poly-genetic hypothesis. There are many examples showing that one and the same grammaticalization process has taken place independently in different parts of the world and in different historical periods (Heine and Kuteva 2002a). Since the

conceptual and pragmatic principles underlying such processes are the same across languages, it is conceivable that grammatical structures as we know them from modern languages arose more than once in human evolution. But this is no more than a possibility; there simply are no appropriate data to answer (d).

(e) Were the forms and structures characterizing early language iconic or arbitrary? Another widely discussed issue is whether the lexicon and grammar at the early stages of language evolution were motivated in some way, for example by semiotic or pragmatic principles, or whether they were arbitrary (unmotivated) creations, showing no systematic correlations between form and meaning. Givón (2002a: 4, 2005) assumes that language evolution led from more iconic, more indexical, less arbitrary to more arbitrary linguistic codings (see also Newmeyer 1991). One piece of argument can be seen in the rise of some restricted linguistic systems. Goldin-Meadow (2002: 348) discusses the results of a study of ten homesigning children, born to hearing parents who did not educate the children using an oral method. The gesture systems that these deaf children created consisted of pairings between gesture forms and meanings which, although having arbitrary aspects, were based on an iconic framework. And Givón (2002b: 29) finds evidence for this directionality, for example, in the fact that frequent use of well-coded signals inevitably leads to automation, speed-up, signal reduction and ritualization, and he draws attention to the development of writing systems, "whose time-course in all five known centers where literacy is known to have arisen independently—China, India, Mesopotamia, Egypt, Maya—followed the same gradual move from early iconicity to later abstraction, arbitrariness and symbolization". That the coding of sequences of events in early language was largely iconic, in that sequences of events were presented in speech in the order they happened, would in fact seem to be a plausible assumption. Accordingly, with the emergence of morphology for clause combining, iconicity is likely to have lost in significance.

But it would seem that the role played by iconicity in early language requires much further research; whether the first lexical and functional forms to arise in human language were indexical, iconic or symbolic is an issue that must remain unresolved at the present state of research. Quite possibly, the use of non-iconic form–meaning units was not a human invention. As we observed in Chapter 3, the alarm calls of vervet monkeys and other animal species are arbitrary forms (see, e.g., Cheney and Seyfarth 1990, 1992), and with appropriate training, a number of non-human animals have been found both to comprehend and produce what can be called elementary arbitrary symbols (see, e.g., Pepperberg 1999b: 43; Savage-Rumbaugh *et al.* 1980; Savage-Rumbaugh and Lewin 1994: 160).

A better understanding of this general issue depends to quite some extent on how the next question is to be answered.

(f) Did language originate as a vocal or a gestural system? Gestural and signed languages are generally characterized by a higher amount of indexical and iconic coding than spoken languages, and the present question falls, in principle, within the scope of our methodology. But our reconstructions were overwhelmingly restricted to speech, which is the modality that is most readily accessible to analysis and diachronic reconstruction; hence, we have little to say about the evolution of gestures and signing. Still, as we saw in Chapter 2, grammaticalization theory has also been applied successfully to sign languages (e.g. Janzen 1995, 1998, 1999; Janzen and Shaffer 2002; Wilbur 1999; Shaffer 2000; Morford 2002; Sexton 1999; Pfau 2004; Pfau and Steinbach 2005a, 2005b). What surfaces from this research is that signed languages appear to develop in accordance with the same parameters of grammaticalization as languages in the speech modality do. But more importantly, this research shows that grammatical development may by-pass the general pathway from lexicon to grammar in that there is an additional pathway leading straight from manual or non-manual gesture to functional category (see Section 2.3). In fact, a hypothesis to the effect that human language started out—primarily or exclusively—as a gestural system would be supported in particular by the following observations:

(a) Our closest relatives in the animal kingdom, the great apes, are good at gesturing but clearly less good at vocalization. Chimpanzees and other non-human primates, for example, have little voluntary control over their vocal signals but fairly good voluntary control over their manual gestures, and a pre-adaptation that was necessary for the emergence of modern spoken languages would simply have involved the extension of voluntary control from the hands to the vocal tract (Hurford 2003: 42–3). Note also that chimpanzee gestures are predominantly dyadic, that is, they involve reciprocal interactions between individuals,[30] whereas their vocalizations are generally not directed to specific others (Tomasello and Call 1997; Corballis 2002b: 167).

(b) Being more indexical and iconic than speech, gesturing provided a more readily available modality of comprehension and production than speech. In modern languages, certain concepts rely more heavily on gestural than on vocal expressions. Such concepts relate most of all to shape and spatial orientation; for example, when asked what a spiral is, people tend to resort to a manual demonstration (Corballis 2002b: 161), and a question for information on some location is likely to trigger a response that includes some pointing gesture.

(c) Gestures have been claimed to be implicitly syntactic (Armstrong, Stokoe, and Wilcox 1995), and if people are prevented from speaking and asked to communicate with gestures, their gestures tend to take on some syntactic format (Goldin-Meadow, McNeill, and Singleton 1996).

(d) That language evolved from manual gestures would also be supported by recent research on mirror neurons (Stamenov and Gallese 2002; Corballis 2002a, 2002b: 165).

To conclude, grammaticalization theory would be compatible with a gesturing hypothesis, or with a hypothesis arguing for a co-evolution of gesturing and vocalization, but so far lacks appropriate evidence to substantiate such a hypothesis—or else a hypothesis according to which language origin involved simultaneously gesturing and vocalization, followed by a gradual transition from the former to the latter (Auel 1980). Gesturing and signing provide promising lines of future research which could provide additional access to and new insights into the evolution of early language; but at the present stage of research, no conclusive answer to the present question seems possible.

(g) Can language genesis be related to behavior of non-human animals? This question has been controversially discussed in the literature on animal communication; nevertheless, we argued when relating our grammaticalization scenario (Table 7.1) to the behavior to be found in non-human

---

[30] Reports on inter-animal gesturing indicate, however, that gesturing is not the most common means of communication, and when it occurs it tends to be limited to begging, taking, and embracing motions (Rumbaugh *et al.* 1978: 140).

animals (Chapter 3) that there is a plausible link between the two: The language-related cognitive abilities observed in some animals can immediately be linked to early stages in the evolution of human language (Section 7.1.2).

(h) Was language evolution abrupt or gradual? This question was the subject of Section 7.1.7, where we showed that there is strong evidence to argue that language evolution proceeded gradually from one layer to another. Accordingly, a hypothesis in terms of an abrupt shift from non-language to language, or from early language to modern languages would seem implausible.

(i) Which is older—the lexicon or grammar? This question was the subject of section 7.1.3, and the answer that we proposed is that there must have been some lexicon before morphology and syntax could evolve. Our findings on grammaticalization are thus in full agreement with the following observation made by Comrie (2000: 1000, 2002): If there is lexicon, then it seems that, at least in the presence of a community of potential speakers, language will develop, and will develop rapidly.

(j) What was the structure of language like when it first evolved? The answer to this question was given in Section 7.1: Based on our scenario of evolution that was summarized in Table 7.1, we hypothesized that at the earliest stage, language consisted of one-word utterances being noun-like in character, followed at the second layer by the rise of noun–verb combinations allowing for mono-clausal verb-argument constructions.

(k) How did language change from its genesis to now? This question was answered in detail in Section 7.1.1, where we hypothesized that language evolution must have proceeded incrementally from one-word utterances to propositional structures having some basic argument structure, subsequently to phrase structure, and eventually to clause subordination. And it also must have led from an isolating morphosyntax to one that was characterized by a gradual increase in agglutinating and inflectional properties.

(l) How long did it take to develop a structure that corresponds to what we find in modern languages? What we said about the preceding question also applies to this question. Grammaticalization theory offers a means to reconstruct relative chronologies, providing information, for example, on whether process X preceded or followed Y in time; but it does not

allow for absolute dating of processes. Still, there is at least a partial answer that we volunteered in Section 7.1.7.

Since the development from concrete lexical forms to abstract functional categories, from basic sentences to complex structures of clause subordination, and from free forms to derivational and inflectional affixes must have taken considerable time to be accomplished, requiring a number of intermediate stages, the development from the beginning of human language to the structures characterizing modern languages is likely to have taken many millennia, perhaps tens of millennia.

(m) How did phonology evolve? The method that we applied in this book was restricted to morphosyntax. This means that we had nothing to say about other grammatical phenomena, especially about phonology. Note that according to some, phonology was already there in early language, perhaps before syntax and other properties of language (Lieberman 1984, 1991, 1998; Carstairs-McCarthy 1999). Nevertheless, in the absence of any sound empirical data we follow others (e.g., Jackendoff 2002;[31] Bickerton 2005) in tentatively assuming that at the initial stages of early language there was no distinct phonological level and that it took some time for phonology to acquire a degree of internal structure—this is at least a position that is most readily compatible with findings on grammatical evolution. How exactly this may have happened is open to question.

But we suspect that the same general principles of evolution that we found in morphosyntax may also have been at work in phonology. For example, observations on phonological development suggest that some properties arose earlier in human languages than others (see Comrie 1992: 206–8, 2002; Newmeyer 2003: 72 for more details): (i) Since loss of earlier vowels or nasalizing consonants appears to be the ultimate origin of nasalized vowels, there must have been a stage in the evolution of human language in which a distinction between nasalized and non-nasalized vowels was nonexistent; thus, there may have been oral vowels and nasal consonants, but no nasalized vowels. (ii) Rounded front vowels

---

[31] In Jackendoff's scenario of evolution (2002: 244), the innovation of phonological structure is located at stage III, that is, after there was already a large lexicon and after the beginnings of syntax. While we are unable to assess this hypothesis, it is at least not incompatible with findings on grammaticalization.

can frequently be shown to go back to other vowels, which could mean that at some earlier stage there may have been no rounded front vowels. (iii) Since many tonal oppositions in language have been shown to have non-tonal origins, tone might not have been a distinguishing feature in early language. (iv) Since morphophonemic alternation is assumed to have developed out of a situation where there was no such alternation, it is likely that there also was an earlier situation without alternation, at least not of the many kinds that are found today.

Such observations could suggest that at some earlier stage in the evolution of human languages there were no nasalized vowels, no rounded front vowels, no tone system, and no morphophonemic alternations. But since these are observations that are beyond the scope of grammaticalization theory, we will not attempt to draw any conclusions from them.

(n) How did the properties believed to be restricted to modern human languages arise, in particular syntax and the recursive use of language structures? We saw in the preceding chapters how syntax gradually evolved from one layer of grammaticalization to the next. But the main issue associated with this question is the rise of the recursive use of language structures. As we saw in Section 7.1.1, recursive structure was not present at the earliest layers of language evolution but must have emerged subsequently after the appearance of hierarchical noun phrase structure. It is only considerably later that recursive clause subordination came into being. Thus, our scenario suggests a development as sketched in (5).

(5) The rise of recursive structures
Layer   I II III   IV V   VI
Kind of recursion None Noun modification Clause subordination

The preceding discussion may have shown that there are answers to some of our questions, and at least partial answers to others. Still, there remain a number of questions that we had to leave unanswered. These questions have been the subject of a wealth of hypotheses and debates (see, e.g., the contributions in Hurford, Studdert-Kennedy, and Knight 1998; Givón and Malle 2002; Wray 2002; Botha 2003a; Christiansen and Kirby 2003). But unfortunately, the methodology used in the present book does not allow for any non-conjectural answers; these are issues that cannot be resolved without further research across disciplines.

7.4 Conclusions

We observed in Chapter 1 that this book is about the use of old means for novel purposes, and in subsequent chapters we showed how people constantly create novel usage patterns and functional categories by using existing linguistic material, extending it to new contexts and endowing it with new grammatical functions. In subsequent chapters we showed how this general strategy can be held responsible for human language developing into an increasingly more refined tool of communication and conceptualization. It is hoped that our discussion has demonstrated that there is nothing mysterious about complex structural phenomena such as syntactic recursion or displacement, which fall out naturally from a long sequence of applications of that very strategy, as we saw in Chapters 5 and 6.

A number of the hypotheses discussed here have been proposed in some form by other authors (especially Givón 2002a, 2005; see also Jackendoff 2002; Tallerman 2007); what distinguishes our presentation from those works is that it is based strictly on the application of grammaticalization theory. This restriction has the advantage that our hypotheses can be falsified with reference to this theory; at the same time it also had enormous disadvantages. Language evolution has become a paradigm field of interdisciplinary research, where new findings made in areas such as evolutionary biology, psychology, psycholinguistics, neurology, genetics, palaeo-anthropology, the computer sciences, etc. have been or can be combined in order to reconstruct what a few decades ago still seemed to be beyond the scope of empirical research. In an attempt to be consistent in our methodology, we had to ignore all this work except for occasional cross-references.

For example, recent studies in computer simulation show independently that complex syntactic rules can emerge out of quite simple systems, such as neural nets, which have a small number of initial assumptions and learn from imperfect inputs, with numerous words acquired by observational learning (e.g. Batali 1998; Kirby 2000; Steels *et al.* 2002; Tonkes and Wiles 2002; Davidson 2003). Some of these simulations exhibit striking similarities to our grammaticalization scenario, and it would be tempting to relate the two to one another. However, as long as it remains unclear how exactly these similarities are to be defined, we refrain from pursuing this issue—hoping that future research will be able to draw on a more comprehensive basis of analysis.

In the preceding section we tried to find answers to questions that we consider to be relevant for understanding language genesis and evolution. But there were many other questions in addition that have been raised by students of language evolution. Suffice it, in concluding, to mention one question that was ignored throughout the book, namely to what extent language is, or is part of, an innate human faculty, and whether language genesis should not be accounted for with reference to such a faculty. This is in fact a much discussed issue, but it is also an unresolved one. For our purposes it was of secondary import since our account of the genesis of grammar did not require any assumptions on innateness. Accordingly, irrespective of whether future research will establish that language evolution is, at least to some extent, determined by innate mechanisms, this would not seem to affect the findings presented here in any dramatic way.

References

Adone, D. (1994). Creolization and language change in Mauritian creole. In Adone and Plag (eds.), 23–44.
—— and Plag, I. (eds.) (1994). *Creolization and language change*. Tübingen: Niemeyer.
Ahlqvist, A. (ed.) (1982). *Papers from the 5th International Conference on Historical Linguistics*. Amsterdam, Philadelphia: John Benjamins.
Aikhenvald, A. Y. (2000). *Classifiers: a typology of noun categorization devices*. (Oxford Studies in Typology and Linguistic Theory.) Oxford: Oxford University Press.
—— and Dixon, R. M. W. (2006). *Serial verb constructions: a cross-linguistic typology*. Oxford: Oxford University Press.
Aitchison, J. (1989). *The articulate mammal*. London, New York: Routledge.
—— (1996). *The seeds of speech: Language origin and evolution*. Cambridge: Cambridge University Press.
—— (1998). On discontinuing the continuity–discontinuity debate. In Hurford *et al.* (eds.), 17–28.
—— (2003). Psycholinguistic perspectives on language change. In Janda and Joseph (eds.), 736–43.
Akimoto, M. (ed.) (2004). *Linguistic studies based on corpora*. Tokyo: Hituzi Syobo Publishing Co.
Amha, A. (2001). The Maale language. Ph.D. dissertation, University of Leiden.
Andersen, H. (ed.) (1995). *Historical linguistics 1993*. (Current Issues in Linguistic Theory, 124.) Amsterdam, Philadelphia: John Benjamins.
—— (ed.) (2001). *Actualization: Papers from a workshop held at the 14th International Conference on Historical Linguistics, Vancouver, BC, 14 August 1999*. Amsterdam, Philadelphia: John Benjamins.
Andersson, P. (2005). Swedish *må* and the (de)grammaticalization debate. Typescript, Gothenburg University.
Arbib, M. A. (2005). From monkey-like action recognition to human language: an evolutionary framework for neurolinguistics. *Behavioral and Brain Sciences* 28(2): 105–24.
—— (2005). An action-oriented neurolinguistic framework for the evolution of protolanguage. Typescript, University of Southern California.
—— (ed.) (2006). *Action to language via the mirror neuron system*. Cambridge: Cambridge University Press.

Arends, J. (1986). Genesis and development of the equative copula in Sranan. In Muysken and Smith (eds.), 103–27.
—— (ed.) (1995). *The early stages of creolization.* (Creole Language Library, 13.) Amsterdam, Philadelphia: Benjamins.
—— Muysken, P. and Smith, N. (1995). *Pidgins and creoles: An introduction.* (Creole Language Library, 15.) Amsterdam, Philadelphia: Benjamins.
Arieti, S. (1976). *Creativity: The magic synthesis.* New York: Basic Books, Inc.
Armstrong, D. F., Stokoe, W. C. and Wilcox, S. E. (1995). *Gesture and the nature of language.* Cambridge: Cambridge University Press.
Askedal, J. O. (1997). *drohen* und *versprechen* als sog. 'Modalitätsverben' in der deutschen Gegenwartssprache. *Deutsch als Fremdsprache* 34: 12–19.
Auel, J. M. (1980). *The clan of the cave bear.* London: Hodder & Stoughton.
Auwera, J. van der (ed.) (1998). *Adverbial constructions in the languages of Europe.* (Empirical approaches to language typology/EUROTYP, 20-3.) Berlin: Mouton de Gruyter.
Bach, E. and Harms, R. T. (eds.) (1968). *Universals in linguistic theory.* New York: Holt, Rinehart & Winston.
Bailey, B. L. (1966). *Jamaican creole syntax: a transformational approach.* Cambridge: Cambridge University Press.
Baker, P. (1995). Some developmental inferences from the historical studies of pidgins and creoles. In Arends (ed.), 1–24.
—— (2006). Productive bimorphemic structures and the concept of gradual creolization. Typescript.
—— and Syea, A. (eds.) (1996). *Changing meanings, changing functions: Papers relating to grammaticalization in contact languages.* (Westminster Creolistics Series, 2.) London: University of Westminster Press.
Bakker, P. (1987a). *Autonomous languages: Signed and spoken languages created by children in the light of Bickerton's bioprogram hypothesis.* (Publikaties van het Instituut voor Algemene Taalwetenschap, 53.) Amsterdam: University of Amsterdam.
—— (1987b). Autonomous languages of twins. *Acta Genet Med Gemellol* (The Mendel Institute, Rome) 36: 233–8.
—— (1990). The structure of autonomous languages: natural or not? In Koch (ed., 1990b), 74–95.
—— (1995). Pidgins. In Arends, Muysken, and Smith (eds.), 25–39.
—— (2003). Pidgins as a window on the genesis of language. Paper presented at the Seminar on Windows on Language Genesis, The Netherlands Institute for Advanced Study, 7–8 November, 2003.
—— (2006). Twins' languages as windows on language genesis. Paper presented at the Netherlands Institute for Advanced Study (NIAS), Wassenaar, 13 June, 2006.
Baron-Cohen, S. (1992). How monkeys do things with 'words'. *Behavioral and Brain Sciences* 15(1): 148–9.

Batali, J. (1998). Computational simulations of the emergence of grammar. In Hurford *et al.* (eds.), 405–26.

Bates, E. and Goodman, J. C. (1999). On the emergence of grammar from the lexicon. In MacWhinney (ed.), 29–79.

Bauer, B. L. M. and Pinault, G.-J. (eds.) (2003). *Language in time and space: a festschrift for Werner Winter on the occasion of his 80th birthday.* Berlin, New York: Mouton de Gruyter.

Bauer, L. (1978). *The grammar of nominal compounding, with special reference to Danish, English and French.* Odense: Odense University Press.

—— and Renouf, A. (2001). A corpus-based study of compounding in English. *Journal of English Linguistics* 29(2): 101–23.

Bechert, J., Bernini, G. and Buridant, C. (eds.) (1990). *Toward a typology of European languages.* (Empirical Approaches to Language Typology, 8.) Berlin: Mouton de Gruyter.

Behaghel, O. (1923–32). *Deutsche Syntax: eine geschichtliche Darstellung.* Four volumes. Heidelberg: Carl Winter.

Benazzo, S. (2006). A framework for comparing restricted linguistic systems: preliminary remarks. Typescript, Wassenaar, Netherlands Institute for Advanced Study (NIAS).

Benson, T. G. (1964). *Kikuyu-English dictionary.* Oxford: Clarendon.

Bergs, A. and Stein, D. (2001). The role of markedness in the actuation and actualization of linguistic change. In Andersen (ed.), 79–93.

Berwick, R. C. (1998). Language evolution and the Minimalist Program: the origins of syntax. In Hurford *et al.* (eds.), 320–40.

Bichakjian, B. H. (1999). Language diversity and the straight flush pattern of language evolution. *The Journal of the Linguistic Society of St. Petersburg* 2: 18–44.

Bickerton, D. (1981). *Roots of language.* Ann Arbor, MI: Karoma.

—— (1984). The language bioprogram hypothesis. *Behavioral and Brain Sciences* 7(2): 173–222.

—— (1990). *Language and species.* Chicago: University of Chicago Press.

—— (1995). *Language and human behavior.* Seattle: University of Washington Press.

—— (1998). Catastrophic evolution: the case for a single step from protolanguage to full human language. In Hurford *et al.* (eds.), 341–58.

—— (2003). Symbol and structure: a comprehensive framework for language evolution. In Christiansen and Kirby (eds.), 77–93.

—— (2005). Language evolution: A brief guide for linguists. *Lingua* 2005 (to appear).

Bierwisch, M. (2001). The apparent paradox of language evolution: can Universal Grammar be explained by adaptive selection? In Trabant and Ward (eds.), 55–79.

Bisang, W. (1996). Areal typology and grammaticalization: Processes of grammaticalization based on nouns and verbs in East and Mainland South East Asian languages. *Studies in Language* 20(3): 519–97.

Bisang, W. (1998). Adverbiality: The view from the Far East. In van der Auwera (ed.), pp. 643–812.

—— and Rinderknecht, P. (eds.) (1991). *Von Europa bis Ozeanien—von der Antonymie zum Relativsatz: Gedenkschrift für Meinrad Scheller*. Zurich: Arbeiten des Seminars für Allgemeine Sprachwissenschaft der Universität Zürich.

Blackings, M. and Fabb, N. (2003). *A grammar of Ma'di*. (Mouton Grammar Library, 32.) Berlin, New York: Mouton de Gruyter.

Blackmore, S. (1999). *The meme machine*. New York: Oxford University Press.

Blake, B. J. (1994). *Case*. Cambridge: Cambridge University Press.

——, Burridge, K. and Taylor, J. (eds.) (2003). *Historical Linguistics 2001*. Selected papers from the 15th International Conference on Historical Linguistics, Melbourne, 13–17 August 2001. (Amsterdam Studies in the Theory and History of Linguistic Science. Series IV: Current Issues in Linguistic Theory, 237.) Amsterdam, Philadelphia: John Benjamins.

Blake, F. R. (1934). The origin of pronouns of the first and second persons. *The American Journal of Philology* 55: 244–8.

Blanche-Benveniste, C. (1985). Coexistence de deux usages de la syntaxe du français parlé. In *Actes du XVIIe Congrès International de Linguistique et Philologie Romanes, 1983*, 7), 203–14.

—— (1997). *Approches de la langue parlée*. Paris, Gap: Ophrys.

Blecke, T. (1996). *Lexikalische Kategorien und grammatische Strukturen im Tigemaxo (Bozo, Mande)*. (Mande Languages and Linguistics, 1.) Cologne: Köppe.

Bloom, L., Lahey, M., Hood, L., Lifter, K., and Fiess, K. (1980). Complex sentences: acquisition of syntactic connectives and the semantic relations they encode. *Journal of Child Language* 7: 235–61.

Bloomfield, L. (1933). *Language*. New York. (Quoted from the English edition, London: Allen & Unwin (1935).

Borer, H. and Wexler, K. (1987). The maturation of syntax. In Roeper and Williams (eds.), 123–72.

Boretzky, N. (1983). *Kreolsprachen, Substrate und Sprachwandel*. Wiesbaden: Harrassowitz.

——, Enninger, W. and Stolz, T. (eds.) (1991). *Beiträge zum 6. Essener Kolloquium über „Kontakt und Simplifikation" vom 18. – 19.11.1989 an der Universität Essen*. (Bochum–Essener Beiträge zur Sprachwandelforschung, 11.) Bochum: Brockmeyer.

Borgman, D. (1990). Sanuma. In Derbyshire and Pullum (eds.), 15–248.

Botha, R. (2003a). *Unravelling the evolution of language*. (Language and Communication Library.) Amsterdam: Elsevier.

—— (2003b). Windows on language genesis: What are they and wherein lies their virtue? Paper presented at the Seminar on Windows on Language Genesis, The Netherlands Institute for Advanced Study, 7–8 November, 2003.

—— (2005). Pidgin languages as a putative window on language evolution. *Language and Communication* (in print).

—— (2005/6). On homesign systems as a potential window on language evolution. *Language and Communication* (in print).

Bowden, J. (1992). *Behind the preposition: Grammaticalization of locatives in Oceanic languages.* (Pacific Linguistics Series B, 107.) Canberra: Australian National University.

Bower, B. (2005). The Piraha challenge: an Amazonian tribe takes language to a strange place. *Science News* 10 December, 2005.

Bowerman, M. (1973). *Early syntactic development.* Cambridge: Cambridge University Press.

—— and Levinson, S. C. (eds.) (2001). *Language acquisition and conceptual development.* Cambridge: Cambridge University Press.

Boysen, S. T. and Berntson, G. G. (1989). Numerical competence in a chimpanzee. *Journal of Comparative Psychology* 103(1): 23–31.

Brannon, E. M. and Terrace, H. S. (1998). Ordering of the numerosities 1 to 9 by monkeys. *Science* 282: 746–9.

Briscoe, T. (ed.) (2002). *Linguistic evolution through language acquisition: formal and computational models.* Cambridge: Cambridge University Press.

Brown, K. (ed.) (2005). *Encyclopedia of language and linguistics*, 2nd edn. 14 volumes. Oxford: Elsevier.

Brown, P. and Levinson, S. C. (1987). *Politeness: Some universals in language usage.* (Studies in Interactional Sociolinguistics, 4.) Cambridge: Cambridge University Press.

Brown, R. (1965). *Social psychology.* New York: The Free Press.

—— and Gilman, A. (1968). The pronouns of power and solidarity. In Fishman (ed.), 252–81.

Bruyn, A. (1995a). Relative clauses in early Sranan. In Arends (ed.), 149–202.

—— (1995b). *Grammaticalization in creoles: The development of determiners and relative clauses in Sranan.* Amsterdam: Institute for Functional Research into Language and Language Use (IFOTT).

—— (1996). On identifying instances of grammaticalization in creole languages. In Baker and Syea (eds.), 29–46.

—— (1998). What can *this* be? In Schmid, Austin, and Stein (eds.), 25–40.

Buchholz, O. and Fiedler, W. (1987). *Albanische Grammatik.* Leipzig: Enzyklopädie.

——, ——, and Uhlisch, G. (1993). *Wörterbuch Albanisch–Deutsch.* Leipzig, Berlin, München: Langenscheidt Verlag Enzyklopädie.

Budwig, N. (1989). The linguistic marking of agentivity and control in child language. *Journal of Child Language* 16: 263–84.

Burch, L. C. (1980). *Ute reference grammar.* Ignacio, Colorado: Ute Press.

Burdyn, L. E. and Thomas, R. K. (1984). Conditional discrimination with conceptual simultaneous and successive cues in the squirrel monkey (*Saimiri sciureus*). *Journal of Comparative Psychology* 98(4): 405–13.

Burling, R. (2002). The slow growth of language in children. In Wray (ed.), 297–313.
—— (2005). *The talking ape: how language evolved.* Oxford: Oxford University Press.
Burridge, K. (1995). From modal auxiliary to lexical verb: The curious case of Pennsylvania German *wotte*. In Hogg and van Bergen (eds.), 19–33.
Bybee, J. L. (1985). *Morphology: A study of the relation between meaning and form.* Amsterdam: John Benjamins.
—— (2002). Sequentiality as the basis of constituent structure. In Givón and Malle (eds.), 109–34.
—— and Fleischman, S. (eds.) (1995). *Modality in grammar and discourse.* (Typological Studies in Language, 32.) Amsterdam, Philadelphia: Benjamins.
—— and Hopper, P. J. (eds.) (2001). *Frequency and the emergence of linguistic structure.* (Typological Studies in Language, 45.) Amsterdam, Philadelphia: Benjamins.
—— and Slobin, D. I. (1982). Why small children cannot change language on their own: suggestions from the English past tense. In Ahlqvist (ed.), 29–37.
—— and Noonan, M. (eds.) (2001). *Complex sentences in grammar and discourse: Studies presented to Sandra Thompson.* Amsterdam, Philadelphia: Benjamins.
——, Perkins, R. D. and Pagliuca, W. (1994). *The evolution of grammar: Tense, aspect, and modality in the languages of the world.* Chicago: University of Chicago Press.
Byrne, F. X. (1984). *Fi* and *fu*: origins and functions in some Caribbean English-based creoles. *Lingua* 62: 97–120.
—— (1988). Deixis as a noncomplementizer strategy for creole subordination marking. *Linguistics* 26(3): 335–64.
—— and Holm, J. A. (eds.). (1993). *Atlantic meets Pacific: A global view of pidginization and creolization.* (Creole Language Library, 11.) Amsterdam, Philadelphia: Benjamins.
Callanan, S. (2006). The pragmatics of protolanguage. Paper presented at the Cradle of Language Conference, Spier Estate, Stellenbosch, 6–10 November, 2006.
Calvin, W. H. (2005). The creative explosion. Typescript.
—— and Bickerton, D. (2000). *Lingua ex machina: reconciling Darwin and Chomsky with the human brain.* Cambridge, MA: MIT Press.
Campbell, L. (1991). Some grammaticalization changes in Estonian and their implications. In Traugott and Heine (eds.), 285–99.
—— (2001). What's wrong with grammaticalization? *Language Sciences* 23(2–3): 113–61.
—— and Janda, R. (2001). Introduction: conceptions of grammaticalization and their problems. *Language Sciences* 23(2–3): 93–112.
Candland, D. K. (1993). *Feral children and clever animals: reflections on human nature.* Oxford: Oxford University Press.

Cangelosi, A., Smith, A. D. M., and Smith, K. (eds.) (to appear). *The Evolution of language: Proceedings of the 6th International Conference (EVOLANG6), Rome, Italy, 12–15 April 2006.* Singapore: World Scientific Publishing.

Carden, G. and Stewart, W. A. (1988). Binding theory, bioprogram, and creolization: evidence from Haitian Creole. *Journal of Pidgin and Creole Languages* 3: 1–67.

—— (1989). Mauritian Creole reflexives: a reply to Corne. *Journal of Pidgin and Creole Languages* 4(1): 65–101.

Carlson, R. (1994). *A grammar of Supyire.* (Mouton Grammar Library.) Berlin, New York: Mouton de Gruyter.

Carpenter, K. (1987). How children learn to classify nouns in Thai. Ph.D. dissertation, Stanford University.

Carstairs-McCarthy, A. (1999). *The origins of complex language: an inquiry into the evolutionary beginnings of sentences, syllables, and truth.* Oxford: Oxford University Press.

Carstens, V. and Parkinson, F. (eds.) (2002). *Advances in African linguistics.* (Trends in African Linguistics, 4.) Trenton, NJ, Asmara: Africa World Press.

Casad, E. (1984). Cora. In Langacker (ed.), 151–459.

Casey, S. and Kluender, R. (1998). Evidence for intermediate forms in the evolution of language. *Chicago Linguistic Society* 31: 66–81.

Chafe, W. L. (1982). Integration and involvement in speaking, writing, and oral literature. In Tannen (ed., 1982a), 35–53.

Chapman, S. and Derbyshire, D. C. (1991). Paumarí. In Derbyshire and Pullum (eds.), 161–352.

Cheney, D. L. and Seyfarth, R. M. (1980). Vocal recognition in free-ranging vervet monkeys. *Animal Behaviour* 28: 362–7.

—— and —— (1990). *How monkeys see the world.* Chicago: University of Chicago Press.

—— and —— (1992). Précis of "How monkeys see the world". *Behavioral and Brain Sciences* 15(1): 135–47.

Chomsky, N. (1957). *Syntactic structure.* The Hague: Mouton.

—— (1965). *Aspects of the theory of syntax.* Cambridge, MA: MIT Press.

—— (1980). *Rules and representations.* Oxford: Blackwell.

—— (1988). *Language and problems of knowledge—the Managua lectures.* Cambridge, MA: MIT Press.

—— (2002). *On nature and language.* Cambridge: Cambridge University Press.

Christiansen, M. H. and Chater, N. (1999). Toward a connectionist model of recursion in human linguistic performance. *Cognitive Science* 23: 157–205.

—— and Kirby, S. (2003). *Language evolution.* (Studies in the Evolution of Language.) Oxford: Oxford University Press.

Christy, T. C. (1983). *Uniformitarianism in linguistics.* (Amsterdam Studies in the Theory and History of Linguistic Science, 31.) Amsterdam, Philadelphia: Benjamins.

Clark, E. V. (ed.) (1997). *Proceedings of the 28th Annual Child Language Research Forum*. Stanford Linguistics Association, Center for the Study of Language and Information.

——(2001). Emergent categories in first language acquisition. In Bowerman and Levinson (eds.), 379–405.

——(2003). *First language acquisition*. Cambridge: Cambridge University Press.

Claudi, U. (1998). Intensifiers of adjectives in German. Typescript, University of Cologne.

——(2003). *First language acquisition*. Cambridge: Cambridge University Press.

Cole, P. (1982). *Imbabura Quechua*. Amsterdam: North-Holland.

Coleman, H. (2006). Can we conceive of art without language? A neuroscientific answer. Paper presented at the Cradle of Language Conference, Spier Estate, Stellenbosch, 6–10 November, 2006.

Comrie, B. (1975). Polite plurals and predicate agreement. *Language* 51: 406–41.

——(1981). *The languages of the Soviet Union*. Cambridge: Cambridge University Press.

——(1988). Topics, grammaticalized topics, and subjects. *Berkeley Linguistics Society 14*: 265–79.

——(1992). Before complexity. In Hawkins and Gell-Mann (eds.), 193–211.

——(2000). From potential to realization: an episode in the origin of language. *Linguistics* 38(5): 989–1004.

——(2002). Reconstruction, typology and reality. In Hickey (ed.), 243–57.

——and Kuteva, T. (2005). Relativization strategies in the languages of the world. In Haspelmath *et al.* (eds.).

——and Thompson, S. A. (1985). Lexical nominalization. In Shopen (ed., 1985b), 349–98.

Conard, N. J. (2006). Is the archaeological evidence for early symbolic communication and music consistent with an African origin of language? Paper presented at the Cradle of Language Conference, Spier Estate, Stellenbosch, 6–10 November, 2006.

Connolly, K. J. and Bruner, J. S. (eds.) (1974). *The growth of competence*. London, New York: Academic Press.

Cooke, J. R. (1968). *Pronominal reference in Thai, Burmese, and Vietnamese*. (University of California Publications in Linguistics, 52.) Berkeley: University of California Press.

Coolidge, F. L. and Wynn, T. (2006). Recursion, pragmatics, and the evolution of modern speech. Paper presented at the Cradle of Language Conference, Spier Estate, Stellenbosch, 6–10 November, 2006.

Corballis, M. C. (2002a). *From hand to mouth*. Princeton, NJ: Princeton University Press.

——(2002b). Did language evolve from manual gestures? In Wray (eds.), 161–79.

Corne, C. (1977). *Seychelles Creole grammar: Elements for Indian Ocean Proto-Creole reconstruction.* (Tübinger Beiträge zur Linguistik, 91.) Tübingen: Narr.
—— (1978). A note on "passives" in Indian Ocean creole dialects. *Journal of Creole Studies* 1(1): 33–57.
—— (1988a). Patterns of subject-object coreference in Seychelles Creole. *Te Reo* 31: 73–84.
—— (1988b). Mauritian Creole reflexives. *Journal of Pidgin and Creole Languages* 3: 69–94.
Court, C. (1998). Untitled posting on the SEALTEACH list, 30 April 1998.
Coveney, A. (2000). Vestiges of *nous* and the 1st person plural verb in informal spoken French. *Language Sciences* 22: 447–81.
Cowan, J. R. and Schuh, R. G. (1976). *Spoken Hausa.* Part 1: *Hausa language-grammar.* Ithaca, New York: Spoken Language Services.
Craig, C. G. (1977). *The structure of Jacaltec.* Austin, London: University of Texas Press.
—— (ed.) (1986). *Noun classes and categorization: Proceedings of a symposium on categorization and noun classification, Eugene, Oregon, October 1983.* Amsterdam, Philadelphia: John Benjamins.
—— (1991). Ways to go in Rama: a case study in polygrammaticalization. In Traugott and Heine (eds., 1991b), 455–92.
Crass, J. (2002). Die Grammatikalisierung des Verbes "sagen" im Beria. Typescript, University of Mainz.
—— Dehnhard, B., Meyer, R., and Wetter, A. (2001). Von "sagen" zum Verbbildungsmorphem: die Grammatikalisierung des Verbs "sagen" einmal anders. *Afrikanistische Arbeitspapiere (AAP)* 65: 129–41.
Crazzolara, P. (1978). *A study of the Pokot (Suk) language: grammar and vocabulary.* Bologna: Editrice Missionaria Italiana.
Cristofaro, S. (1998). Grammaticalization and clause linkage strategies: a typological approach with particular reference to Ancient Greek. In Giacalone Ramat and Hopper (eds.), 59–88.
Croft, W. (1991). The evolution of negation. *Journal of Linguistics* 27: 1–27.
—— (2000). *Explaining language change: an evolutionary approach.* London: Longman.
——, Denning, K., and Kemmer, S. (eds.) (1990). *Studies in typology and diachrony. Papers presented to Joseph H. Greenberg on his 75th birthday.* (Typological Studies in Language, 20.) Amsterdam, Philadelphia: Benjamins.
Crow, T. J. (2002). ProtocadherinXY: a candidate gene for cerebral asymmetry and language. In Wray (ed.), 93–112.
Crowley, T. (1990). *Beach-la-Mar to Bislama: The emergence of a national language in Vanuatu.* Oxford: Clarendon Press.
Cruse, D. A. (1986). *Lexical semantics.* Cambridge: Cambridge University Press.

Csikszentmihályi, M. (1990). The domain of creativity. In Runco and Albert (eds.), 190–202.
Curtiss, S. (1977). *Genie: a linguistic study of a modern-day "wild child"*. New York: Academic Press.
——(1994). Language as a cognitive system: its independence and selective vulnerability. In Otero (ed.), 211–55.
Dahl, Ö. (2004). *The growth and maintenance of linguistic complexity*. (Studies in Language Companion Series, 71.) Amsterdam, Philadelphia: Benjamins.
——and Koptjevskaja-Tamm, M. (eds.) (2001). *Circum-Baltic languages: Typology and contact*. Volume 2: *Grammar and typology*. (Studies in Language and Companion Series, 55.) Amsterdam, Philadelphia: Benjamins.
Darwin, C. ([1859] 1964). *On the origin of species*. London: John Murray.
——(1871). *The descent of man and selection in relation to sex*. 2 volumes. London: J. Murray.
Dasser, V. (1988). A social concept in Java monkeys. *Animal Behaviour* 36(1): 225–30.
Davidson, D. (2004). *Inquiries into truth & interpretation*. Oxford: Oxford University Press.
Davidson, I. (2003). The archaeological evidence of language origins: states of the art. In Christiansen and Kirby (eds.), 140–57.
——and Noble, W. (1993). Tools and language in human evolution. In Gibson and Ingold (eds.), 125–55.
Dayley, J. P. (1985). *Tzutujil grammar*. (University of California Publications in Linguistics, 107.) Berkeley, Los Angeles, London: University of California Press.
de Luce, J. and Wilder, H. T. (eds.) (1983). *Language in primates*. New York: Springer-Verlag.
de Rooij, V. (1995). Shaba Swahili. In Arends, Muysken, and Smith (eds.), 179–90.
de Vries, M. (2002). The syntax of relativization. Ph.D. dissertation, University of Amsterdam.
de Waal, F. (2001). *The ape and the sushi master: cultural reflections by a primatologist*. New York: Basic Books.
Deacon, T. W. (1997). *The symbolic species: the co-evolution of language and the brain*. New York: W. W. Norton.
——(2003). Universal grammar and semiotic constraints. In Christiansen and Kirby (eds.), 111–39.
DeCamp, D., Hancock, I. F. et al. (eds.) (1974). *Pidgins and creoles: Current trends and prospects*. Washington, DC: Georgetown University Press.
Dechmann, D. K. N. (2005). Studying communication in bats. *Cognition, Brain, Behavior* 9(3): 479–96.
Dediu, D. (2006). Mostly out of Africa, but what did the others have to say? Paper presented at the Sixth International Conference on "Evolution of Language", Rome, 12–15 April, 2006.

DeGraff, M. (ed.) (1997). *Language creation and language change: creolization, diachrony, and development.* Cambridge, MA, London: MIT Press.
—— (1999). *Language creation and language change: Creolization, diachrony, and development.* Cambridge, MA: MIT Press.
—— (2000). À propos des pronoms objets dans le créole d'Haïti: Regards croisés de la morphologie et de la diachronie. *Langages* 138: 89–113.
Demopoulos, W. and Marras, A. (eds.) (1986). *Language learning and concept acquisition: foundational issues.* Norwood, MA: Ablex.
Dennell, R. and Roebroeks, W. (2005). An Asian perspective on early human dispersal from Africa. *Nature* 438, 22, 29 December, 2005: 1099–104.
Derbyshire, D. C. and Pullum, G. K. (eds.) (1986). *Handbook of Amazonian languages.* Volume 1. Berlin, New York: Mouton de Gruyter.
—— and —— (eds.) (1990). *Handbook of Amazonian languages.* Volume 2. Berlin, New York: Mouton de Gruyter.
—— and —— (eds.) (1991). *Handbook of Amazonian languages.* Volume 3. Berlin, New York: Mouton de Gruyter.
Deutscher, G. (2000). *Syntactic change in Akkadian: the evolution of sentential complementation.* New York: Oxford University Press.
Dickey, E. (1997). Forms of address and terms of reference. *Journal of Linguistics* 33: 255–74.
Diessel, H. (1997). The diachronic reanalysis of demonstratives in crosslinguistic perspective. *Chicago Linguistic Society* 33: 83–98.
—— (1998). Discourse-deictics in Oneida narratives. Conference on Iroquois Research. Rensselaerville, USA.
—— (1999a). The morphosyntax of demonstratives in synchrony and diachrony. *Linguistic Typology* 3: 1–49.
—— (1999b). *Demonstratives: Form, function, and grammaticalization.* (Typological Studies in Language, 42.) Amsterdam, Philadelphia: Benjamins.
—— (2005). *The acquisition of complex sentences.* (Cambridge Studies in Linguistics, 105.) Cambridge: Cambridge University Press.
Dixon, R. M. W. (1972). *The Dyirbal language of North Queensland.* (Cambridge Studies in Linguistics, 9.) Cambridge: Cambridge University Press.
—— (1980). *The languages of Australia.* Cambridge: Cambridge University Press.
—— (1994). *Ergativity.* (Cambridge Studies in Linguistics, 69.) Cambridge: Cambridge University Press.
Donohue, M. (2003). Agreement in the Skou language: a historical account. *Oceanic Linguistics* 42(2): 479–98.
Dougherty, R. (2004). G61.2820: Introduction to computational modeling of recursion: coordination, subordination, and embeddings. Typescript, New York University.
Downing, P. (1996). *Numeral classifier systems: the case of Japanese.* Amsterdam, Philadelphia: Benjamins.
Du Bois, J. W. (1987). The discourse basis of ergativity. *Language* 63: 805–55.

Duden Grammatik der deutschen Gegenwartssprache (1998). (Duden, 4.) 6th edn. Mannheim, Leipzig, Vienna, Zurich: Dudenverlag.

Dunbar, R. (1996). *Grooming, Gossip and the Evolution of Language.* London: Faber and Harvard University Press.

—— (1998). Theory of mind and the evolution of language. In Hurford *et al.* (eds.), 92–110.

—— (2004). *The human story.* London: Faber.

—— (2006). Why is language unique to humans? Paper presented at the Cradle of Language Conference, Spier Estate, Stellenbosch, 6–10 November.

Duran, J. J. (1979). Non-standard forms of Swahili in west-central Kenya. In Hancock *et al.* (eds.), 129–51.

d'Errico, F., Henshilwood, C., Vanhaeren, M., and van Niekerk, K. (2005). *Nassarius kraussianus* shell beads from Blombos Cave: evidence for symbolic behaviour in the Middle Stone Age. *Journal of Human Evolution* 48(1): 3–24.

Ebert, K. (1991). Vom Verbum dicendi zur Konjunktion: ein Kapitel universaler Grammatikentwicklung. In Bisang and Rinderknecht (eds.), 77–95.

—— (ed.) (1994). *The structure of Kiranti languages.* (Arbeiten des Seminars für Allgemeine Sprachwissenschaft, 13.) Zurich: University of Zurich.

Egli, H. (1990). *Paiwangrammatik.* Wiesbaden: Otto Harrassowitz.

Eldrege, N. (1995). *Reinventing Darwin: the great evolutionary debate.* London: Weidenfeld & Nicolson.

Emery, N. J. and Clayton, N. S. (2001). Effects of experience and social context on prospective caching strategies by scrub jays. *Nature* 414: 443–6.

Emlen, S. T. and Wrege, P. H. (1988). The role of kinship in helping decisions among white-fronted bee-eaters. *Behavioral Ecology & Sociobiology* 23: 305–15.

Epps, P. (2007). From "wood" to future tense: Nominal origins of the future construction in Hup. Typescript, University of Texas.

Erbaugh, M. (1986). Taking stock: the development of Chinese noun classifiers historically and in young children. In Craig (ed.), 399–436.

Ervin-Tripp, S. (1976). Is Sybil there? The structure of some American English directives. *Language in Society* 5: 25–66.

Evans, N. (2004). Reciprocal constructions: towards a fuller typology. Paper presented at the workshop in Reciprocity and Reflexivity—Description, Typology and Theory, Freie Universität Berlin, 1–2 October 2004.

Everaert, M. (2000). Types of anaphoric expressions: reflexives and reciprocals. In Frajzyngier and Curl (eds., 2000b), 63–83.

Everett, D. L. (1986). Pirahã. In Derbyshire and Pullum (eds.), 200–325.

—— (2005). Cultural constraints on grammar and cognition in Pirahã: another look at the design features of human language. *Current Anthropology* 46(4): 621–46.

Fant, L. J. (1994). *The American Sign Language phrase book.* Chicago: Contemporary Books.

Faraclas, N. (1996). *Nigerian Pidgin.* London: Routledge.

Fauquenoy, M. St-J. (1974). Guyanese: A French creole. In DeCamp and Hancock (eds.), 27–37.
Ferraz, L. I. (1979). *The creole of São Tomé*. Johannesburg: Witwatersrand University Press.
Fishman, J. A. (ed.) (1968). *Readings in the sociology of language*. The Hague, Paris: Mouton.
Fitch, W. T. (2002). Comparative vocal production and the evolution of speech: reinterpreting the descent of the larynx. In Wray (ed.), 21–45.
—— and Hauser, M. D. (2004). Computational constraints on syntactic processing in a nonhuman primate. *Science* 303: 377–80.
——, —— and Chomsky, N. (in press). The evolution of the language faculty: clarifications and implications. *Cognition*.
Fleischer, M. (1990). Die Sprache des Hundes oder die "Sprache" des Hundes. In Koch (ed., 1990a), 44–76.
Fónagy, I. (1988). Live speech and preverbal communication. In Landsberg (ed.), 183–203.
Fortescue, M. (1984). *West Greenlandic*. (Croom Helm Descriptive Grammars.) London, Sydney, Dover: Croom Helm.
Fouts, R. S. (1987). Chimpanzee signing and emergent levels. In Greenberg and Tobach (eds.), 57–84.
—— and Mills, S. T. (1997). *Next to kin: what chimpanzees have taught me about who we are*. New York: Morrow.
Fox, B. (ed.) (1995). *Studies in anaphora*. (Typological Studies in Language, 33.) Amsterdam, Philadelphia: Benjamins.
Frajzyngier, Z. (1987). From verb to anaphora. *Lingua* 72: 155–68.
—— (1993). *A grammar of Mupun*. (Sprache und Oralität in Afrika, 14.) Berlin: Dietrich Reimer.
—— (1995). On sources of demonstratives and anaphors. In Fox (ed.), 169–203.
—— (1996). *Grammaticalization of the complex sentence: A case study in Chadic*. (Studies in Language Companion Series, 32.) Amsterdam, Philadelphia: Benjamins.
—— (2000). Coding the reciprocal function: two solutions. In Frajzyngier and Curl (eds., 2000b), 179–94.
—— and Curl, T. S. (eds.) (2000a). *Reflexives: Forms and functions*. (Typological Studies in Language, 40.) Amsterdam, Philadelphia: Benjamins.
—— and —— (2000b). *Reciprocals: Forms and functions*. (Typological Studies in Language, 41.) Amsterdam, Philadelphia: Benjamins.
Friedländer, M. (1974). *Lehrbuch des Susu*. Leipzig: Enzyklopädie.
—— (1992). *Lehrbuch des Malinke*. Leipzig, Berlin, München: Langenscheidt Verlag Enzyklopädie.
Froger, F. (1910). *Étude sur la langue des Mossi, suivie d'un vocabulaire & de textes*. Paris.

Gage, W. W. (ed.) (1974). *Language in its social setting*. Washington, DC: Anthropological Society.

Gardiner, A. (1957). *Egyptian grammar: being an introduction to the study of hieroglyphs*. 3rd edn. Oxford: Oxford University Press.

Gardner, R. A. and Gardner, B. T. (1969). Teaching sign language to a chimpanzee. *Science* 165: 664–72.

—— and —— (1975). Early signs of language in child and chimpanzee. *Science* 187: 752–3.

—— and —— (1978). Comparative psychology and language acquisition. *Annals of the New York Academy of Sciences* 309: 37–76.

——, ——, and Van Cantfort, T. E. (eds.) (1989). *Teaching sign language to chimpanzees*. Albany: State University of New York Press.

Gast, V. and Haas, F. (2004). Why are there no reciprocal uses of German *sich* in PPs? Paper presented at the workshop in Reciprocity and Reflexivity—Description, Typology and Theory, Freie Universität Berlin, 1–2 October 2004.

Gelman, R. and Butterworth, B. (2005). Number and language: how are they related? *Trends in Cognitive Sciences* 9(1): 6–10.

—— and Gallistel, C. R. (2004). Language and the origin of numerical concepts. *Science* 306: 441–3.

Genetti, C. (1986). The development of subordinators from postpositions in Bodic languages. *Berkeley Linguistics Society* 12: 387–400.

—— (1991). From postposition to subordinator in Newari. In Traugott and Heine (eds., 1991b), 227–55.

Geniušiene, E. (1987). *The typology of reflexives*. (Empirical Approaches to Language Typology, 2.) Berlin, New York, Amsterdam: Mouton de Gruyter.

—— (1993). The development of tense markers from demonstrative pronouns in Panare (Cariban). *Studies in Language* 17: 53–73.

Giacalone Ramat, A. (1998). Testing the boundaries of grammaticalization. In Giacalone Ramat and Hopper (eds.), pp. 107–28.

—— and Hopper, P. J. (eds.) (1998). *The limits of grammaticalization*. Amsterdam, Philadelphia: Benjamins.

—— Carruba, O. and Bernini, G. (eds.) (1987). *Papers from the 7th International Conference on Historical Linguistics*. (Amsterdam Studies in the Theory and History of Linguistic Science, 48.) Amsterdam, Philadelphia: Benjamins.

Gibson, K. R. and Ingold, T. (eds.) (1993). *Tools, language and cognition in human evolution*. Cambridge: Cambridge University Press.

Gil, D. (2001). Creoles, complexity, and Riau Indonesian. *Linguistic Typology* 5: 325–71.

Gildea, S. (1993). The development of tense markers from demonstrative pronouns in Panare (Cariban). *Studies in Language* 17: 53–73.

—— (ed.) (2000). *Reconstructing grammar: comparative linguistics and grammaticalization*. (Typological Studies in Language, 41.) Amsterdam, Philadelphia: Benjamins.

Givón, T. (ed.) (1979a). *Syntax and semantics, 12: Discourse and syntax.* New York: Academic Press.
Givón, T. (1979b). From discourse to syntax: grammar as a processing strategy. In Givón (ed., 1979a), 81–111.
—— (1979c). *On understanding grammar.* New York: Academic Press.
—— (1984). *Syntax: A functional-typological introduction.* Volume I. Amsterdam, Philadelphia: John Benjamins.
—— (1995). *Functionalism and grammar.* Amsterdam, Philadelphia: Benjamins.
—— (2000). Internal reconstruction: as method, as theory. In Gildea (ed.), 107–59.
—— (2002a). *Bio-linguistics: the Santa Barbara lectures.* Amsterdam, Philadelphia: John Benjamins.
—— (2002b). The visual information-processing system as an evolutionary precursor of human language. In Givón and Malle (eds.), 3–50.
—— (2004). Grammatical relations in passive clauses: The impact of diachrony. Typescript, University of Oregon.
—— (2005). *Context as other minds.* Amsterdam, Philadelphia: Benjamins.
—— (2006). Multiple routes to clause union: the diachrony of syntactic complexity. Typescript, Eugene, University of Oregon.
—— and Savage-Rumbaugh, S. (2005). Can apes learn grammar? A short detour into language evolution. Typescript.
—— and Malle, B. F. (eds.) (2002). *The evolution of language out of pre-language.* (Typological Studies in Language, 53.) Amsterdam, Philadelphia: John Benjamins.
Glinert, L. (1989). *The grammar of Modern Hebrew.* Cambridge: Cambridge University Press.
Goldenberg, G. (1991). "Oneself"; "One's own" and "one another" in Amharic. In Kaye (ed.), 531–49.
Goldin-Meadow, S. (1982). The resilience of recursion: a study of a communication system developed without a conventional language model. In Wanner and Gleitman (eds.), 51–77.
—— (2002). Getting a handle on language creation. In Givón and Malle (eds.), 343–74.
—— (2003). *The resilience of language: What gesture creation in deaf children can tell us about how all children learn language.* New York, Hove: Psychology Press.
—— and Mylander, C. (1984). Gestural communication in deaf children: The effects and non-effects of parental input on early language development. *Monographs of the Society for Research in Child Development* 49: 1–121.
—— and —— (1990). Beyond the input given: the child's role in the acquisition of language. *Language* 66(2): 323–55.
——, ——, and Butcher, C. (1995). The resilience of combinatorial structure at the word level: morphology in self-styled gesture systems. *Cognition* 56: 195–262.
——, McNeill, D., and Singleton, J. (1996). Silence is liberating: Removing the handcuffs on grammatical expression and speech. *Psychological Review* 103: 34–55.

Goodman, M. (1984). Are creole structures innate? *The Behavioral and Brain Sciences* 7: 193–4.

Görlach, M. and Holm, J. A. (eds.) (1986). *Focus on the Caribbean*. (Varieties of English around the world, General Series, 8.) Amsterdam, Philadelphia: Benjamins.

Gould, S. J. (1991). Exaptation: a crucial tool for evolutionary psychology. *Journal of Social Issues* 47: 43–65.

——(1997). The exaptive excellence of spandrels as a term and prototype. *Proceedings of the National Academy of Science (USA)* 94: 10750–5.

——and Vrba, E. S. (1982). Exaptation—a missing term in the science of form. *Paleobiology* 8: 4–15.

Grant, A. P. (1996). The evolution of functional categories in Grand Ronde Chinook Jargon: ethnolinguistic and grammatical considerations. In Baker and Syea (eds.), 225–42.

Greenberg, G. and Tobach, E. (eds.) (1987). *Cognition, language, and consciousness: integrative levels*. (The T. C. Schneirla Conference Series, 2.) Hillsdale, NJ: Erlbaum.

Greenberg, J. H. (1959). The origin of the Maasai passive. *Africa* 29, 2: 171–6.

——(1966). Synchronic and diachronic universals on phonology. *Language* 42: 508–17.

——(1978). How does a language acquire gender markers? In Greenberg, Ferguson, and Moravcsik (eds., 1978a), 47–82.

——(1991). The last stages of grammatical elements: contractive and expansive desemanticization. In Traugott and Heine (eds., 1991a), 301–14.

——(1992). Preliminaries to a systematic comparison between biological and linguistic evolution. In Hawkins and Gell-Mann (eds.), 139–58.

——, Ferguson, C. A., and Moravcsik, E. (eds.) (1978a). *Universals of human language*. Volume 3: *Word structure*. Stanford: Stanford University Press.

——,——, and —— (1978b). *Universals of human language*. Volume 4: *Syntax*. Stanford: Stanford University Press.

Greenfield, P. M. and Smith, J. (1976). *The structure of communication in early language development*. New York: Academic Press.

——and Savage-Rumbaugh, E. S. (1990). Grammatical combinations in *Pan paniscus*: processes of learning and invention in the evolution and development of language. In Parker and Gibson (eds.), 540–78.

Gregersen, E. A. (1974). The signalling of social distance in African languages. In Gage (ed.), 47–55.

Grinstead, J., MacSwan, J., Curtiss, S., and Gelman, R. (n.d.). The independence of language and number. Typescript.

Grüßner, K.-H. (1978). *Arleng Alam: die Sprache der Mikir, Grammatik und Texte*. (Beiträge zur Südasienforschung, 39.) Wiesbaden: Franz Steiner.

Güldemann, T. (2001). *Quotative constructions in African languages: a synchronic and diachronic survey*. Habilitationsschrift, University of Leipzig.

Guy, J. B. M. (1974). *Handbook of Bichelamar—Manuel de Bichelamar.* (Pacific Linguistics, Series C, 34.) Canberra: The Australian National University.

Haase, M. (1992). *Sprachkontakt und Sprachwandel im Baskenland: die Einflüsse des Gaskognischen und Französischen auf das Baskische.* Hamburg: Buske.

—— (1994). *Respekt: die Grammatikalisierung von Höflichkeit.* (Edition Linguistik, 44.) Munich, Newcastle: LINCOM Europa.

—— (1997). Gascon et basque: bilinguisme et substrat. *Sprachtypologie und Universalienforschung* 50(3): 189–228.

—— and Nau, N. (eds.) (1996). *Sprachkontakt und Grammatikalisierung. Sprachtypologie und Universalienforschung.* (Special issue, 49, 1.) Berlin: Akademie-Verlag.

Hackstein, O. (2003). Apposition and word-order typology in Indo-European. In Bauer and Pinault (eds.), 131–52.

Hagège, C. (1995). *La structure des langues.* 4th edn. Paris: Presses Universitaires de France.

Hagman, R. S. (1977). *Nama Hottentot Grammar.* (Language Science Monographs; 15.) Bloomington, IN: Indiana University Press.

Haiman, J. (1978). Conditionals are topics. *Language* 54: 564–89.

—— (ed.) (1985a). *Iconicity in Syntax.* (Typological Studies in Language, 6.) Amsterdam: Benjamins.

—— (1985b). *Natural syntax: Iconicity and erosion.* Cambridge: Cambridge University Press.

—— (1987). On some origins of medial verb morphology in Papuan languages. *Studies in Language* 11(2): 347–64.

—— (1994). The divided self in a Papuan language. In Reesink (ed.), 42–9.

—— (1998). *Talk is cheap. Sarcasm, alienation and the evolution of language.* Oxford: Oxford University Press.

—— (1999). Auxiliation in Khmer. The case of *BAAN. Studies in Language* 23(1): 149–72.

——, Heine, B. Kuteva, T. (forthc.). The origins of the generic 2p. sg. pronoun.

—— and Thompson, S. A. (eds.) (1988). *Clause combining in grammar and discourse.* (Typological Studies in Language, 18.) Amsterdam, Philadelphia: Benjamins.

Hammond, M. and Noonan, M. (eds.) (1988). *Theoretical Morphology: Approaches in Modern Linguistics.* New York: Academic Press.

Hancock, I. F., Polomé, E., Goodman, M., and Heine, B. (eds.) (1979). *Readings in creole studies.* Ghent: E. Story-Scientia P.V.B.A.

Hare, B., Call, J., Agnetta, B., and Tomasello, M. (2000). Chimpanzees know what conspecifics do and do not see. *Animal Behavior* 59: 771–85.

Harnard, S. R., Steklis, S. D., and Lancaster, J. (eds.) (1976). *Origins and evolution of language and speech.* New York: New York Academy of Sciences.

Harris, Alice C. and Campbell, L. (1995). *Historical syntax in cross-linguistic perspective.* Cambridge: Cambridge University Press.

Hartmann, J. (1980). *Amharische Grammatik.* (Äthiopische Forschungen, 3.) Wiesbaden: Franz Steiner.

Haspelmath, M. (1990). The grammaticization of passive morphology. *Studies in Language* 14(1): 25–72.

—— (1993). *A grammar of Lezgian.* Berlin, New York: Mouton de Gruyter.

—— (1997a). *Indefinite pronouns.* Oxford: Clarendon Press.

—— (1997b). *From space to time. Temporal adverbs in the world's languages.* Munich, Newcastle: Lincom Europa.

—— (1998a). Does grammaticalization need reanalysis? *Studies in Language* 22(2): 315–51.

—— (1998b). The semantic development of old presents: new futures and subjunctives without grammaticalization. *Diachronica* 15: 29–62.

—— (1999a). Why is grammaticalization irreversible? *Linguistics* 37(6): 1043–68.

—— (1999b). Are there principles of grammatical change? *Journal of Linguistics* 35: 579–95.

—— (1999c). Optimality and diachronic adaptation. *Zeitschrift für Sprachwissenschaft* 18(2): 180–205.

—— (ed.) (2004a). *Coordinating constructions.* (Typological Studies in Language, 58.) Amsterdam, Philadelphia: Benjamins.

—— (2004b). On directionality in language change with particular reference to grammaticalization. In Norde and Perriodon (eds.), 17–44.

——, Dryer, M., Gill, D., and Comrie, B. (eds.) (2005). *World Atlas of Linguistic Structures (WALS).* Oxford: Oxford University Press.

Hauser, M. D. (1997). *The evolution of communication.* Cambridge, MA: MIT Press.

—— (2000). *Wild minds: what animals really think.* New York: Henry Holt.

——, Chomsky, N., and Fitch, W. T. (2002). The faculty of language: What is it, who has it, and how did it evolve? *Science* 298: 1569–79.

—— and Fitch, W. T. (2003). What are the uniquely human components of the language faculty? In Christiansen and Kirby (eds.), 158–81.

Hawkins, J. A. (ed.) (1988). *Explaining language universals.* Oxford: Blackwell.

—— and Gell-Mann, M. (eds.) (1992). *The evolution of human language. Proceedings of the Workshop on the Evolution of Human Languages, held August, 1989 in Santa Fe, New Mexico.* Santa Fe: Addison-Wesley Publishing Company.

Head, B. F. (1978). Respect degrees in pronominal reference. In Greenberg, Ferguson, and Moravcsik (eds.), 151–211.

Heath, J. (1999a). *A grammar of Koyra Chiini: the Songhay of Timbuktu.* (Mouton Grammar Library, 19.) Berlin, New York: Mouton de Gruyter.

—— (1999b). *A grammar of Koyraboro (Koroboro) Senni: the Songhay of Gao, Mali.* (Westafrikanische Studien, 19.). Cologne: Köppe.

Heikkinen, T. (1987). An outline of the grammar of the !Xũ language spoken in Ovamboland and West Kavango. *South African Journal of African Languages* (Pretoria) 7, Supplement 1 (African Languages).

Heine, B. (1973). *Pidgin-Sprachen im Bantu-Bereich.* (Kölner Beiträge zur Afrikanistik, 3.) Berlin: Dietrich Reimer.
—— (1979). Some generalizations on African-based pidgins. In Hancock *et al.* (eds.), 89–98.
—— (1983). Eine Bemerkung zur Gliederung der Swahili-Dialekte. *Afrika und Übersee* 66: 57–65.
—— (1990). The dative in Ik and Kanuri. In Croft, Denning, and Kemmer (eds.), 129–49.
—— (1991). On the development of Kenya Pidgin Swahili. In Boretzky, Enninger, and Stolz (eds.), 29–54.
—— (1993). *Auxiliaries: cognitive forces and grammaticalization.* New York, Oxford: Oxford University Press.
—— (1997a). *Possession: Sources, forces, and grammaticalization.* Cambridge: Cambridge University Press.
—— (1997b). *Cognitive foundations of grammar.* Oxford, New York: Oxford University Press.
—— (1999). *The ‖Ani: Grammatical notes and texts.* (Khoisan Forum, 11.) Cologne: Institut für Afrikanistik.
—— (2000). Polysemy involving reflexive and reciprocal markers in African languages. In Frajzyngier and Curl (eds., 2000b), 1–29.
—— (2002a). Personal deixis: Some rhetorical aspects of grammaticalization. Typescript, University of Cologne.
—— (2002b). On the role of context in grammaticalization. In Wischer and Diewald (eds.), 83–101.
—— (2003). On degrammaticalization. In Blake, Burridge, and Taylor (eds.), 163–79.
—— (2005). On reflexive forms in creoles. *Lingua* 115: 201–57.
—— and Claudi, U. (1986). *On the rise of grammatical categories: Some examples from Maa.* (Kölner Beiträge zur Afrikanistik, 13.) Berlin: Reimer.
—— and Kuteva, T. (2002a). *World lexicon of grammaticalization.* Cambridge: Cambridge University Press.
—— and —— (2002b). On the evolution of grammatical forms. In Wray (ed.), 376–97.
—— and —— (2003). Contact-induced grammaticalization. *Studies in Language* 27(3): 529–72.
—— and —— (2005). *Language contact and grammatical change.* (Cambridge Approaches to Language Contact, 3.) Cambridge: Cambridge University Press.
—— and —— (2006). *The changing languages of Europe.* Oxford: Oxford University Press.
—— and Miyashita, H. (2006). Accounting for a functional category: German *drohen.* Typescript, University of Cologne.
—— and Reh, M. (1984). *Grammaticalization and reanalysis in African languages.* Hamburg: Buske.

Heine, B., Claudi, U. and Hünnemeyer, F. (1991). *Grammaticalization: a conceptual framework.* Chicago: University of Chicago Press.
Helbig, G. and Buscha, J. (1988). *Deutsche Grammatik: Ein Handbuch für den Ausländerunterricht.* 11th edn. Leipzig: VEB Verlag Enzyklopädie.
Held, G. (1999). Submission strategies as an expression of the ideology of politeness: reflections on the verbalisation of social power relations. *Pragmatics* 9: 21–36.
Henshilwood, C. (2006). Reading the artefacts: gleaning language skills from the Middle Stone Age in Southern Africa. Paper presented at the Cradle of Language Conference, Spier Estate, Stellenbosch, 6–10 November, 2006.
Herman, L. M. (1987). Receptive competencies of language-trained animals. In Rosenblatt *et al.* (ed.), 1–60.
—— (1989). In which procrustean bed does the sea lion sleep tonight? *The Psychological Record* 39: 19–50.
—— and Forestell, P. H. (1985). Reporting presence or absence of named objects by a language-trained dolphin. *Neuroscience and Biobehavioral Reviews* 9: 667–81.
——, Richards, D. G. and Wolz, J. P. (1984). Comprehension of sentences by bottlenosed dolphins. *Cognition* 16: 129–219.
Herrnstein, R. J., Loveland, D. H. and Cable, C. (1976). Natural concepts in pigeons. *Journal of Experimental Psychology: Animal Behavior Processes* 2: 285–302.
Herring, S. C. (1988). Aspect as a discourse category in Tamil. *Berkeley Linguistics Society* 14: 280–92.
—— (1991). The grammaticalization of rhetorical questions in Tamil. In Traugott and Heine (eds, 1991a), 253–84.
Hewitt, B. G. (1995). *Georgian: a structural reference grammar.* (London Oriental and African Language Library, 2.) Amsterdam, Philadelphia: John Benjamins.
Heyes, C. and Huber, L. (eds.) (2000). *The evolution of cognition.* (The Vienna Series in Theoretical Biology.) Cambridge, MA: MIT Press.
Hickey, R. (ed.) (2002). *Motives for language change.* Cambridge: Cambridge University Press.
Hill, K. C. (ed.) (1979). *The genesis of language.* (The First Michigan Colloquium, 1979.) Ann Arbor: Karoma Publishers.
Hiraga, M. K. (1995). DEFERENCE as DISTANCE: Metaphorical base of honorific verb constructions in Japanese. Paper presented at the 4th Conference of the International Cognitive Linguistics Association, Albuquerque, 17–21 July, 1995.
Hobbs, J. R. (2006). The origin and evolution of language: a plausible, strong AI account. In Arbib (ed.), 48–88.
Hockett, C. (1960). The origin of speech. *Scientific American* 203(3): 89–96.
Hoffmann, C. (1963). *A grammar of the Margi language.* London: Oxford University Press.
Hofmann, I. (1983). *Einführung in den nubischen Kenzi Dialekt.* (Veröffentlichungen der Institute für Afrikanistik und Ägyptologie der Universität Wien, 27.) Vienna: University of Vienna.

Hogg, R. M. and van Bergen, L. (eds.) (1995). *Historical linguistics 1995*. Vol. 2: *Germanic linguistics*. (Amsterdam Studies in the Theory and History of Linguistic Science, 162.) Amsterdam, Philadelphia: Benjamins.

Holden, C. (2004). The origin of speech. *Science* 303: 1316–19.

Holm, J. A. (1988). *Pidgins and creoles*. Volume 1: *Theory and structure*. Cambridge: Cambridge University Press.

—— (2000). *An introduction to pidgins and creoles*. (Cambridge Textbooks in Linguistics.) Cambridge: Cambridge University Press.

Hopper, P. J. (1991). On some principles of grammaticization. In Traugott and Heine (eds., 1991a), 17–35.

—— and Thompson, S. A. (eds.) (1982). *Studies in transitivity*. (Syntax and Semantics, 15.). New York: Academic Press.

—— and Traugott, E. C. (1993). *Grammaticalization*. Cambridge: Cambridge University Press.

—— and —— (2003). *Grammaticalization*. 2nd edn. Cambridge: Cambridge University Press.

Howell, S. C., Fish, S. A., and Keith-Lucas, T. (eds.) (2000). *The proceedings of the Boston University Conference on Language Development*, 24. Boston: Cascadilla Press.

Huang, C.-T. J. and Li, Y.-H. (eds.) (1996). *New horizons in Chinese linguistics*. Dordrecht et alibi: Kluwer.

Huber, M. (1996). The grammaticalization of aspect markers in Ghanaian Pidgin English. In Baker and Syea (eds.), 53–70.

—— (1999). *Ghanaian Pidgin English in its West African context: a sociohistorical and structural analysis*. (Varieties of English Around the World, 24.) Amsterdam, Philadelphia: Benjamins.

Hull, D. L. (1988). *Science as a process: an evolutionary account of the social and conceptual development of science*. Chicago: University of Chicago Press.

Hünnemeyer, F. (1985). *Die serielle Verbkonstruktion im Ewe: Eine Bestandsaufnahme und Beschreibung der Veränderungstendenzen funktional-spezialisierter Serialisierungen*. MA Thesis, University of Cologne.

Hurford, J. R. (1975). *The linguistic theory of numerals*. Cambridge: Cambridge University Press.

—— (1987). *Language and number: the emergence of a cognitive system*. Oxford: Basil Blackwell.

—— (2002). The role of expression and representation in language evolution. In Wray (eds.), 311–34.

—— (2003). The language mosaic and its evolution. In Christiansen and Kirby (eds.), 38–57.

—— (2004). Human uniqueness, learned symbols and recursive thought. Preliminary version, Language Evolution and Computation Research Unit, Linguistics Department, University of Edinburgh.

Hurford, J. R., Studdert-Kennedy, M., and Knight, C. (eds.) (1998). *Approaches to the evolution of language.* Cambridge: Cambridge University Press.

Huttar, G. L. and Huttar, M. L. (1994). *Ndyuka.* London, New York: Routledge.

Ittmann, J. (1939). *Grammatik des Duala (Kamerun).* With the collaboration of Carl Meinhof. (Zeitschrift für Eingeborenen-Sprachen, Beihefte, 20.) Berlin: Reimer; Hamburg: Friederichsen, de Gruyter.

Jackendoff, R. (1990). *Semantic structures.* Cambridge, MA: MIT Press.

—— (1999). Possible stages in the evolution of the language capacity. *Trends in Cognitive Science* 3(7): 272–9.

—— (2002). *Foundations of language: brain, meaning, grammar, evolution.* Oxford: Oxford University Press.

—— and Pinker, S. (2005). The nature of the language faculty and its implications for the evolution of language (reply to Fitch, Hauser, and Chomsky). *Cognition* 2005: 1–36.

Jacobs, J. (2001). The dimensions of topic-comment. *Linguistics* 39: 641–81.

——, von Stechow, A., Sternefeld, W., and Vennemann, T. (eds.) (1995). *Syntax: Ein internationales Handbuch.* Berlin: Mouton de Gruyter.

Jäger, G. (2005). Evolutionary game theory and typology. *Unpublished ms, University of Bern* http://www.ling.uni-potsdam.de/~jaeger/games_dcm.pdf.

Janda, R. D. and Joseph, B. D. (eds.) (2003). *The handbook of historical linguistics.* Oxford: Blackwell.

Janson, T. (1979). *Mechanisms of language change in Latin.* (Studia Latina Stockholmiensia, 23.) Stockholm: Almqvist & Wiksell.

Janzen, T. (1995). *The polygrammaticalization of FINISH in ASL.* Masters Thesis, University of Manitoba, Winnipeg.

—— (1998). *Topicality in ASL: information ordering, constituent structure, and the function of topic marking.* Ph.D. dissertation, University of New Mexico, Albuquerque.

—— (1999). The grammaticization of topics in American Sign Language. *Studies in Language* 23: 271–306.

—— and Shaffer, B. (2002). Gesture as the substrate in the process of ASL grammaticization. In Meier *et al.* (eds.), 199–223.

Jenkins, L. (2000). *Biolinguistics: exploring the biology of language.* Cambridge: Cambridge University Press.

Jenkins, R. S. (2002). *Language contact and composite structures in New Ireland, Papua New Guinea.* Ann Arbor: UMI Dissertation Services.

Johansson, S. (2001). *Animal communication, animal minds and animal language.* BA Thesis, University of Lund.

—— (2002). *The evolution of the human language capacity.* MA Thesis, University of Lund.

—— (2005). *Origins of language—constraints on hypotheses.* (Converging Evidence in Language and Communication Research, 5.) Amsterdam: Benjamins.

—— (to appear). Working backwards from modern language to proto-grammar. In Cangelosi, Smith and Smith (to appear).
Joseph, B. D. and Janda, R. D. (1988). The how and why of diachronic morphologization and demorphologization. In Hammond and Noonan (eds.), 193–210.
—— and Zwicky, A. M. (eds.) (1990). *When verbs collide: Papers from the 1990 Ohio State Mini-Conference on Serial Verbs.* (Working Papers in Linguistics, 39.) Columbus: The Ohio State University, Department of Linguistics.
Jurafsky, D., Bell, A., Gregory, M., and Raymond, W. D. (2001). Probabilistic relations between words: evidence from reduction in lexical production. In Bybee and Hopper (eds.), 229–54.
Just, M. A. and Carpenter, P. A. (1992). A capacity theory of comprehension: individual differences in working memory. *Psychological Review* 98: 122–49.
Kahr, J. C. (1975). Adpositions and locationals: typology and diachronic development. *Working Papers on Language Universals* (Stanford) 19: 21–54.
—— (1976). The renewal of case morphology: sources and constraints. *Working Papers on Language Universals* (Stanford) 20: 107–51.
Kako, E. (1999). Elements of syntax in the systems of three language-trained animals. *Animal Learning & Behavior* 27: 1–14.
Kaye, A. S. (ed.) (1991). *Semitic studies, in honor of Wolf Leslau.* Volume 1. Wiesbaden: Otto Harrassowitz.
Keesing, R. M. (1988). *Melanesian Pidgin and the Oceanic substrate.* Stanford: Stanford University Press.
—— (1991). 'Substrates, calquing and grammaticalization in Melanesian Pidgin'. In Traugott and Heine (eds., 1991a), 315–42.
Kegl, J. and McWhorter, J. H (1997). Perspectives on an emerging language. In Clark (ed.), 15–38.
——, Senghas, A., and Coppola, M. (1999). Creation through contact: Sign language emergence and sign language change in Nicaragua. In DeGraff (ed.), 179–237.
Keller, R. (1994). *Language change: the invisible hand in language.* London: Routledge.
Kellog, W. N. (1968). Communication and language in the home-raised chimpanzee. *Science* 162: 423–7.
Kemmer, S. E. (1993). *The middle voice.* (Typological Studies in Language, 23.) Amsterdam, Philadelphia: Benjamins.
Kilian-Hatz, C. (1995). *Das Baka: Grundzüge einer Grammatik aus der Grammatikalisierungsperspektive.* (Afrikanistische Monographien, 6.) Cologne: Institut für Afrikanistik, Universität zu Köln.
—— (2003). *Khwe dictionary. With a supplement on Khwe place-names of West Caprivi by Matthias Brenzinger.* (Namibian African Studies, 7.) Cologne: Köppe.
Kim, H.-S. (2000). Indefinite subject marking (alias "impersonals") in Chadic. Paper presented at *3e Congrès Mondial de Linguistique Africaine*, Université du Bénin, Lome, 21–26 August, 2000.

Kiparsky, P. (1968). Linguistic universals and linguistic change. In Bach and Harms (eds.), 171–202.
—— (2005). Grammaticalization as optimization. Typescript, Stanford University.
Kirby, S. (2000). Syntax without natural selection: how compositionality emerges from vocabulary in a population of learners. In Knight *et al.* (eds.), 303–23.
—— (2002). Learning, bottlenecks and the evolution of recursive syntax. In Briscoe (ed.), 173–203.
Kitching, A. L. (1915). *A handbook of the Ateso language.* London: Society for Promoting Christian Knowledge.
Klamer, M. (1994). Kambera, a language of Eastern Indonesia. Ph.D. dissertation, University of Amsterdam.
—— (2000). How report verbs become quote markers and complementisers. *Lingua* 110: 69–98.
Klausenburger, J. (2000). *Grammaticalization: studies in Latin and Romance morphosyntax.* (Amsterdam Studies in the Theory and History of Linguistic Science, 193.) Amsterdam, Philadelphia: Benjamins.
Klein, R. G. and Edgar, B. (2002). *The dawn of human culture.* New York: John Wiley & Sons.
Klein, W. and Perdue, C. (1992). *Utterance Structure: Developing Grammars.* Amsterdam, Philadelphia: Benjamins.
—— and —— (1997). The Basic Variety (or: Couldn't language be much simpler?). *Second Language Research* 13: 301–47.
Knight, C., Studdert-Kennedy, M., and Hurford, J. R. (eds.) (2000). *The evolutionary emergence of language: social function and the origins of linguistic form.* Cambridge: Cambridge University Press.
Koch, P. and Oesterreicher, W. (1990). *Gesprochene Sprache in der Romania: Französisch, Italienisch, Spanisch.* (Romanistische Arbeitshefte, 31.) Tübingen: Niemeyer.
Koch, W. A. (ed.) (1990a). *Geneses of language.* Bochum: Universitätsverlag Brockmeyer.
—— (ed.) (1990b). *Natürlichkeit der Sprache und der Kultur: Acta Colloquii.* (Bochumer Beiträge zur Semiotik.) Bochum: Universitätsverlag Brockmeyer.
Kochanski, G. (2004). Is a phrase structure grammar the important difference between humans and monkeys? (Modified comment on Fitch and Hauser 2004 in *Science*). http://kochanski.org/gpk/papers/2004/FitchHauser
Koehler, O. (1943). "Zähl"-Versuche an einem Kolkraben und Vergleichsversuche an Menschen. *Zeitschrift für Tierpsychologie* 5: 575–712.
—— (1950). The ability of birds to "count". *Bulletin of Animal Behavior* 9: 41–5.
Köhler, O. (1973a). Grundzüge der Grammatik der Kxoe-Sprache. Typescript, Cologne, University of Cologne.
—— (1973b). Grundzüge der Grammatik der !Xū-Sprache. Typescript, Cologne, University of Cologne.

König, C. (2002). *Kasus im Ik*. (Nilo-Saharan Studies, 16.) Cologne: Köppe.
—— (2006). Marked nominative in Africa. *Studies in Language* 30(4): 705–82.
—— (forthc.) *Case in Africa*. Oxford: Oxford University Press.
—— (in prep). *A grammar of !Xun (W2 dialect)*. Typescript, University of Cologne.
König, E. (1988). Concessive connectives and concessive sentences: cross-linguistic regularities and pragmatic principles. In Hawkins (ed.), 145–66.
—— (2004). Towards a typology of reciprocity. Paper presented at the workshop in Reciprocity and Reflexivity—Description, Typology and Theory, Freie Universität Berlin, 1–2 October, 2004.
—— and Kortmann, B. (1991). On the reanalysis of verbs as prepositions. In Rauh (ed.), 109–25.
—— and Siemund, P. (2000). Intensifiers and reflexives: a typological perspective. In Frajzyngier and Curl (eds., 2000a), 41–74.
Koopman, H. (1986). The genesis of Haitian: Implications of a comparison of some features of the syntax of Haitian, French, and West African languages. In Muysken and Smith (eds.), 231–58.
Kortmann, B. (1998). The evolution of adverbial subordinators in Europe. In Schmid, Austin, and Stein (eds.), 213–27.
—— and König, E. (1992). Categorial reanalysis: the case of deverbal prepositions. *Linguistics* 30: 671–97.
Kouwenberg, S. (1994). *A grammar of Berbice Dutch Creole*. (Mouton Grammar Library, 12.) Berlin, New York: Mouton de Gruyter.
—— and Murray, E. (1994). *Papiamentu*. (Languages of the World Materials, 83.) Munich, Newcastle: Lincom Europa.
Krantz, G. (1980). Sapientization and speech. *Current Anthropology* 21: 773–92.
Krifka, M. (2005). A functional similarity between bimanual coordination and topic/comment structure. Paper presented at the Blankensee Colloquium on "Language Evolution: Cognitive and Cultural Factors", 14–16 July, 2005.
Krönlein, J. G. (1889). *Wortschatz der Khoi-khoin*. Berlin: Deutsche Kolonialgesellschaft.
Kuno, S. (1973). *The structure of the Japanese language*. Cambridge, MA: Massachusetts Institute of Technology.
Kunze, J. (1997). Typen der reflexiven Verbverwendung im Deutschen und ihre Herkunft. *Zeitschrift für Sprachwissenschaft* 16: 83–180.
Kuteva, T. (1998). On identifying an evasive gram: action narrowly averted. *Studies in Language* 22(1): 113–60.
—— (1999). On 'sit'/'stand'/'lie' auxiliation. *Linguistics* 37(2): 191–213.
—— (2001). *Auxiliation: An enquiry into the nature of grammaticalization*. Oxford, New York: Oxford University Press.
—— and Comrie, B. (2005). The typology of relative clause formation in African languages. In Voeltz (ed.), 209–28.
Kutsch Lojenga, C. (1994). *Ngiti: A Central-Sudanic language of Zaire*. (Nilo-Saharan, 9.) Cologne: Köppe.

Laberge, S. (1977). Étude de la variation des pronoms sujets définis dans le français parlé à Montréal. Ph.D. dissertation, University of Montreal.

Lakoff, G. (1987). *Women, fire, and dangerous things: What categories reveal about the mind.* Chicago: University of Chicago Press.

—— and Johnson, M. (1980). *Metaphors we live by.* Chicago: University of Chicago Press.

Lambrecht, K. (1981). *Topic, antitopic and verb agreement in non-standard French.* Amsterdam: Benjamins.

—— (1994). *Information structure and sentence form: topic, focus and the mental representation of discourse referents.* (Cambridge Studies in Linguistics, 71.) Cambridge: Cambridge University Press.

Landsberg, M. E. (ed.) (1988). *The genesis of language: a different judgement of evidence.* (Studies in Anthropological Linguistics, 3.) Berlin: Mouton de Gruyter.

Langacker, R. W. (1977). Syntactic reanalysis. In Li (ed.), 59–139.

—— (ed.) (1984). *Studies in Uto-Aztecan grammar.* Volume 4: *Southern Uto-Aztecan grammatical sketches.* (Summer Institute of Linguistics Publications in Linguistics, 56, IV.) Arlington: The Summer Institute of Linguistics & the University of Texas at Arlington.

Lass, R. (1990). How to do things with junk: exaptation in language evolution. *Journal of Linguistics* 26: 79–102.

—— (1997). *Historical linguistics and language change.* Cambridge: Cambridge University Press.

Lee, N. and Schumann, J. H. (2003). The evolution of language and of the symbolosphere as complex adaptive systems. Paper presented at the conference of the American Association for Applied Linguistics, Arlington VA, 22–25 March, 2003.

Leeman, D. (1991). On thème. *Lingvisticae Investigationes* 15: 101–13.

Lefebvre, C. and Brousseau, A.-M. (2001). *A grammar of Fongbe.* (Mouton Grammar Library, 25.) Berlin, New York: Mouton de Gruyter.

Leger, R. (1994). *Eine Grammatik der Kwami-Sprache (Nordostnigeria).* (Westafrikanische Studien, 8.). Cologne: Rüdiger Köppe.

Lehmann, C. (1982). *Thoughts on Grammaticalization. A programmatic Sketch.* Volume 1. *AKUP* 48 (Arbeiten des Kölner Universalien-Projekts, 48.). Cologne: University of Cologne, Institut für Sprachwissenschaft.

—— (1984). *Der Relativsatz: Typologie seiner Strukturen, Theorie seiner Funktionen, Kompendium seiner Grammatik.* Tübingen: Narr.

—— (1985). Grammaticalization: synchronic variation and diachronic change. *Lingua e Stile* 20(3): 303–18.

—— (1988). Towards a typology of clause linkage. In Haiman and Thompson (eds.), 181–225.

—— (1995). Synsemantika. In Jacobs *et al.* (eds.), 1251–66.

Leslau, W. (1995). *Reference grammar of Amharic.* Wiesbaden: Harrassowitz.

Levelt, W. J. M. (1974). *Formal grammars in linguistics and psycholinguistics.* Volume 2: *Applications in linguistic theory.* (Janua Linguarum, Series Minor, 192/2.) The Hague, Paris: Mouton.

Lewis, G. L. ([1967] 1985). *Turkish grammar.* Oxford: Oxford University Press.
Li, C. N. (ed.) (1975). *Word order and word order change.* Austin/London: University of Texas Press.
—— (ed.) (1976). *Subject and topic.* New York: Academic Press.
—— (1977). *Mechanisms of syntactic change.* Austin: University of Texas Press.
—— (2002). Missing links, issues and hypotheses in the evolutionary origin of language. In Givón and Malle (eds.), 83–106.
—— and Hombert, J.-M. (2002). On the evolutionary origin of language. In Stamenov and Gallese (eds.), 175–205.
—— and Thompson, S. A. (1974). Co-verbs in Mandarin Chinese: verbs or prepositions? *Journal of Chinese Linguistics* 2(3): 257–78.
—— and —— (1977). A mechanism for the development of copula morphemes. In Li (1977), 419–44.
—— and —— (1980). Synchrony and diachrony: the Mandarin comparative. *Folia Linguistica Historica* 1(2): 231–50.
—— and —— (1981). *Mandarin Chinese. A functional reference grammar.* Berkeley, Los Angeles: University of California Press.
Lichtenberk, F. (1985). Multiple uses of reciprocal constructions. *Australian Journal of Linguistics* 5: 19–41.
—— (1991). Semantic change and heterosemy in grammaticalization. *Language* 67(3): 475–509.
—— (1994). Reflexives and reciprocals. *The Encyclopedia of Language and Linguistics.* Volume 7. Oxford: Pergamon Press, 3504–9.
—— (2000). Reciprocals without reflexives. In Frajzyngier and Curl (eds., 2000b), 31–62.
Lieberman, P. (1984). *The biology and evolution of language.* Cambridge, MA: Harvard University Press.
—— (1991). Uniquely human: the evolution of speech, thought, and selfless behavior. Cambridge, MA: Harvard University Press.
—— (1998). *Eve spoke: human language and human evolution.* New York: Norton.
Lightfoot, D. (1979). *Principles of diachronic syntax.* Cambridge: Cambridge University Press.
—— (1991). *How to set parameters: arguments from language change.* Cambridge, MA: MIT Press.
—— (1999a). *The development of language: acquisition, change, and evolution.* Oxford: Blackwell.
—— (1999b). *The development of language: acquisition, change, and evolution.* (Maryland Lectures in Language and Cognition, 1.) Malden, MA & Oxford: Blackwell.
—— (2003). Grammatical approaches to syntactic change. In Janda and Joseph (eds.), 495–508.
Liu, M. (2000). Reciprocal marking with deictic verbs *come* and *go* in Mandarin. In Frajzyngier and Curl (eds., 2000b), 123–32.

Lock, A. (1988). Implication and the evolution of language. In Landsberg (ed.), 89–100.
Lockwood, W. B. (1968). *Historical German syntax*. Oxford: Clarendon Press.
Lord, C. D. (1973). Serial verbs in transition. *Studies in African Linguistics* 4(3): 269–96.
—— (1975). Igbo verb compounds and the lexicon. *Studies in African Linguistics (Los Angeles)* 6(1): 23–48.
—— (1976). Evidence for syntactic reanalysis: from verb to complementizer in Kwa. In Steever, Walker, and Mufwene (eds.), 179–91.
—— (1982). The development of object markers in serial verb languages. In Hopper and Thompson (eds.), 277–99.
—— (1993). *Historical change in serial verb constructions*. (Typological Studies in Language, 26.) Amsterdam, Philadelphia: Benjamins.
Luce, P., Bush, R., and Galanter, E. (eds.) (1963). *Handbook of mathematical psychology, Volume 2*. New York: Wiley.
Lüpke, F. (2005). A grammar of Jalonke argument structure. Ph.D. dissertation, Katholieke Universiteit Nijmegen, Nijmegen.
Luraghi, S. (2001). Some remarks on instrument, comitative, and agent in Indo-European. *Sprachtypologie und Universalienforschung* (STUF) 54(4): 385–401.
Lyell, Sir C. (1830–33). *Principles of geology; being an attempt to explain the former changes of the earth's surface by reference to causes now in operation*. 3 volumes. London: John Murray.
Lyons, J. (1977). *Semantics*. Cambridge: Cambridge University Press.
Macaulay, M. (1996). *A grammar of Chalcatongo Mixtec*. Berkeley, Los Angeles, London: University of California Press.
McBrearty, S. and Brooks, A. (2000). The revolution that wasn't: a new interpretation of the origin of modern human behavior. *Journal of Human Evolution* 39: 453–563.
McNeill, D. (1974). Sentence structure in chimpanzee communication. In Connolly and Bruner (eds.), 75–94.
MacSwan, J. and Rolstad, K. (2005). Modularity and the facilitation effect: Psychological mechanisms of transfer. *Hispanic Journal of Behavioral Science* 27(2): 224–243.
MacWhinney, B. (ed.) (1999). *The emergence of language*. Mahwah, NJ, London: Erlbaum Associates.
—— (2002). The gradual emergence of language. In Givón and Malle (eds.), 233–63.
McWhorter, J. H. (1997). *Towards a new model of creole genesis*. New York: Peter Lang.
—— (1998). Identifying the creole prototype: Vindicating a typological class. *Language* 74(4): 788–818.
Malle, B. F. (2002). The relation between language and theory of mind in development and evolution. In Givón and Malle (eds.), 265–84.

Malsch, D. L. (1987). The grammaticalization of social relationship: the origin of number to encode deference. In Giacalone Ramat *et al.* (eds.), 407–18.
Marchese, L. (1984). Tense innovation in the Kru language family. *Studies in African Linguistics* 15(2):189–213.
—— (1986). *Tense/aspect and the development of auxiliaries in Kru languages.* (Summer Institute of Linguistics. Publications in Linguistics, 78.) Arlington, Texas: Summer Institute of Linguistics & The University of Texas at Arlington.
Mason, W. A. and Hollis, J. H. (1962). Communication between young rhesus monkeys. *Animal Behaviour* 10: 211–21.
Matsumoto, Y. (1998). Semantic change in the grammaticalization of verbs into postpositions in Japanese. In Ohori (ed.), 25–60.
Maurer, P. (1995). *L'angolar: Un créole afro-portugais parlé à São Tomé. Notes de grammaire, textes, vocabulaires.* (Krcolische Bibliothek, 16.) Hamburg: Buske.
Meier, R. P., Cormier, K., and Quinto-Pozos, D. (eds.) (2002). *Modality and structure in signed and spoken languages.* Cambridge: Cambridge University Press.
Mellars, P. (2006). Patterns of cognitive evolution in modern human origins. Paper presented at the Cradle of Language Conference, Spier Estate, Stellenbosch, 6–10 November, 2006.
Meyer, C. F. (1992). *Apposition in contemporary English.* (Studies in English Language.) Cambridge, New York, Port Chester, Melbourne, Sydney: Cambridge University Press.
Meyerhoff, M. (2001). Another look at the typology of serial verb constructions: The grammaticalization of temporal relations in Bislama (Vanuatu). *Oceanic Linguistics* 40(2): 247–68.
Miles, H. L. (1978). Language acquisition in apes and children. In Peng (ed.), 103–20.
—— (1983). Apes and language: the search for communicative competence. In de Luce and Wilder (eds.), 43–61.
—— (1990). The cognitive foundations for reference in a signing orangutan. In Parker and Gibson (eds.), 511–39.
Miles, L. W. and Harper, S. E. (1994). "Ape language" studies and the study of human language origins. In Quiatt and Itani (eds.), 253–78.
Miller, G. and Chomsky, N. (1963). Finitary models of language users. In Luce, Bush, and Galanter (eds.), 419–92.
Milroy, J. and Milroy, L. (1985). Linguistic change, social network and speaker innovation. *Journal of Linguistics* 21: 339–84.
Mitchell, P. and Riggs, K. J. (eds.) (2000). *Children's reasoning and the mind.* Hove: Psychology Press.
Mithen, S. (2005). *The singing Neanderthals: the origins of music, language, mind and body.* London: Weidenfeld & Nicholson.
Mithun, M. (1984). The evolution of noun incorporation. *Language* 60: 847–94.
Mkhatshwa, S. N. L. (1991). *Metaphorical extensions as a basis for grammaticalization. With special reference to Zulu auxiliary verbs.* MA Thesis, University of South Africa, Pretoria.

Möhlig, W. J. G. (1967). Die Sprache der Dciriku: Phonologie, Prosodologie und Morphologie. Ph.D. dissertation, Universität zu Köln, Cologne.
Moravcsik, E. A. (1972). Some crosslinguistic generalizations about intensifier constructions. *Chicago Linguistic Society* 8: 271–7.
Morford, J. P. (1996). Insights to language from the study of gesture: A review of research on the gestural communication of non-signing deaf people. *Language and Communication* 16: 165–78.
—— (2002). Why does exposure to language matter? In Givón and Malle (eds.), 329–41.
—— and Goldin-Meadow, S. (1997). From here and now to there and then: the development of displaced reference in homesign and English. *Child Development* 68(3): 420–35.
Mous, M. (1992). *A grammar of Iraqw*. Ph.D. dissertation, *Rijksuniversiteit te Leiden*, Leiden.
Mufwene, S. S. (1986). Notes on durative constructions in Jamaican and Guyanese creoles. In Görlach and Holm (eds.), 167–82.
—— (1990). Serialization and subordination in Gullah: Toward a definition of serialization. In Joseph and Zwicky (eds.), 91–108.
—— (1996). Creolization and grammaticization: what creolistics could contribute to research on grammaticization. In Baker and Syea (eds.), 5–28.
—— (2001). *The ecology of language evolution*. (Cambridge Approaches to Language Contact.) Cambridge: Cambridge University Press.
Mühlhäusler, P. (1986). *Pidgin & creole linguistics*. Oxford: Blackwell.
—— and Harré, R. (1990). *Pronouns and people: the linguistic construction of social and personal identity*. (Language in Society, 15.) Oxford: Blackwell.
Mustanoja, T. (1960). *A Middle English syntax. Part I: Parts of speech*. (Mémoires de la Société Néophilologique, 23.) Helsinki.
Muthiani, J. (1974). Kisetla. *Studies in Linguistics* 24: 31–44.
Muysken, P. and Smith, N. (eds.) (1986). *Substrata versus universals in Creole genesis*. Amsterdam, Philadelphia: Benjamins.
Newman, J. (1996). *Give: A cognitive linguistic study*. Berlin & New York: Mouton de Gruyter.
—— (ed.) (1997). *The linguistics of giving*. (Typological Studies in Language, 36.) Amsterdam, Philadelphia: Benjamins.
Newman, P. (2000). *The Hausa language: an encyclopedic reference grammar*. New Haven, London: Yale University Press.
Newmeyer, F. J. (1991). Functional explanation in linguistics and the origins of language. *Language and Communication* 11: 3–28.
—— (1998a). *Language form and language function*. Cambridge, MA: MIT Press.
—— (1998b). On the supposed "counter-functionality" of Universal Grammar: some evolutionary implications. In Hurford *et al.* (eds.), 305–19.
—— (2000). Three book-length studies of language evolution. *Journal of Linguistics* 36: 383–95.

—— (2002). Uniformitarian assumptions and language evolution research. In Wray (eds.), 359–75.

—— (2003). 'What can the field of linguistics tell us about the origins of language?' In Christiansen and Kirby (eds.), 58–76.

—— (2004). Cognitive and functional factors in the evolution of grammar. Paper presented at the conference "Coevolution of Language and Theory of Mind" (www.interdisciplines.org/coevolution/papers/3/1).

—— (2006). What can grammaticalization tell us about the origins of language? Paper presented at the Sixth International Conference on "Evolution of Language", Rome, 12–15 April, 2006.

Nichols, J. (1992). *Linguistic diversity in space and time.* Chicago, London: University of Chicago Press.

Noonan, M. (1985). Complementation. In Shopen (ed., 1985a), 42–140.

—— (1992). *A grammar of Lango.* (Mouton Grammar Library, 7). Berlin, New York: Mouton de Gruyter.

Norde, M. (2001). Deflexion as a counterdirectional factor in grammatical change. *Language Sciences* 23(2–3): 231–64.

—— (2002). The final stages of grammaticalization: affixhood and beyond. In Wischer and Diewald (eds.), 45–81.

—— and Perriodon, H. (eds.) (2004). *Up and down the cline.* Amsterdam, Philadelphia: Benjamins.

Odling-Smee, F. J., Laland, K. N., and Feldman, M. W. (2003). *Niche construction: the neglected process in evolution.* Princeton: Princeton University Press.

Ohori, T. (ed.) (1998). *Studies in Japanese grammaticalization: cognitive and discourse perspectives.* Tokyo: Kurosio Publishers.

O'Neil, W. (1977). Clause adjunction in Old English. *General Linguistics* 17: 199–211.

Osumi, M. (1995). *Tinrin grammar.* (Oceanic Linguistics Special Publication, 25.) Honolulu: University of Hawai'i Press.

Otero, C. P. (ed.) (1994). *Noam Chomsky: critical assessments.* Volume 4: *From artificial intelligence to theology: Chomsky's impact on contemporary thought: Tome I.* London, New York: Routledge.

Oxford English Dictionary. 2nd edn. (1989). (ed. J. A. Simpson and E. S. C. Weiner), Additions 1993–7 (ed. John Simpson and Edmund Weiner; Michael Proffitt) & 3rd edn. (in progress) Mar. 2000– (ed. John Simpson). *OED Online.* Oxford University Press (http://oed.com).

Pagliuca, W. (ed.) (1994). *Perspectives on grammaticalization.* (Amsterdam Studies in the Theory and History of Linguistic Science, 109.) Amsterdam, Philadelphia: Benjamins.

Papen, R. A. (1978). The French-based creoles of the Indian Ocean: an analysis and comparison. Ph.D. dissertation, University of California, San Diego. Ann Arbor, MI: University Microfilms.

Paprotté, W. and Dirven, R. (eds.) (1985). *The ubiquity of metaphor.* Amsterdam, Philadelphia: Benjamins.

Parker, A. R. (2005). Evolving the narrow language faculty: was recursion the pivotal step? Typescript, Language Evolution and Computation Research Unit, University of Edinburgh.

Parker, S. T. and Gibson, K. R. (eds.) (1990). *"Language" and intelligence in monkeys and apes: comparative developmental perspectives.* New York: Cambridge University Press.

Patterson, F. P. (1978a). The gestures of a gorilla: language acquisition in another pongid. *Brain and Language* 5: 72–97.

—— (1978b). Linguistic capabilities of a lowland gorilla. In Peng (ed.), 161–201.

Paul, H. ([1880] 1920). *Prinzipien der Sprachgeschichte.* 5th edn. Tübingen: Niemeyer.

Payne, D. L. and Payne, T. E. (1990). Yagua. In Derbyshire and Pullum (eds.), 249–474.

Peng, F. C. C. (ed.) (1978). *Sign language and language acquisition in man and ape: new dimensions in comparative pedlinguistics.* (AAAS Selected Symposia Series, 16.) Boulder, CO: Westview Press.

Pepperberg, I. M. (1987a). Acquisition of the same/different concept by an African Grey parrot (*Psittacus erithacus*): Learning with respect to categories of color, shape, and material. *Animal Learning & Behavior* 15(4): 423–32.

—— (1987b). Evidence for conceptual quantitative abilities in the African Grey parrot: labeling of cardinal sets. *Ethology* 75: 37–61.

—— (1987c). Interspecies communication: a tool for assessing conceptual abilities in the African Grey parrot. In Greenberg and Tobach (eds.), 31–56.

—— (1992). Proficient performance of conjunctive, recursive task by an African Grey parrot (*Psittacus erithacus*). *Journal of Comparative Psychology* 106: 295–305.

—— (1994). Evidence for numerical competence in an African Grey parrot (*Psittacus erithacus*). *Journal of Comparative Psychology* 108: 36–44.

—— (1999a). Rethinking syntax: a commentary on E. Kako's "Elements of syntax of three language-trained animals". *Animal Learning & Behavior* 27(1): 15–7.

—— (1999b). *The Alex studies.* Cambridge, MA: Harvard University Press.

Perdue, C. (1996). Pre-basic varieties: the first stages of second language acquisition. *Toegepaste Taalwetenschap in Artikelen* 55: 135–50.

Peyraube, A. (1996). Recent issues in Chinese historical syntax. In Huang and Li (eds.), 161–213.

Pfau, R. (2004). The grammar of headshake: Sentential negation in German Sign Language. Manuscript, University of Amsterdam.

—— and Steinbach, M. (2005). Grammaticalization of auxiliaries in sign languages. Typescript, University of Amsterdam and University of Mainz.

—— and —— (2006). Modality-independent and modality-specific aspects of grammaticalization in sign languages. Typescript, University of Amsterdam and University of Mainz.

Piattelli-Palmarini, M. (1989). Evolution, selection and cognition: From "learning" to parameter setting in biology and the study of language. *Cognition* 31: 1–44.

—— (1990). An ideological battle over modals and quantifiers. *Behavioral and Brain Sciences* 13: 752–4.

Pine, J. M. and Lieven, E. V. M. (1997). Slot and frame patterns and the development of the determiner category. *Applied Psycholinguistics* 18: 123–38.

Pinker, S. (1994). *The language instinct.* New York: Harper/Collins.

—— (2003). Language as an adaptation to the cognitive niche. In Christiansen and Kirby (eds.), 16–37.

—— and Bloom, P. (1990). Natural language and natural selection. *Behavioral and Brain Sciences* 13: 707–27, 765–84.

—— and Jackendoff, R. (2005). The faculty of language: what's special about it? *Cognition* 95: 201 36.

Plag, I. (1993). *Sentential complementation in Sranan: on the formation of an English-based creole language.* Tübingen: Niemeyer.

—— (1994a). Creolization and language change: a comparison. In Adone and Plag (eds.), 3–21.

—— (1994b). On the diachrony of creole complementizers: the development of Sranan *taki* and *dati*. *Amsterdam Creole Studies* 11: 40–65.

—— (1995). The emergence of *taki* as a complementizer in Sranan: on substrate influence, universals, and gradual creolization. In Arends (ed.), 113–48.

Plank, F. (1979). Exklusivierung, Reflexivierung, Identifizierung, relationale Auszeichnung: Variationen zu einem semantisch-pragmatischen Thema. In Rosengren (ed.), 330–54.

—— (2001). Thoughts on the origin, progress and pro status of reciprocal forms, occasioned by those of Bavarian. Typescript, University of Konstanz.

Praetorius, F. (1879). *Die amharische Sprache.* Halle: Verlag der Buchhandlung des Waisenhauses.

Premack, D. (1971). Language in chimpanzee? *Science* 172: 808–22.

—— (1976). *Intelligence in ape and man.* Hillsdale, NJ: Erlbaum.

—— and Premack, A. J. (1983). *The mind of an ape.* New York: Norton.

Quiatt, D. and Itani, J. (eds.) (1994). *Hominid culture in primate perspective.* Boulder: University Press of Colorado.

Radden, G. (1985). Spatial metaphors underlying prepositions of causality. In Paprotté and Dirven (eds.), 177–207.

Ragir, S. (2002). Constraints on communities with indigenous sign languages: Clues to the dynamics of language genesis. In Wray (ed.), 272–94.

Ramat, P. (1992). Thoughts on degrammaticalization. *Linguistics* 30: 549–60.

—— and Bernini, G. (1990). Area influence versus typological drift in western Europe: the case of negation. In Bechert *et al.* (eds.), 25–46.

—— and —— (1992). Thoughts on degrammaticalization. *Linguistics* 30: 549–60.

Rauh, G. (ed.) (1991). *Approaches to prepositions.* (Tübinger Beiträge zur Linguistik, 358.) Tübingen: Narr.

Reesink, G. P. (ed.) (1994). *Topics in descriptive Papuan linguistics.* (Semaian, 10.) Leiden: Vakgroep Talen en Culturen van Zuidoost-Azië en Oceanië, Rijksuniversiteit te Leiden.

Reeve, H. K. and Sherman, P. W. (1993). Adaptation and the goals of evolutionary theory. *Quarterly Review of Biology* 68: 1–32.

Reh, M. (1985). *Die Krongo-Sprache (niino mo-di). Beschreibung, Texte, Wörterverzeichnis.* (Kölner Beiträge zur Afrikanistik, 12.) Berlin: Dietrich Reimer.

—— (1993). *Anywa language: description and internal reconstructions.* (Nilo-Saharan, 11.) Cologne: Rüdiger Köppe.

—— (1994). A grammatical sketch of Deiga. *Afrika und Übersee* 77: 197–261.

—— (1996). *Anywa language. Description and internal reconstructions.* Cologne: Köppe.

Reiter, S. (2000). *On the genesis and development of intensifiers and reflexive anaphors in creole languages.* MA dissertation, Free University, Berlin.

Renck, G. L. (1975). *A grammar of Yagaria.* (Pacific Linguistics, Series B, 40.) Canberra: The Australian National University.

Rennison, J. (1996). *Koromfe.* London, New York: Routledge.

Rettler, J. (1991). Infinitiv-complementizer in Englisch und Französisch lexifizierten Kreols. In Boretzky, Enninger, and Stolz (eds.), 139–60.

Rice, K. (1989). *A grammar of Slave.* (Mouton Grammar Library, 5.) Berlin, New York: Mouton de Gruyter.

Rimpau, J. B., Gardner, R. A. and Gardner, B. T. (1989). Expression of person, place, and instrument in ASL utterances of children and chimpanzees. In Gardner, Gardner, and Van Cantfort (eds.), 240–68.

Rivas, E. (2005). Recent use of signs by chimpanzees (*Pan troglodytes*) in interactions with humans. *Journal of Comparative Psychology* 119(4) (2005, to appear).

Roberts, I. and Roussou, A. (1999). A formal approach to "grammaticalization". *Linguistics* 37: 1011–41.

—— and —— (2003). *Syntactic change.* Cambridge: Cambridge University Press.

Robinson, L. W. and Armagost, J. (1990). *Comanche dictionary and grammar.* Arlington, TX: The Summer Institute of Linguistics and The University of Texas at Arlington.

Roeper, T. and Williams, E. (eds.) (1987). *Parameter setting.* Dordrecht: Reidel.

Roitblat, H. L., Bever, T. G. and Terrace, H. S. (eds.) (1984). *Animal cognition.* Hillsdale, NJ: Erlbaum.

Romaine, S. (1982). *Socio-historical linguistics: its status and methodology.* Cambridge: Cambridge University Press.

—— (1984). Relative clauses in child language, pidgins and creoles. *Australian Journal of Linguistics* 4: 257–81.

—— (1988). *Pidgin and creole languages.* London, New York: Longman.

—— (1989). *Bilingualism.* Oxford: Blackwell.

—— (1992a). The evolution of complexity in a creole language: Acquisition of relative clauses in Tok Pisin. *Studies in Language* 16: 139–82.

—— (1992b). The evolution of linguistic complexity in pidgin and creole languages. In Hawkins and Gell-Mann (eds.), 213–38.

—— (1995). The grammaticalization of irrealis in Tok Pisin. In Bybee and Fleischman (eds.), 389–427.

—— (1999). The grammaticalization of the proximative in Tok Pisin. *Language* 75(2): 322–46.

—— and Lange, D. (1991). The use of *like* as a marker of reported speech and thought: A case of grammaticalization in progress. *American Speech* (1991): 227–79.

Rosenblatt, J. S., Beer, C., Busnel, M.-C., and Slater, P. J. B. (eds.) (1987). *Advances in the study of behavior.* Volume 17. Orlando, FL: Academic Press.

Rosengren, I. (ed.) (1979). *Sprache und Pragmatik. Lunder Symposium 1978.* (Lunder germanistische Forschungen, 48.) Lund: CWK Gleerup.

Rumbaugh, D. M., Savage-Rumbaugh, E. S., and Gill, T. V. (1978). Language skills, cognition, and the chimpanzee. In Peng (ed.), 137–59.

Runco, M. A. and Albert, R. S. (eds.) (1990). *Theories of creativity.* Newbury Park, London, New Delhi: Sage Publications.

Samarin, W. J. (1984). Socioprogrammed linguistics. *The Behavioral and Brain Sciences* 7: 206–7.

—— (2002). Plurality and deference in urban Sango. In Carstens and Parkinson (eds.), 299–311.

Sankoff, G. (1979). The genesis of a language. In Hill (ed.), 23–47.

—— and Brown, P. (1976). The origins of syntax in discourse: a case study of Tok Pisin relatives. *Language* 52: 631–66.

—— and Laberge, S. (1973). On the acquisition of native speakers by a language. *Kivung* 6: 32–47.

—— and —— (1974). The acquisition of native speakers by a language. In DeCamp *et al.* (eds.), 73–84.

Santandrea, S. (1961). *Comparative outline-grammar of Ndogo—Sere—Tagbu—Bai—Bviri.* (Museum Combonianum, 13.) Bologna: Editrice Nigrizia.

—— (1965). *Languages of the Banda and Zande groups: a contribution to a comparative study.* Naples: Istituto Universitario Orientale.

—— (1966). *The Birri language: Brief elementary notes.* (Museum Combonianum, 20) Verona: Editrice Nigrizia. [Also published in *Afrika und Übersee* 49 (1965).]

Savage-Rumbaugh, E. S. (1984). Acquisition of functional symbol usage in apes and children. In Roitblat, Bever, and Terrace (eds.), 291–310.

—— (1986). *Ape language: from conditioned response to symbol.* New York: Columbia University Press.

—— and Lewin, R. (1994). *Kanzi: the ape at the brink of the human mind.* New York: John Wiley & Sons.

Savage-Rumbaugh, E. S., Rumbaugh, D. M., Smith, S. T., and Lawson, J. (1980). Reference: the linguistic essential. *Science* 210: 922–4.

Schladt, M. (2000). The typology and grammaticalization of reflexives. In Frajzyngier and Curl (eds., 2000a), 103–24.

Schmid, M. S., Austin, J. R., and Stein, D. (eds.) (1998). *Historical Linguistics 1997*: Selected papers from the 13th International Conference on Historical Linguistics, Düsseldorf, 10–17 August 1997. (Amsterdam Studies in the Theory and History of Linguistic Science, 164.) Amsterdam, Philadelphia: Benjamins.

Schultze-Berndt, E. F. (2000). Simple and complex verbs in Jaminjung: a study of event categorisation in an Australian language. Ph.D. dissertation, Katholieke Universiteit Nijmegen, Nijmegen.

Schusterman, R. J. and Gisiner, R. (1988). Artificial language comprehension in dolphins and sea lions: the essential cognitive skills. *The Psychological Record* 38: 311–48.

—— and Krieger, K. (1984). California sea lions are capable of semantic comprehension. *The Psychological Record* 34: 3–23.

Schwegler, A. (1988). Word-order changes in predicate negation strategies in Romance Languages. *Diachronica* 5: 21–58.

—— (1993). Subject pronouns and person/number in Palenquero. In Byrne and Holm (eds.), 145–61.

Searle, J. R. (ed.) (1974). *The philosophy of language.* (Oxford Readings in Philosophy.) Oxford: Oxford University Press.

Seidenberg, M. S. (1986). Evidence from great apes concerning the biological bases of language. In Demopoulos and Marras (eds.), 29–53.

Seiler, W. (1985). *Imonda, a Papuan language.* (Pacific Linguistics Series B, 93) Canberra: Department of Linguistics Research School of Pacific Studies. The Australian University.

Senft, G. (1996). *Classificatory particles in Kilivila.* New York: Oxford University Press.

—— (ed.) (2000). *Systems of nominal classification.* Cambridge: Cambridge University Press.

Senghas, A. (1995). Children's contribution to the birth of Nicaraguan Sign Language. Ph.D. dissertation, Massachusetts Institute of Technology.

—— (2000). The development of early spatial morphology in Nicaraguan Sign Language. In Howell *et al.* (eds.), 696–707.

—— and Coppola, M. (2001). Children creating language: how Nicaraguan sign language acquired a spatial grammar. *Psychological Science* 12: 323–8.

——,——, Newport, E. L., and Suppalla, T. (1997). Argument structure in Nicaraguan Sign Language: the emergence of grammatical devices. *Proceedings of Boston University Child Language Development* 21: 550–61.

——, Kita, S., and Özyürek, A. (2004). Children creating core properties of language: evidence from an emerging sign language in Nicaragua. *Science* 305: 1779–82.

Seuren, P. A. M. (1984). The bioprogram hypothesis: facts and fancy. *The Behavioral and Brain Sciences* 7: 208–9.
Sexton, A. L. (1999). Grammaticalization in American sign language. *Language Sciences* 21: 105–41.
Seyfarth, R. M. and Cheney, D. (1990). *How monkeys see the world*. Chicago: University of Chicago Press.
—— (1992). Inside the mind of a monkey. *New Scientist* 4 January, 1992: 25–9.
Shaffer, B. (2000). A syntactic, pragmatic analysis of the expression of necessity and possibility in American Sign Language. Ph.D. dissertation, University of New Mexico, Albuquerque.
Shi, Y. and Li, C. N. (2002). The establishment of the classifier system and the grammaticalization of the morphosyntactic particle *de* in Chinese. *Language Sciences* 24: 1–15.
Shnukal, A. and Marchese, L. (1983). Creolization of Nigerian Pidgin English: a progress report. *English World-wide* 4: 17–26.
Shopen, T. (ed.) (1985a). *Complex constructions.* (Language typology and syntactic description, 2) Cambridge: Cambridge University Press.
—— (ed.) (1985b). *Grammatical categories and the lexicon.* (Language typology and syntactic description, 3) Cambridge: Cambridge University Press.
Simeoni, A. (1978). *Päri. A Luo language of Southern Sudan. Small grammar and vocabulary.* Bologna: E.M.I.
Simon, H. J. (1997). Die Diachronie der deutschen Anredepronomina aus Sicht der Universalienforschung. *Sprachtypologie und Universalienforschung* (STUF, Berlin) 50(3): 267–81.
Sinha, N. K. s.a. *Mundari grammar.* (Central Institute of Indian Languages, Series-2) Mysore: Central Institute of Indian Languages.
Slobin, D. I. (ed.) (1985a). *The crosslinguistic study of language acquisition.* Volume 2. Hillsdale, NJ: Lawrence Erlbaum.
—— (1985b). Crosslinguistic evidence for the Language-Making Capacity. In Slobin (ed., 1985a), 1157–249.
—— (1994). Talking perfectly: discourse origins of the present perfect. In Pagliuca (ed.), 119–33.
—— (2002). Language evolution, acquisition and diachrony: probing the parallels. In Givón and Malle (eds.), 375–92
Smeets, C. J. M. A. (1989). A Mapuche grammar. Ph.D. dissertation, University of Leiden, Leiden.
Smith, A. D. M. (2006). Uncertainty and the origin of linguistic complexity. Paper presented at the Cradle of Language Conference, Spier Estate, Stellenbosch, 6–10 November, 2006.
Sohn, H.-M. (1994). *Korean.* (Descriptive Grammars.) London, New York: Routledge.
Song, J. J. (1997). On the development of MANNER from GIVE. In J. Newman (ed.), 327–48.

Song, K.-A. (2002). *Pronominals, addressing forms, and grammaticalization: A speaker–hearer dynamic approach with reference to Korean.* Typescript, University of Cologne, Institut für Afrikanistik.

—— (forthc.). Korean reflexives and grammaticalization: A speaker–hearer dynamic approach. *Sprachtypologie und Universalienforschung* (STUF, Berlin).

Stamenov, M. I. and Gallese, V. (eds.) (2002). *Mirror neurons and the evolution of brain and language.* (Advances in Consciousness Research, 42) Amsterdam, Philadelphia: John Benjamins.

Stassen, L. (2000). AND-languages and WITH-languages. *Typological Linguistics* 4(1): 1–54.

Steels, L. (2003). Evolving grounded communication for robots. *Trends in Cognitive Science* 7(7): 308–12.

—— and Belpaeme, T. (2005). Coordinating perceptually grounded categories through language: a case study for colour. *Behavioral and Brain Sciences* 28(4): 469–529.

—— Kaplan, F., McIntyre, A., and Van Loveren, J. (2002). Crucial factors in the origins of word-meaning. In Wray (ed.), 252–71.

Steever, S. B., Walker, C. A., and Mufwene, S. S. (eds.) (1976). *Papers from the parasession on diachronic syntax.* Chicago: Chicago Linguistic Society.

Steklis, H. D. (1988). Primate communication, comparative neurology, and the origin of language re-examined. In Landsberg (eds.), 37–63.

Stolz, T. (1991a). Forschungen zu den Interrelationen von Grammatikalisierung und Metaphorisierung: Von der Grammatikalisierbarkeit des Körpers. I: Vorbereitung (= ProPrinS 2). Essen: University of Essen.

—— (1991b). *Sprachbund im Baltikum? Estnisch und Lettisch im Zentrum einer sprachlichen Konvergenzlandschaft.* Bochum: Brockmeyer.

—— (1992a). *Von der Grammatikalisierbarkeit des Körpers.* II: Einleitung. 1. Kritik der "Grammatik mit Augen und Ohren, Händen und Füßen" (Arbeitspapiere des Projekts "Prinzipien des Sprachwandels", 7). Essen: University of Essen.

—— (1992b). *Sekundäre Flexionsbildung: Eine Polemik zur Zielgerichtetheit im Sprachwandel.* Bochum: Brockmeyer.

—— (1994). *Sprachdynamik: Auf dem Weg zu einer Typologie sprachlichen Wandels.* Volume 2: *Grammatikalisierung und Metaphorisierung.* Bochum: Brockmeyer.

—— (1996a). Komitativ-Typologie: MIT- und OHNE-Relationen im crosslinguistischen Überblick. *Papiere zur Linguistik* 51(1): 3–65.

—— (1996b). Some instruments are really good companions—some are not: on syncretism and the typology of instrumentals and comitatives. *Theoretical Linguistics* 23(1/2): 113–200.

—— (1998a). UND, MIT und/oder UND/MIT? Koordination, Instrumental und Komitativ—kymrisch, typologisch und universell. *STUF* (Sprachtypologie und Universalienforschung, Berlin) 51(2): 107–30.

—— (1998b). Komitative sind mehr als Instrumentale sind mehr als Komitative: Merkmalhaftigkeit und Markiertheit in der Typologie der mit-Relationen. In Teržan-Kopecky, K. (ed.) (1998, *Sammelband des II. Internationalen Symposions zur Natürlichkeitstheorie, 23. bis 25. Mai 1996.* Maribor), 83–100.

—— (2001). On Circum-Baltic instrumentals and comitatives. In Dahl and Koptjevskaja-Tamm (eds.), 591–612.

—— and Gugeler, T. (2000). Comitative typology: Nothing about the ape, but something about king-size samples, the European Community, and the little prince. *STUF* (Sprachtypologie und Universalienforschung, Berlin) 53(1):53–61.

Sudlow, D. (2001). *The Tamasheq of north-east Burkina Faso: Notes on grammar and syntax including a key vocabulary.* (Berber Studies, 1.) Cologne: Köppe.

Sun, C. (1996). *Word order change and grammaticalization in the history of Chinese.* Stanford: Stanford University Press.

Sutcliffe, E. F. (1936). *A grammar of the Maltese language, with chrestomathy and vocabulary.* London: Oxford University Press.

Svorou, S. (1986). On the evolutionary paths of locative expressions. *Berkeley Linguistics Society* 12: 515–27.

—— (1994). *The grammar of space.* (Typological Studies in Language, 25.) Amsterdam, Philadelphia: Benjamins.

Syea, A. (1994). The development of genitives in Mauritian Creole. In Adone and Plag (eds.), 85–97.

—— (1996). The development of a marker of definiteness in Mauritian Creole. In Baker and Syea (eds.), 171–86.

Sylvain, S. (1936). *Le créole haitien: morphologie et syntaxe.* Wetteren: Imprimerie de Meester.

Tallerman, M. (ed.) (2005). *Language origins: perspectives on evolution.* Oxford: Oxford University Press.

—— (2007). Did our ancestors speak a holistic protolanguage? *Lingua* 117, 3: 579–604.

Tannen, D. (ed.) (1982a). *Spoken and written language: exploring orality and literacy.* Norwood, NJ: Ablex.

—— (1982b). Oral and literate strategies in spoken and written narratives. *Language* 58(1): 1–21.

Taylor, J. R. (2003). *Linguistic categorization.* (Oxford Textbooks in Linguistics.) Oxford: Oxford University Press.

Ternes, E. (1999). Ist Bretonisch SVO oder VSO? Typologische Überlegungen zu einer umstrittenen Frage. In Zimmer, Ködderitzsch, and Wigger (eds.), 236–53.

Terrace, H. S. (1979). *Nim.* New York: Knopf.

—— (1980). More on monkey talk. Response to Patterson's rejoinder to Martin Gardner's review of *Nim* and *Speaking of Apes. New York Review of Books,* December 4, 1980: 59.

Terrace, H. S. (1983). Apes who "talk": language or projection of language by their teachers? In de Luce and Wilder (eds.), 19–42.
—— (1985). In the beginning was the "name." *American Psychologist* 40: 1011–28.
Thomason, S. G. and Kaufman, T. (1988). *Language contact, creolization, and genetic linguistics.* Berkeley, Los Angeles, London: University of California Press.
Thompson, S. A. and Longacre, R. (1985). Adverbial clauses. In Shopen (ed., 1985a), 171–234.
—— and Mulac, A. (1991). A quantitative perspective on the grammaticization of epistemic parentheticals in English. In Traugott and Heine (eds., 1991b), 313–29.
Todd, L. (1974). *Pidgins and creoles.* London, New York: Routledge & Kegan Paul.
—— (1979). Cameroonian: a consideration of "what's in a name?" In Hancock et al. (eds.), 281–94.
Tomasello, M. (1997). One child's early talk about possession. In Newman (ed.), 349–73.
—— (1999). *The cultural origins of human communication.* Cambridge: Cambridge University Press.
—— (2000). Do young children have adult syntactic competence? *Cognition* 74: 209–53.
—— (2000a). Two hypotheses about primate cognition. In Heyes and Huber (eds.), 165–83.
—— (2002). The emergence of grammar in early child language. In Givón and Malle (eds.), 309–28.
—— (2003a). On the different origins of symbols and grammar. In Christiansen and Kirby (eds.), 94–110.
—— (2003b). *Constructing a language: a usage-based theory of language acquisition.* Cambridge, MA & London: Harvard University Press.
—— and Call, J. (1994). Social cognition of monkeys and apes. *Yearbook of Physical Anthropology,* 37: 273–305.
—— and —— (1997). *Primate cognition.* Oxford: Oxford University Press.
——, Call, J., and Gluckman, A. (1997). Comprehension of novel communicative signs by apes and human children. *Child Development* 68: 1067–80.
Tonkes, B. and Wiles, J. (2002). Methodological issues in simulating the emergence of language. In Wray (ed.), 226–51.
Tosco, M. (2001). *The Dhaasanac language: grammar, texts, vocabulary of a Cushitic language of Ethiopia.* (Cushitic Language Studies, 17) Cologne: Köppe.
—— and Owens, J. (1993). Turku: A descriptive and comparative study. *SUGIA (Sprache und Geschichte in Afrika)* 14: 177–267.
Trabant, J. and Ward, S. (eds.) (2001). *New essays on the origin of language.* (Trends in Linguistics, Studies and Monographs, 133) Berlin, New York: Mouton de Gruyter.
Traugott, E. C. (1975). Spatial expressions of tense and temporal sequencing: a contribution to the study of semantic fields. *Semiotica* 15(3): 207–30.

—— (1980). Meaning-change in the development of grammatical markers. *Language Science* 2: 44–61.
—— (1985). Conditional markers. In Haiman (ed., 1985a), 289–307.
—— (1986). On the origins of and and but connectives in English. *Studies in Language* 10(1): 137–50.
—— (2003). Constructions in grammaticalization. In Janda and Joseph (eds.), 624–47.
—— (2004). Exaptation and grammaticalization. In Akimoto (ed.), 133–56.
—— and Dasher, R. B. (2002). *Regularity in semantic change*. (Cambridge Studies in Linguistics, 96.) Cambridge University Press.
—— and Heine, B. (eds.) (1991a). *Approaches to grammaticalization*. Volume 1. Amsterdam, Philadelphia: Benjamins.
—— and —— (eds.) (1991b). *Approaches to grammaticalization*. Volume 2. Amsterdam, Philadelphia: Benjamins.
—— and König, E. (1991). The semantics-pragmatics of grammaticalization revisited. In Traugott and Heine (eds., 1991a), 189–218.
Tröbs, H. (1998). *Funktionale Sprachbeschreibung des Jeli (West-Mande)*. (Mande Languages and Linguistics, 3) Cologne: Köppe.
Tucker, A. N. (1940). *The Eastern Sudanic languages*. Volume 1. London, New York, Toronto: Oxford University Press.
—— (1994). *A grammar of Kenya Luo (Dholuo)*, edited by C. A. Creider. Two volumes. (Nilo-Saharan Linguistic Analyses and Documentation, 8.1, 8.2) Cologne: Köppe.
—— and Tompo ole Mpaayei, J. (1955). *A Maasai grammar with vocabulary*. (Publications of the African Institute, Leyden, 2) London, New York, Toronto: Longmans, Green & Co.
Valian, V. (1986). Syntactic categories in the speech of young children. *Developmental Psychology* 22: 562–79.
van den Berg, R. (1989). A grammar of the Muna language. Ph.D. dissertation, University of Leiden.
van den Bogaerde, B. (2004). Are there syntactic constraints in the signing of the deaf people of Kosindo, Surinam? Project description, University of Amsterdam.
van Driem, G. (1987). *A grammar of Limbu*. (Mouton Grammar Library, 4) Berlin, New York, Amsterdam: Mouton de Gruyter.
van Gelderen, E. (2004). *Grammaticalization as economy*. (Linguistik Aktuell, 71) Amsterdam, Philadelphia: Benjamins.
Vincent, N. (1995). Exaptation and grammaticalization. In Andersen (ed.), 433–45.
Visser, L. E. and Voorhoeve, C. L. (1987). *Sahu-Indonesian-English dictionary and Sahu grammar sketch*. (Verhandelingen van het Koninklijk Instituut voor Taal-, Land- en Volkenkunde, 126) Dordrecht, Providence: Foris Publications.
Voeltz, E. (ed.) (2005). *Studies in African linguistic typology*. (Typological Studies in Language, 64) Amsterdam, Philadelphia: Benjamins.

Wälchli, B. (2005). *Co-compounds and natural coordination.* (Oxford Studies in Typology and Linguistic Theory.) Oxford: Oxford University Press.

Waltereit, R. (2001). Modal particles and their functional correlates: a speech-act theoretic approach. *Journal of Pragmatics* 33: 1391–417.

Wanner, E. and Gleitman, L. R. (eds.) (1982). *Language acquisition: the state of the art.* Cambridge: Cambridge University Press.

Warner, A. R. (1993). *English auxiliaries: structure and history.* Cambridge: Cambridge University Press.

Watahomigie, L. J., Bender, J., Watahomigie, P., Sr., and Yamamoto, A. Y. (2001). *Hualapai reference grammar (revised and expanded edition).* (Endangered Languages of the Pacific Rim, A2–003) Kyoto: Nakanishi Printing Co.

Waters, B. E. (1989). *Djinang and Djinba—a grammatical and historical perspective.* (Pacific Linguistics, Series C, 114.) Canberra: Department of Linguistics Research School of Pacific Studies. The Australian University.

Weber, D. J. (1989). *A grammar of Huallaga (Huánuco) Quechua.* (University of California Publications in Linguistics, 112) Berkeley, Los Angeles, London: University of California Press.

Wehr, B. (1984). *Diskurs-Strategien im Romanischen: ein Beitrag zur romanischen Syntax.* Tübingen: Narr.

Welke, K. (1997). Eine funktionalgrammatische Betrachtung zum Reflexivum: das Reflexivum als Metapher. *Deutsche Sprache* 1997: 209–31.

Wells, M. A. (1979). *Siroi grammar.* (Pacific Linguistics, Series B, 51) Canberra: The Australian National University.

Westermann, D. (1905). *Wörterbuch der Ewe-Sprache.* Berlin: Reimer.

——(1921). *Die Gola-Sprache in Liberia: Grammatik, Texte und Wörterbuch.* Hamburg: L. Friederichsen & Co.

——(1930). *A study of the Ewe language.* London: Oxford University Press.

Wierzbicka, A. (2004). Conceptual primes in human languages and their analogues in animal communication and cognition. *Language Sciences* 26: 413–41.

Wiese, H. (2003). *Numbers, language, and the human mind.* Cambridge: Cambridge University Press.

Wilbur, R. B. (1999). Metrical structure, morphological gaps, and possible grammaticalization in ASL. *Sign Language & Linguistics* 2: 217–44.

Wilcox, S. (2004a). Language from gesture. *Behavioral and Brain Sciences* 27(4): 525–6.

——(2004b). Gesture and language: cross-linguistic and historical data from signed languages. *Gesture* 4: 43–73.

——and Wilcox, P. (1995). The gestural expression of modality in ASL. In Bybee and Fleischman (eds.), 135–62.

Wildgen, W. (2004). *The evolution of human language: scenarios, principles, and cultural dynamics.* Amsterdam, Philadelphia: Benjamins.

Wiliam, U. (1960). *A short Welsh grammar.* Llandybie: Llyfrau'r Dryw.

Wilkins, W. K. and Wakefield, J. (1995). Brain evolution and neurolinguistic preconditions. *Behavioral and Brain Sciences* 18: 161–82, 205–26.
—— (1996). Further issues in neurolinguistic preconditions. *Behavioral and Brain Sciences* 19: 793–8.
Williams, G. C. (1992). *Natural selection: domains, levels and challenges.* New York: Oxford University Press.
Wilson, W. A. A. (1962). *The Crioulo of Guiné.* Johannesburg: Witwatersrand University Press.
Wischer, I. and Diewald, G. (eds.) (2002). *New reflections on grammaticalization.* (Typological Studies in Language, 49) Amsterdam, Philadelphia: Benjamins.
Woll, B. (1981). Question structure in British Sign Language. In Woll *et al.* (eds.), 136–49.
——, Kyle, J., and Deuchar, M. (eds.) (1981). *Perspectives on British Sign Language and deafness.* London: Croom Helm.
Woolford, E. (1979). The developing complementizer system of Tok Pisin: syntactic change in progress. In Hill (ed.), 108–24.
Wordick, F. J. F. (1982). *The Yindjibarndi language.* (Pacific Linguistics, Series C, 71) Canberra: The Australian National University, Research School of Pacific Studies.
Wray, A. (1998). Protolanguage as a holistic system for social interaction. *Language and Communication* 18: 47–67.
—— (2000). Holistic utterances in protolanguage. In Knight *et al.* (eds.), 285–302.
—— (ed.) (2002). *The transition to language.* Oxford: Oxford University Press.
Wu, Z. (2004). *Grammaticalization and language change in Chinese.* London: RoutledgeCurzon.
Wurz, S. (2006). Stone, bone and ochre—inferring symbolic behaviour at Klasies River Main Site, South Africa. Paper presented at the Cradle of Language Conference, Spier Estate, Stellenbosch, 6–10 November, 2006.
Wynn, T. (1979). The intelligence of later Acheulian hominids. *Man* 14: 371–91.
Zavala, R. (2000). Multiple classifier systems in Akatek (Mayan). In Senft (eds.), 114–46.
Zeevaert, L. (2006). Variation und kontaktinduzierter Wandel im Altschwedischen. (Arbeiten zur Mehrsprachigkeit, Folge B, 74) Hamburg: Universität Hamburg.
Ziegeler, D. (2004). The grammaticalisation of constructions in the history of English causatives. Paper presented at the conference on Current Trends in Cognitive Linguistics, University of Hamburg, 9–10 December, 2004.
Zimmer, S., Ködderitzsch, R. and Wigger, A. (eds.) (1999). *Akten des Zweiten Deutschen Keltologen-Symposiums.* Tübingen: Niemeyer.
Zuberbühler, K. (2002). A syntactic rule in forest monkey communication. *Animal Behaviour* 63: 293–9.
——, Cheney, D. L., and Seyfarth, R. M. (1999). Conceptual semantics in a non-human primate. *Journal of Comparative Psychology* 113(1): 33–42.

Author index

Aikhenvald, Alexandra Y. 62, 149, 335
Aitchison, Jean 9, 12, 24, 167, 187, 195, 203, 302, 337
Amha, Azeb 103
Andersson, Peter 47
Arbib, Michael A. 26
Armstrong, D. F. 350
Askedal, John Ole 39
Auel, J. M. 350
Austin, Jennifer R. 123, 131, 134, 152, 153, 159, 278

Baker, Philip 21, 166, 167, 183, 187, 197
Bakker, Peter 166, 167, 168, 169, 183, 194, 195, 200, 201, 202, 203, 207
Baron-Cohen, Simon 126
Batali, J. 13, 354
Bauer, Laurie 268
Belpaeme, T. 13
Benazzo, Sandra 301
Benson, T. G. 249
Bergs, Alexander 256
Bernini, Giuliano 77
Berntson, G. G. 137
Berwick, Robert C. 11, 260, 275, 338
Bichakjian, Bernard H. 10, 338
Bickerton, Derek 4, 5, 8, 10, 11, 12, 14, 15, 27, 32, 135, 164, 166, 167, 192, 193, 195, 205, 210, 249, 260, 262, 300, 301, 303, 308, 309, 312, 315, 316, 331, 338, 340, 345, 346, 347, 352

Bierwisch, Manfred 6, 309
Bisang, Walter 235
Blackmore, Susan 8
Blake, Barry J. 83
Bloom, Paul 6, 10, 20, 27, 164, 319, 338, 340, 346
Boretzky, Norbert 89, 167
Borgman, Donald 75
Botha, Rudie 6, 10, 11, 12, 14, 20, 27, 28, 167, 208, 343, 353
Bowden, John 62
Bower, Bruce 161, 273
Boysen, S. T. 137
Brannon, Elizabeth M. 137, 138
Briscoe, Ted 13
Brooks, A. 13
Brown, Penelope 187, 188
Bruyn, Adrienne 89, 288, 289
Buchholz, Oda 251
Burdyn, Leonhard E. 124, 128, 131
Burling, Robbins 339
Butcher, C. 199
Butterworth, Brian 275
Bybee, Joan L. 19, 22, 33, 34, 36, 38, 43, 45, 56, 57, 74, 75, 82, 90, 178, 210, 335

Cable, C. 128
Call, Josep 124, 150, 152, 156, 157, 161, 350
Callanan, Sam 26
Calvin, William H. 5, 8, 12, 166, 167, 195, 274, 275, 303, 308, 309, 315

Campbell, Lyle 30, 31, 32, 35, 36, 47, 48, 66, 210, 211, 212, 214, 215, 218, 223, 225, 229, 238, 241, 242, 243, 251, 254, 255, 256, 257, 258
Candland, D. K. 203
Carstairs-McCarthy, Andrew 8, 27, 260, 352
Casey, Shannon 145, 207, 303
Chafe, Wallace L. 30
Chater, Nick 266, 269
Cheney, Dorothy L. 125, 126, 127, 128, 156, 157, 159, 349
Chomsky, Noam 3, 4, 9, 10, 12, 34, 128, 133, 135, 138, 156, 158, 160, 161, 163, 262, 263, 264, 266, 267, 269, 271, 274, 276, 277, 318, 319, 321, 338
Christiansen, Morten H. 12, 266, 269, 353
Christy, T. Craig 29
Clark, Eve V. 285
Claudi, Ulrike 32, 34, 38, 39, 40, 57, 62, 98, 106, 236, 241, 242, 253, 294, 314, 334, 335
Cole, Peter 102, 219
Coleman, Helen 13
Comrie, Bernard 5, 24, 25, 29, 30, 66, 167, 218, 225, 302, 313, 314, 351, 352
Conard, Nicholas J. 13
Coolidge, Frederick L. 262
Coppola, Marie 319, 332, 343
Corballis, Michael C. 338, 350
Corne, Chris 81
Craig, Colette Grinevald 81, 218, 249
Crass, Joachim 99, 236, 253
Cristofaro, Sonia 235
Croft, William 6, 19, 23, 30, 35, 78, 79, 80, 106, 179, 331
Crow, Tim J. 11, 340
Cruse, D. A. 151
Curtiss, Susan 203, 204, 205, 208

Dahl, Östen 57
Dasher, Richard B. 36, 37, 45, 337
Darwin, Charles 6, 8, 10, 56, 338, 340, 341, 346
Dasser, Verena 157
Davidson, Donald 270, 271
Davidson, Iain 2, 11, 13, 338, 354
de Luce, J. 160
de Vries, Mark 215
Deacon, Terrence W. 134, 264, 274
Dechmann, Dina K. N. 156
Dennel, Robin 347
Deutscher, Guy 49, 240
Diessel, Holger 38, 84, 88, 148, 210, 305
Dixon, Robert M.W. 101, 149
Donohue, Mark 95, 96
Dougherty, Ray 265, 270
Du Bois, John W. 318
Dunbar, Robin 7, 324, 345
Duran, James J. 170
d'Errico, Francesco 13, 347

Ebert, Karen 236, 253
Edgar, B. 13, 346
Eldrege, Niles 15
Emlen, S. T. 157
Everett, Daniel L. 29, 161, 272, 273

Fant, L. J. 75
Faraclas, N. 192
Fiedler, Wilfried 251
Fitch, W. Tecumseh 2, 4, 9, 12, 128, 133, 135, 138, 148, 156, 158, 160, 161, 163, 262, 263, 264, 267, 274, 276, 277, 318, 321
Fleischer, Michael 162
Fónagy, Ivan 14
Forestell, P. H. 141
Fouts, Roger S. 122, 127, 137, 139

Frajzyngier, Zygmunt 47, 50, 76, 77, 85, 225, 229, 236, 246, 247, 301
Friedländer, Marianne 44, 105, 246

Gallese, Vittorio 350
Gallistel, C. R. 276
Gardiner, A. 88
Gardner, B. T. 122, 124, 127, 130, 135, 136, 137, 139, 141, 145, 148
Gardner, R. A. 122, 124, 127, 130, 135, 136, 137, 139, 141, 145, 148
Gelman, Rochel 275, 276
Genetti, Carol 93, 104, 221, 222, 223, 246
Gibson, K. R. 12
Gil, David 29, 52, 59
Gildea, Spike 94
Gisiner, R. 124
Givón, Talmy 1, 3, 5, 6, 10, 12, 14, 30, 51, 75, 76, 80, 97, 98, 104, 117, 118, 166, 167, 195, 210, 211, 214, 217, 224, 244, 262, 303, 305, 306, 308, 313, 314, 315, 317, 318, 319, 330, 331, 338, 340, 348, 353, 354
Gluckman, A. 124
Goldin-Meadow, Susan 198, 199, 200, 205, 270, 332, 344, 348, 350
Gould, Stephen Jay 8, 10, 338
Grant, Anthony P. 89
Greenberg, Joseph H. 19, 23, 88, 98, 99, 100
Grinstead, John 275
Güldemann, Tom 240

Haase, Martin 327
Hagman, Roy S. 235
Haiman, John 3, 210
Hare, Brian 156
Harper, S. E. 130, 136
Harris, Alice C. 30, 31, 35, 36, 66, 210, 211, 212, 214, 215, 218, 223, 225, 229, 238, 241, 242, 243, 251, 254, 255, 256, 257, 258
Haspelmath, Martin 6, 17, 19, 37, 47, 48, 53, 69, 80, 88, 97, 98, 112, 326, 327, 331, 337
Hauser, Marc D. 4, 9, 12, 128, 133, 134, 135, 138, 148, 156, 158, 160, 161, 163, 203, 208, 262, 263, 264, 267, 274, 276, 277, 294, 318, 321
Head, Brian F. 325
Heine, Bernd 10, 11, 17, 21, 22, 25, 32, 33, 34, 36, 38, 39, 40, 43, 44, 46, 48, 49, 51, 57, 62, 63, 64, 66, 67, 68, 69, 70, 74, 75, 76, 77, 78, 80, 81, 82, 84, 87, 88, 91, 94, 97, 98, 103, 106, 107, 112, 115, 167, 169, 170, 171, 172, 173, 177, 178, 180, 181, 183, 191, 192, 193, 195, 197, 212, 229, 235, 236, 240, 241, 242, 249, 253, 255, 256, 280, 281, 283, 294, 302, 314, 325, 327, 329, 334, 338, 347
Henshilwood, Christopher 13, 347
Herman, Louis M. 123, 124, 125, 131, 135, 139, 141, 143, 146, 147, 149, 158, 161, 164
Herrnstein, R. J. 128
Hockett, Charles 12
Holm, John A. 70, 107, 167, 168, 195
Hombert, Jean-Marie 5, 132, 135, 346
Hopper, Paul J. 33, 34, 37, 38, 39, 40, 41, 45, 48, 50, 57, 66, 77, 89, 90, 96, 104, 210, 212, 214, 215, 220, 225, 227, 241, 242, 244, 255
Huber, Magnus 187, 190
Hünnemeyer, Friederike 33, 34, 38, 39, 40, 57, 62, 73, 74, 106, 236, 241, 242, 253, 294, 314
Hurford, James R. 12, 27, 135, 166, 167, 275, 278, 301, 303, 319, 326, 349, 353
Hurford, Jim 59, 300

Ingold, T. 12

Jackendoff, Ray 5, 6, 9, 10, 11, 14, 30, 135, 140, 160, 164, 167, 195, 208, 262, 267, 274, 275, 303, 308, 309, 314, 315, 316, 318, 319, 321, 324, 331, 338, 340, 352, 354
Janda, Richard D. 32, 47
Janson, Tore 65, 74, 83, 86, 251
Janzen, Terry 25, 75, 108, 109, 110, 349
Johansson, Sverker 5, 8, 12, 13, 30, 134, 135, 160, 303, 309, 310, 345
Johnson, Mark 39, 330
Joseph, Brian D. 47
Jurafsky, Daniel 38

Kako, E. 124, 139, 140, 142, 143, 145, 147, 154, 162, 163
Keesing, Roger M. 187, 191
Kegl, Judy 332, 343
Keller, Rudi 17
Kemmer, Suzanne E. 98
Kilian-Hatz, Christa 68, 90
Kim, Hak-Soo 68
Kiparsky, Paul 45, 48, 331
Kirby, Simon 10, 12, 13, 26, 274, 352, 354
Kita, S. 332, 343
Kitching, A.L. 65
Klamer, Marian 236, 237, 238, 239
Klein, R. G. 13, 346
Klein, Wolfgang 301, 343
Kluender, Robert 145, 207, 303
Knight, Chris 12, 353
Kochanski, Greg 277
Koehler, O. 137
Köhler, Oswin 77
König, Christa 61, 94, 101, 103, 105, 227, 228, 231, 232, 233, 234, 245, 246

König, Ekkehard 31, 40, 63, 72, 79, 82, 106, 248, 299
Kortmann, Bernd 40, 63, 72, 82
Kouwenberg, Silvia 85
Krantz, G. 346
Krifka, Manfred 161
Krieger, K. 125, 155
Krönlein, J. G. 235
Kuno, Susumu 236
Kuteva, Tania 10, 11, 21, 22, 25, 33, 34, 43, 49, 57, 63, 64, 66, 67, 68, 69, 74, 75, 76, 77, 78, 80, 81, 82, 84, 87, 88, 91, 94, 97, 107, 112, 115, 178, 191, 192, 193, 197, 225, 229, 235, 236, 240, 241, 249, 255, 256, 280, 302, 327, 338, 347
Kutsch Lojenga, Constance 69

Laberge, Suzanne 187, 336
Lakoff, George 39, 330
Lambrecht, Knud 96
Lange, Deborah 240
Lass, Roger 100
Lehmann, Christian 33, 34, 57, 63, 66, 67, 69, 91, 92, 97, 112, 235, 236, 272
Lewin, R. 124, 132, 136, 144, 145, 146, 147, 349
Li, Charles N. 5, 80, 132, 135, 228, 229, 244, 259, 307, 317, 341, 346
Lieberman, P. 8, 27, 352
Lightfoot, David 332
Lock, Andrew 346
Lockwood, W. B. 241, 242, 256
Longacre, Robert 215
Lord, Carol Diane 236, 240
Loveland, D. H. 128
Lyons, John 272

Macaulay, Monica 235
MacSwan, J. 203, 204, 205, 208
MacWhinney, Brian 55

Malle, Betram F. 12, 353
Marchese, Lynell 78, 79, 87, 336
Matsumoto, Yo 71, 72, 254
Maurer, Philippe 85
McBrearty, S. 13
McNeill, David 122, 126, 140, 148, 350
McWhorter, John H. 167, 332, 343
Mellars, Paul 13, 347
Meyerhoff, Miriam 187
Miles, H. Lyn 123, 124, 125, 128, 130, 134, 136, 140, 143, 147, 148, 155
Miles, L. W. 130, 136
Miller, George 269
Mills, S. T. 122, 127
Mithen, Stephen 27
Miyashita, Hiroyuki 17, 39, 98
Mkhatshwa, Simon Nyana Leon 81
Morford, Jill P. 75, 108, 109, 199, 200, 314, 332, 343, 349
Mpaayei, J. Tompo ole 65
Mufwene, Salikoko S. 6, 86, 167
Mühlhäusler, Peter 197
Mulac, Anthony 212
Murray, Eric 85
Mustanoja, T. 256
Muthiani, Joseph 170, 172
Mylander, Carolyn 199, 344

Newman, Paul 76
Newmeyer, Frederick J. (Fritz) 2, 10, 29, 30, 36, 40, 46, 47, 48, 49, 51, 52, 53, 116, 164, 206, 216, 226, 230, 235, 254, 255, 260, 265, 269, 272, 306, 310, 312, 316, 319, 321, 322, 328, 329, 332, 338, 339, 340, 348, 352
Nichols, Johanna 56, 320
Noble, W. 13, 338
Noonan, Michael 210, 215, 219, 230, 243

Odling-Smee, F. J. 345
Owens, Jonathan 167, 169
O'Neil, Wayne 89
Özyürek, A. 332, 343

Pagliuca, William 19, 22, 33, 34, 36, 43, 56, 57, 74, 75, 82, 90, 178
Parker, Anna R. 272, 279, 282
Patterson, Francine P. 124, 128, 130, 131, 136, 138, 139, 140, 142, 147, 153, 155, 158
Paul, Herrmann 37
Pepperberg, Irene M. 123, 124, 129, 130, 131, 132, 134, 136, 137, 138, 139, 141, 142, 144, 145, 154, 161, 277, 278, 349
Perdue, Clive 301, 343
Perkins, Revere D. 19, 22, 33, 34, 36, 43, 56, 57, 74, 75, 82, 90, 178
Peyraube, Alain 79, 80
Pfau, Roland 25, 57, 69, 75, 82, 97, 108, 109, 110, 248, 249, 349
Piattelli-Palmarini, M. 6, 9, 338, 347
Pinker, Stephen 6, 10, 20, 27, 135, 160, 164, 262, 267, 274, 275, 309, 314, 319, 321, 324, 331, 338, 340, 346
Plag, Ingo 291, 292
Plank, Frans 84
Premack, A. J. 141
Premack, David 122, 124, 125, 133, 134, 138, 140, 141, 142, 145, 146, 148, 149, 150, 153, 154, 155, 156, 161, 162, 278

Ramat, Paolo 47, 77
Reh, Mechthild 43, 51, 57, 69, 98, 212
Renck, G. L. 70
Rettler, Josef 249
Rimpau, J. B. 145
Rivas, Esteban 122, 127, 135, 137, 139, 140, 145, 152

Roberts, Ian 45, 328
Roebroeks, Wil 347
Rolstad, Kellie 203, 204, 205, 208
Romaine, Suzanne 30, 167, 168, 187, 188, 189, 192, 195, 240, 256
Roussou, Anna 45
Rumbaugh, Duane M. 312, 350

Samarin, William J. 333
Sankoff, Gillian 24
Savage-Rumbaugh, E. Susan 122, 123, 124, 125, 127, 132, 133, 134, 135, 136, 141, 144, 145, 146, 147, 152, 153, 278, 349
Schladt, Mathias 70
Schultze-Berndt, Eva Friederike 102, 223
Schusterman, R. J. 124, 125, 155
Schwegler, Armin 70, 77
Seidenberg, M. S. 122
Seiler, Walter 254
Senft, Gunter 335
Senghas, Ann 319, 332, 343
Seuren, Pieter A. M.
Sexton, A. L. 25, 57, 75, 87, 108, 349
Seyfarth, Robert M. 125, 126, 127, 128, 156, 157, 158, 159, 349
Shaffer, Barbara 25
Shnukal, Anna 336
Shi, Yuzhi 228, 229
Singleton, J. 350
Slobin, Dan I. 6, 332, 335, 336, 337
Smith, Andrew D. M. 33, 308
Song, Jae Jung 249
Stamenov, Maxim I. 350
Steels, Luc 13, 354
Steklis, Horst D. 156
Stein, Dieter 256
Steinbach, Markus 25, 57, 69, 75, 82, 97, 108, 109, 110, 248, 249, 349
Stolz, Thomas 66, 68, 83

Stokoe, W. C. 350
Studdert-Kennedy, Michael 12, 353
Svorou, Soteria 62
Sylvain, Suzanne 73

Tallerman, Maggie 12, 26, 59, 300, 354
Tannen, Deborah 30
Terrace, H. S. 122, 124, 125, 127, 136, 137, 138, 148, 160
Thomas, Roger K. 124, 128, 131
Thompson, Sandra A. 80, 212, 215, 244
Tomasello, Michael 10, 124, 148, 150, 152, 156, 157, 161, 287, 350
Tonkes, Bradley 13, 354
Tosco, Mauro 167, 169
Traugott, Elizabeth Closs 32, 33, 34, 36, 37, 39, 40, 41, 45, 48, 57, 66, 77, 89, 90, 96, 104, 210, 212, 214, 215, 220, 224, 225, 227, 241, 242, 244, 248, 255, 337
Tröbs, Holger 44
Tucker, Archibald N. 65

Uhlisch, Gerda 251

Valian, Virginia 287
van den Bogaerde, Beppie 199, 200
van Gelderen, Elly 34, 45, 92, 221, 225, 226, 227, 328
Vincent, Nigel 42, 88, 89
Vrba, Elisabeth S. 8,

Wakefield, Jenny 8, 9, 164
Waltereit, Richard 26
Westermann, Diedrich 231
Wiese, Heike 275
Wilbur, Ronnie B. 25, 108, 349
Wilcox, Phyllis 109, 350
Wilcox, Sherman 109, 110, 350
Wilder, H. T. 160

Wildgen, W. 345
Wiles, Janet 13, 354
Wiliam, Urien 78
Wilkins, Wendie K. 8, 9, 164
Woll, Bencie 109, 198
Wray, Alison 12, 26, 353

Wrege, P. H. 157
Wurz, Sarah 13
Wynn, T. 262, 346

Zeevaert, Ludger 226
Zuberbühler, Klaus 125, 128, 141, 143

Subject index

adaptation 6, 8, 10, 20, 43, 48, 128, 150, 264, 321, 331, 338, 346, 349, 374, 389, 390
additive combining, *see* combining
adjective 41, 43, 58, 60–1, 82–3, 85, 99, 109, 111, 114, 116, 118, 147, 153, 154, 155, 198, 200, 205, 267, 273, 286–7, 294, 299, 303–4, 307 n., 313, 326, 334–5
adposition 20, 39, 41, 62–4, 65, 66, 71, 72–3, 75–6, 82, 83, 87, 91, 92, 93, 94, 104, 111, 114, 115–16, 117, 119, 139, 164, 199, 202, 216, 220–1, 222–4, 331, 246, 247, 253, 282, 293–4, 299, 304, 324
adverb 41, 64–5, 70–1, 72, 73–4, 81, 82, 83–7, 90, 91, 92, 94, 95, 104, 109, 111, 114, 116, 117, 182, 183, 184 n., 185, 187–9, 191, 204, 212, 214 n. 2, 215, 216, 220–1, 224, 231, 240, 244–54, 267, 273, 281–3, 288–90, 291, 292, 293, 299, 303, 304, 306, 326 n. 22, 328–9, 334 n. 24
alternative combining, *see* combining
appositive combining, *see* combining
argument structure 93, 144–6, 163, 192, 244, 306, 311, 318, 351
article:
 definite article 19, 20, 34 n. 24, 42, 45, 46, 88, 89, 98–9, 111, 220, 300, 320

 indefinite article 19, 20, 45–6, 82, 320, 327
aspect 2, 16, 17–20, 21, 22, 30, 33, 34, 39–40, 41–4, 45, 50–2, 59, 73–5, 79, 81–2, 87, 90, 93, 94, 105, 107, 111, 112, 114, 117, 118, 149, 158, 168, 171, 172 n., 173, 176–8, 191, 192, 199, 202, 203, 215, 217, 218, 237, 238, 272, 299, 305, 307, 317, 322, 323, 332, 341, 348
autonomous language 200 n., 201; *see also* twins' language

Basic Variety 301, 314, 316, 343
Bonobo 124, 125, 127, 131, 133, 136, 140, 143, 146

Campbell monkey, *see* monkey
case function 66, 145
case marker 62–3, 65, 66, 67, 75–6, 91–2, 94, 101–4, 107, 111, 114, 202, 217, 219, 220, 221, 223–4, 253, 267, 280, 282–3, 299–300
Chelsea 203, 205, 208
chimpanzee 122–5, 127–8, 130, 131, 133–7, 138–42, 143, 145, 146–8, 149–50, 152–3, 154–6, 159, 160–1, 162, 278, 311, 349
classifier:
 noun classifier 41, 45–6, 52, 58, 60–71, 82–6, 88, 89, 91–2, 93, 94, 98, 99–102, 104–5, 106–8, 111–14, 116–17, 119, 155, 159,

Subject index   409

161, 169, 170–2, 175, 176 n., 179,
    181, 194, 199, 200, 205, 215,
    219–20, 222, 224, 227, 230–5,
    240–1, 245–8, 250, 252, 262, 273,
    278, 279, 282–7, 294–6, 299, 300,
    301–5, 313, 335, 342, 353
  numeral classifier 20, 45–6, 61 n.,
    192, 227, 327, 335
  verbal classifier 18, 40, 42, 50
coding:
  iconic coding 132, 136, 147, 166,
    308, 348–50
  indexical coding 132, 135, 348–50
  symbolic coding 133
combinatorics, discrete 143
combining:
  additive combining 270, 283 n.
  alternative combining 64, 151, 270,
    283 n., 284, 317
  appositive combining 283 n.
  modifying combining 16, 152, 280,
    283, 284, 296
compound 26, 74, 113, 152, 202, 267,
    268–9, 278, 280, 283, 284–5, 294,
    296, 344
  modifying compound 152, 280,
    283–4, 296
  noun-noun compound 152, 268,
    283–5
  noun-verb compound 302, 351
co-optation 338
coordination 148, 173, 211, 255 n.,
    256, 257, 258
counting algorithm 270, 271 n.
creole 66, 70, 73, 80–1, 82, 85–6, 87,
    89, 106–7, 167, 183, 187, 240, 249,
    254, 288, 313, 331, 333, 336–7, 341
cryptophasia 201

decategorialization 18, 34, 35, 40–2,
    43, 46, 47, 61, 62, 64, 65, 67, 68,
    71, 72, 79, 81, 84, 87, 88, 89, 91,
    103, 113–14, 117, 179, 199, 213,
    216 n., 217, 225, 226, 234, 237,
    238, 242, 243, 248, 283, 286, 325,
    336, 338
definite article, *see* article
degrammaticalization 47–8, 184 n.
deixis 42, 69–70, 85, 87, 89, 91, 96–7,
    139–40, 163, 191, 204, 207, 213,
    225, 226, 242, 311, 320, 325–6,
    344
  personal deixis 42, 70, 96–7, 139,
    163, 191, 204, 207, 320, 326
derivation 11, 40, 43, 52, 75, 82, 113,
    114, 168, 169, 171–3, 175, 194,
    197, 203, 307, 312, 339, 352
desemanticization 18 n. 6, 34, 35, 37,
    39, 40, 43, 46, 47, 60, 62, 64, 65,
    67, 68, 71, 72, 78–9, 84, 86, 87, 88,
    89, 91, 97–8, 113, 114, 115, 117,
    176, 179, 186, 199, 213, 216 n.,
    217, 225, 226, 229, 237, 238, 242,
    248, 282, 283, 286, 325, 335
Diana monkey, *see* monkey
displacement 12, 121, 134, 299,
    310, 354
dog 103, 130, 133, 136, 162, 202, 246,
    248, 273, 281, 282, 285, 313
dolphin, bottle-nosed 124–5, 130,
    134, 135, 139, 141, 142, 143, 146,
    149, 155, 158, 161, 164
downgrading 48

elementary linguistic system 205–8, 303
erosion 21 n., 34–5, 42, 43–44, 46, 47,
    62, 66, 69, 71, 87, 88–9, 91, 100,
    115, 117, 213, 217, 226, 228, 234,
    237, 238, 242, 243, 248, 283, 284,
    290, 325, 336, 338
exaptation 8–9, 10, 20, 48, 264, 338
extended pidgin, *see* pidgin

extension 130, 164, 19, 34, 35–9, 47, 79, 81, 94, 95, 97, 99, 102, 105, 171, 173, 179, 185, 194, 199, 213, 217, 220, 222, 242, 248, 253, 257, 258, 263, 282, 286, 289, 315, 325, 335, 338, 349

feral children 203, 314

Genie 5, 203, 204–5, 207–8
gorilla 124, 128, 130–1, 136, 138, 139, 140, 142, 143, 147, 152, 153, 155, 158, 279

head-dependent relation 151–2
homesign 3, 198 n., 198–200, 206, 207, 208–9, 297, 303, 314, 319, 332, 336, 337, 343, 344, 348
hypernym 152, 159, 278, 279
hyponym 151, 159, 279
hypotaxis 31 n., 172, 254–5

iconic coding, see coding
iconicity 348
idioglossia 201
imperative 72–3, 77–8, 126, 127, 249, 251, 320
impersonal pronoun, see pronoun
implicature, conventional 37
incorporation 121
indefinite article, see article
indefinite pronoun, see pronoun
indexical coding, see coding
inference, invited 36n., 38, 40, 258
inflection 52, 67, 83, 87, 99, 117, 168, 171–3, 175, 179, 182, 197, 202, 225, 282, 312, 339, 342, 351, 352
integration 214–15, 220, 224, 230, 343, 255, 259, 260, 287–8, 292, 295
intensifier 326, 334–5
isolated children 203, 206, 208–9
iteration 161, 200 n. 20, 270, 277, 345 n.

jackdaw 137
jargon 88–89, 166, 167, 170, 172, 184, 186 n.

Kaspar Hauser 203, 208

lexicalization 34 n. 23, 48, 60
linear arrangement 146–8, 163, 175, 302, 303 n. 6, 311, 315

meronymy 151, 280; see also partonomy
modality 17, 20, 22, 42, 50, 51, 82, 99, 108, 112, 168, 171, 178, 197, 217, 238, 320, 337, 341, 342, 349, 350
modifying combining, see combining
modifying compound, see compound
monkey:
    Campbell monkey 144
    Diana monkey 128, 143–4
    Rhesus monkey 138
    Spider monkey 156
    Squirrel monkey 124, 128, 131
    Vervet monkey 125, 126–7, 133, 157, 348
mono-genetic hypothesis 2, 347
morphologization 34, 117, 125, 166, 168–9, 221–2, 307–8, 348, 351

negation 18, 19, 41, 42, 72, 77–9, 87, 100, 105–6, 110, 111, 114, 140, 142, 163, 173, 179, 202, 206, 207, 217, 299, 305, 306–7, 311, 344
noun classifier, see classifier
noun-noun compound, see compound
noun-verb compound, see compound
numeral classifier, see classifier
numerosity 137–8, 271, 275

object permanence 130, 306
obligatorification 34 n. 23

onomatopoetic 201
orangutan 124, 125, 130, 134, 135, 136, 139, 148, 155
over-generalization 131, 204

parataxis 172, 215, 254–5, 271
parrot, (African) Grey 124, 129, 131, 132, 134, 136, 137, 138, 142, 143–4, 145, 153, 159, 160, 161, 276, 278
partonomy (meronymy) 151, 155, 159, 280
passive 80, 93, 97–8, 110–12, 173, 299, 306, 331
personal deixis, *see* deixis
personal pronoun, *see* pronoun
phonetic reduction 34, 44
phrase structure, hierarchical 161
pidgin:
    extended pidgin 167, 168, 170, 187, 191, 192, 197, 208, 343
    stable pidgin 167
pidgin/creole 87, 167, 240, 333, 336
pluralization 325, 334
poly-genetic hypothesis 2, 340, 347
possession, attributive 62, 64, 66, 267, 268, 280–1, 282, 284, 295, 296
postposition 20, 43, 62, 63, 66, 71–2, 92, 93, 104, 221–2, 224, 245, 294
pre-basic variety 301
preposition 20, 36, 38, 40, 44, 51, 52, 62–4, 65, 66, 71–3, 82, 83, 91–2, 104, 111, 116, 138, 176, 192, 204, 220–1, 224, 281–3, 293, 323, 324, 328–9
pronoun:
    impersonal pronoun 68, 70, 207
    indefinite pronoun 58, 68–9
    personal pronoun 48, 68–70, 87–8, 96–8, 119, 139, 136, 187, 191, 204, 207, 250, 320, 325
    reflexive pronoun 70, 98, 110, 112

question marker 109, 154, 202, 244

raven 137
reappropriation 9
reconstruction, internal 14, 23, 45–6
recursion, embedding 161, 200, 265–6, 267, 270–1, 272, 275, 277, 279, 292, 344 n.
reflexive pronoun, *see* pronoun
repeater 62, 189, 222
Rhesus monkey, *see* monkey

sea lion, Californian 124–5, 160
secret language 201
serial verb construction 149, 173, 271, 344
Spider monkey, *see* monkey
Squirrel monkey, *see* monkey
stable pidgin, *see* pidgin
"stripping" process 168, 171, 175, 187, 343
subjectification 34, 40
subordination 30–1, 54, 71, 92, 93, 94, 95, 100, 101, 104, 120, 148, 161, 168, 171, 173, 175, 188, 190, 192, 196, 197, 199, 202, 203, 206, 207, 208–9, 210–17, 219, 225, 231, 234, 247, 251–2, 254, 255–60, 262, 266–7, 268, 269, 270, 272, 273, 279, 287–8, 290–2, 295, 296, 299, 304, 305, 306, 310, 312, 341, 342, 351, 352, 353
subordinator 41, 43, 70–1, 81, 86, 92–3, 94–5, 103–4, 107, 112, 114, 116, 182, 188, 190, 221–4, 226, 230, 233–4, 236, 245, 246–9, 250–3, 255, 257, 289, 291, 293, 299, 305
succession 270–1, 275
suppletive 25 n., 69
symbol 125, 129, 131, 132–5, 146, 152–3, 161, 162, 166, 264, 265, 275, 308, 348–9

symbolic coding, *see* coding
syntacticization 34, 317

tamarin, cotton-top 276
taxonomy, hierarchical 151, 207
tense 17, 19–20, 21–2, 39, 40–2, 50,
 51–2, 59, 72, 73, 75 n., 79, 81–2,
 86–7, 90–1, 94, 99, 107, 109,
 111, 112, 114, 117, 138, 168, 171,
 172 n., 173, 176–8, 185, 187,
 191, 196, 197, 199, 202, 217,
 218, 237, 238, 267, 277, 299,
 305, 306, 307, 323, 328–9,
 335, 341

thematic role 145, 147
twins' language 200–3, 206–9, 297,
 303, 343, 344

uniformitarianism 28–9, 49

valence 95, 244
verbal classifier, *see* classifier
Vervet monkey,
 see monkey
vocalization 136, 349–50

WH-question 117
written language 30–1, 57, 108–9

Language index

Accadian 49
   Old Accadian 240
African languages 44, 55–6, 69
Afrikaans 85
Afroasiatic languages 68, 76, 85, 103, 170, 229, 246–7
Akan 190
Akkadian, *see* Accadian
Albanian 68, 251
American Sign Language (ASL), *see* sign languages
Ancient Egyptian 49, 88, 240
Angolar 85
‖Ani 63–4, 69
Arabic 19, 66, 169, 171, 247
   Cairene Arabic 19
Archaic Chinese, *see* Chinese, Archaic
Armenian, Modern 19
Australian languages 62, 241
Austronesian languages 78, 96, 187, 237–8, 335

Baka 68, 89–90
Balto-Finnic 66
Bantu languages 61, 64, 72, 81, 83–4, 86–7, 98, 170, 172, 194, 249, 251
Bari 87
Basque 64, 66, 97, 327
Bislama 97, 168, 187, 191
Bodic languages 104, 221–2
British Sign Language, *see* sign languages
Bulgarian 46
Bulu 64
Buru 237–9

Carib 94, 101, 104
Catalan 281
Caucasian languages, South, *see* South Caucasian languages
Central African languages 44, 89, 251
Central Malayo-Polynesian, *see* Malayo-Polynesian
Central Sudanic languages, *see* Sudanic languages, Central
Chadic 50, 68, 76–7, 85, 100, 225, 229, 246–7, 301
   East Chadic 246–7
Chalcatongo Mixtec 235
Chamling 253
Chibchan languages 75, 81, 218, 249, 273
Chinese 31, 49, 77, 79–80, 106, 169, 228–9, 243–4, 259, 307
   Archaic Chinese 77, 79
   Classical Chinese 228
   Early Medieval Chinese 80
   Mandarin Chinese 80, 243–4
   Modern Chinese 259
Chinese language family 307
Chinook Jargon 88–9
Coastal Swahili (CS), *see* Swahili, Coastal
Cushitic languages 170

Danish 89, 201–2
Dolakhari Newari, *see* Newari, Dolakhari

Dorobo, *see* Okiek
Duri 239
Dutch 281, 288
  Netherlands Sign Language, *see* sign languages

East Asian languages 80
East Australian Aboriginal Pidgin English 187
East Chadic, *see* Chadic languages
East Nilotic, *see* Nilotic languages
Egyptian, Ancient, *see* Ancient Egyptian
English 21–3, 26, 31–2, 37, 39–46, 50–2, 60, 62–4, 70–2, 74, 80–2, 89–90, 92, 95, 97, 104, 107, 112, 114–16, 119, 168, 180, 200–2, 212–13, 215, 217, 220–1, 225, 227–30, 241–2, 248–9, 252, 254–9, 266–8, 271, 281–2, 284–7, 294–5, 304, 307, 323–4, 328–9, 335, 341
  British Sign Language, *see* sign languages
  colloquial English 43
  Early Middle English 92, 221, 281–2
  Late Old English 248, 281
  Middle English 92, 221, 256, 281–2
  Modern English 23, 44, 51, 115, 282
  Nigerian Pidgin English 192, 273, 336, 343
  Old English 43, 50, 70, 90, 92, 115, 221, 224–5, 227, 242, 248, 256, 281–2, 341
English-based pidgins or creoles 85–6, 89, 97, 168, 180, 187, 189–93, 218, 240–2, 273, 288–91, 301, 336
  Bislama 97, 168, 187, 191
  Ghanaian Pidgin English (GhaPE) 168, 187, 190, 240, 343
  Guayanese 85
  Jamaican 85–6

Nigerian Pidgin English 168, 192, 273, 336, 343
Solomons Pijin 168, 187, 191, 273, 336, 343
Sranan 89, 288–93
Tok Pisin 87, 97, 168, 187–9, 191–2, 240, 273, 336, 343
Estonian 66, 82, 201, 218
European languages 77, 91, 112, 229, 243, 281–2, 325, 327, 334
Evenki 217–18
Ewe 44, 73–4, 95, 106, 231, 237, 285, 296

Fanagalo 168, 194
Faroese 241–2, 282
Finnic language 66, 82
Finnish 66
Finno-Ugric language 104, 218
French 18–19, 22, 42, 52, 68, 77, 84, 88–9, 91–2, 96, 176, 207, 268, 281, 296, 327, 334
  Modern French 44
  non-standard French 96
  Old French 18, 44, 46
French-based creoles 73, 80–2
  Haitian 73
  Seychellois 81
  Tayo 82
Frisian 281

Gascon 327
Georgian 238, 243
German 17–18, 26, 38–40, 42–3, 63, 68, 70–1, 80, 82, 90–2, 201–4, 212–13, 241–2, 247–8, 268–9, 281, 284, 296, 328–9, 334
  Bavarian 204
  Early New High German 17
  German Sign Language (DGS), *see* sign languages

Middle High German 17
Modern High German 90, 248
Old High German 17, 43, 70, 90, 247
Swiss German 201–2
Germanic languages 216, 241–2, 268
Ghanaian Pidgin English
   (GhaPE) 168, 187, 190, 240, 343
Gidar 229
Giziga 246
Gorokan languages 104
Gorze 272
Grand Ronde Chinook Jargon, see
   Chinook Jargon
Greek, Tsakonian 19
Guayanese 85
Gude 68
Guduf 68
Gunwinggu 241
Gur 63–4
Gusii 170

Haitian 73
Hausa 76
Hawaiian Pidgin 193, 301
Hebrew, Early Biblical 235
Hindi 32
Hittite 49
Hungarian 66, 92

Icelandic 63, 68, 201
Idoma 240
Ik 61, 94, 101, 103, 105, 218, 227–8,
   231–4, 245–6, 286
Imbabura Quechua, see Quechua
Imonda 253–4
Indic, Modern 19
Indo-Aryan 253
Indo-European languages 26, 32, 113,
   229, 243
Proto-Indo-European 32
Indonesian, Riau 29, 51–2, 307

Indopakistani Sign Language, see sign
   languages
Irish 282
Israeli Sign Language, see sign languages
Italian 64, 325
   Italian Sign Language, see sign
      languages

Jacaltec 218
Jamaican 85–6
Jaminjung 102, 223
Japanese 67, 71–2, 236, 254
Jordanian Sign Language, see sign
   languages
Ju|'hoan, see !Xun

Kabuverdiano 70
Kalenjin 170, 186
Kamba 170, 186
Kanuri 79, 99–100, 219
Kapsiki 68
Kartvelian 223
Kenya Pidgin Swahili (KPS) 108,
   169–86, 192–197, 290, 295–6, 343
Khoisan languages 31, 63–4, 67, 74,
   77, 79, 106, 112, 212, 219–20, 235,
   239, 241, 245, 250, 252–3
   Central Khoisan 31, 63–4, 67, 69,
      74–5, 212, 219–20, 235, 245
   North Khoisan 77, 79, 106, 112,
      220, 239, 241, 244, 250, 253, 294
Khwe (Kxoe) 74–5, 212, 219–20, 245
Kihindi 170
Kikongo 168, 194
Kikuyu 170, 186, 249
Kilivila 335
Kimvita, see Swahili
Kisetla 170, 172
Kituba 168, 194
Kiunguja 171
Korean 80

KPS, *see* Kenya Pidgin Swahili
Kru 78–9, 86
Kuliak 61, 94, 101, 103, 105, 218,
    227–8, 232–3, 245–6
  Proto-Kuliak 94
Kurdish 32
Kwa 73–4, 106, 231, 237
Kxoé, *see* Khwe

Latin 32, 42, 44, 65, 68, 74, 78, 83, 86,
    88–9, 96, 251, 256, 281, 341
Latvian 83
Laz 102, 223
Lele 246–7, 301
  East Lele 246
Lendu 69
Lezgian 88, 225
  Proto-Lezgian 225
Lingala 87, 251
Luba 98
Luhya 186
Luo 170, 186
Luyia 170

Maa 170
  Maasai 64–5, 98, 170, 174, 186
Maale 103, 223
Maasai, *see* Maa
Macedonian 281
Malayo-Polynesian languages, 238–9;
    see also Austronesian languages
  Central Malayo-Polynesian 238
Maltese 66
Mandarin Chinese, *see* Chinese
Mande 44, 105, 246
Maninka 43–4, 105
Margi 68
Melanesian PE (Pidgin English) 187
Mingrelian 258
Mopun, *see* Mupun
Moré 63–4

Mupun 77, 85
Mura 273

Nahuatl 68
Nama 67, 235–6
Neo-Babylonian 240
Nepali 253
Netherlands Sign Language, *see* sign
    languages
Newari 93, 222–3, 246
  Classical Newari 222–3, 246
  Dolakhari Newari 93
Neyo 87
Ngan'gityemerri 62
Ngiti 69
Nguni 168, 194
Nicaraguan Sign Language, *see* sign
    languages
Niger-Congo languages 44, 61,
    68–9, 72–74, 78, 81, 84, 90, 98,
    105–6, 170, 231, 237, 240, 246,
    249, 251
Nigerian Pidgin English 168, 192, 273,
    336, 343
Nilo-Saharan languages 61, 65, 69, 94,
    99, 101, 103, 105, 170, 218–19,
    227–8, 232–3, 245–6
Nilotic languages 64–5, 170
  East Nilotic 64–5, 87, 98
Norse, Old 89, 225, 226
Norwegian 282–3
Nubi 66

Okiek 56
Ometo 103
Omotic languages 103–4, 223, 272
Oromo 170
Oto-Manguean 235

Palenquero 69
Palu'e 95–6

Panare 94
Papiamentu 85, 107
Papuan 70, 253
Persian 19
Pipil 66
Pirahã 29, 161, 273, 276
Portuguese 42, 68, 325

Quechua 102, 104, 219
   Imbabura Quechua 102, 219

Rama 81, 249
Riau Indonesian, *see* Indonesian
Romance languages 42, 88, 92, 268, 281, 325, 327, 341
Russian 201, 202, 243
Russonorsk 193, 301

Saharan languages 99–100
Sami 66
Sango 333–4
Sanuma 75
Sapo 78
Scottish Gaelic 282
Semitic 66, 240
Seychellois 81
sign languages 207, 313, 332, 349
   British Sign Language 109
     American Sign Language (ASL) 75, 87, 108–9, 249
     German Sign Language (DGS) 69, 108, 248
     Indopakistani Sign Language 110
     Israeli Sign Language 75
     Italian Sign Language 75
     Jordanian Sign Language 110
     Netherlands Sign Language (NGT) 69, 249
     Nicaraguan Sign Language 200, 208, 319, 332, 343–4
Skou 95

Solomons Pijin 168, 187, 191, 273, 336, 343
Somali 170
Sorbian, Upper 281
South Caucasian languages 102, 223, 238, 243
Spanish 46, 70, 92, 325, 327
Sranan 89, 288–293
Sudanic languages, Central 69
Susu 246
Swahili 60–1, 63, 72, 83–4, 86, 108, 169, 170–2, 186, 296
   Coastal Swahili (CS) 170–80, 182–6, 194
   Kenya Pidgin Swahili (KPS) 108, 169–86, 192–7, 290, 295–6, 343
   Kimvita 170
   Standard Swahili 170–1, 182, 194
Swedish, Old 225
Swiss German, *see* German

Tagalog 78
Tayo 82
Teso 64–5
Thai 80, 235, 249
Tibetan, Classical 222
Tibeto-Burman 92, 104, 221–3, 249, 253
Tok Pisin 87, 97, 168, 187–9, 191–2, 240, 273, 336, 343
Tukang Besi 238–9
Tungusic 217–18
Turkana 170
Turkic 104
Turku 169
Twi 240

Ubangi 68, 90
Ute 75–6

Uto-Aztecan 75–6
  Northern Uto-Aztecan 104

Vietnamese 80

Waris 254
Warlpiri 276
Welsh 77–8
West African Pidgin English
  (WAPE) 190

!Xun 77, 79, 106, 110–12,
  220, 235, 239, 241,
  244, 250, 252–3,
  276, 294

Yagaria 70
Yanomam 75
Yoruba 69

Zulu 81, 168, 194